THE MAMMAL IN THE MIRROR

THE MAMMAL
IN THE MIRROR

Understanding Our Place
in the Natural World

David P. Barash

Ilona A. Barash

W.H. Freeman and Company
New York, New York

Text Designer: Victoria Tomaselli

Library of Congress Cataloging-in-Publication Data

Barash, David P.
    The mammal in the mirror: understanding our place in the natural
world/David P. Barash and Ilona A. Barash.
    p.   cm.
    Includes bibliographical references and index.
    ISBN 0-7167-3391-9
    1. Human biology.   I. Barash, I.   II. Title.
QP34.5.B34   1999   99-38972
599.9–dc21   CIP

Printed in the United States of America

First printing 1999

W. H. Freeman and Company
41 Madison Avenue, New York, NY 10010
Houndmills, Basingstoke RG21 6XS, England

CONTENTS

A PERFECTLY GOOD MAMMAL

There is a chilling moment toward the end of Ray Bradbury's science fiction classic, *The Martian Chronicles*. A human family, having escaped to Mars to avoid impending nuclear war, looks eagerly into the "canals" of their new planetary home, expecting to see Martians. They do: they see their own reflections. Similarly, if you want to see a perfectly good mammal, look in the mirror.

But the question isn't "How do we establish a civilization?," as it was for Bradbury's fictional family, or "Who is the fairest?," as it was for the wicked witch in *Snow White and the Seven Dwarfs*. For us, the question to ask is simply "Who—or what—is that creature?"

We all think we know who we are: name, address, social security number, family situation, job or profession, and so forth. We might even venture a self-characterization. And yet, increasingly, many people sense that, deep down, they don't really have a good grasp of who or what they are. We aren't speaking here of existential angst, metaphysical speculation, or religious doctrine, but rather, of the nuts and bolts of everyone's shared biology. Ironically, even as scientists discover more about their own species, many people—even the well educated—find themselves slipping further behind, if only because discoveries are occurring so quickly.

Hence, this book. In it, we offer a core curriculum in life itself, a primer of human biology, a view of and from the human condition that everyone should have. "When you look into the abyss," wrote Nietzsche, "the abyss looks into you." Martians in the canals, mammals in the mirror: when you look into life, you cannot help looking at yourself. Most of the time, therefore, we shall present a human-centered view of life, one that—fortunately—isn't an abyss but a bounty, and an opportunity for greater self-knowledge.

Some people, to be sure, are affronted by their "mammalhood," their creatureliness, the unavoidable fact that whatever else they may be (symbol-using, self-conscious, made-in-the-image-of-God, etc.), they are also organisms that live, breathe, pump blood, masticate and defecate, metabolize, reproduce, and—eventually—die. Composed of regular old mammalian meat and bones, a bubbling brew of interacting chemicals, criss-crossed with a myriad of electric currents, all of us carry the stigmata of our biology, deeply mired in a long, shared organic past.

At the same time, there are a slew of new kids on the block, novel faces that life science has been unearthing for us all, as we enter the twenty-first century: cloning, genetic fingerprinting, AIDS, birth control pills, and sociobiology are just a few. Then there are new insights into old phenomena, such as Ebola virus, neurotransmitters, cholesterol, and biodiversity. Not to mention causes of and possible cures for cancer, or the latest facts and fads on weight loss and exercise. We meet these newcomers regularly—on television, in newspapers, as important parts of our daily lives—whether we recognize them or not.

Much of the time, however, there is a curious "disconnect" between what so many of us hear about, read about, even experience "up close and personal," and what, deep down, we recognize and understand. Never, in fact, have people heard so much about things they know so little about—and yet, which are so important. Recent discoveries in the life sciences have been leaving even scientists trained in related disciplines feeling like strangers in a strange land, detached and perplexed bystanders to what is, ironically, closest to their own existence—the details and complexity of life itself. Paradoxically, even as this complexity is unraveled, it often seems to grow more distant, harder to grasp.

Part of the problem is that the sciences of life have been moving so rapidly, and on so many different fronts, that it is hard not to sympathize with Alice in *Through the Looking Glass*, who finds herself in a high-velocity world in which she must run just to keep in place. Tired and out of breath, Alice is then admonished by the Red Queen that if she wants to get anywhere, she'll have to run twice as fast!

Not only is the pace of discovery increasing but preparedness to start the race, or even to admire it from the sidelines, seems to be diminishing. Too many biology courses—in high school and even college—are mired in tedious dissections of preserved cadavers (especially earthworms and frogs); hence they are not so much biology, the exploration of *life*, as thanatology, an obsession with things deceased. And then there is the bane of all biology students: eternal lists of accumulated facts, such as the various digestive enzymes (ptyalin is secreted by the salivary glands; pepsin, by the stomach; etc., etc., ad nauseum).

No wonder many people in their student years conclude that biology is for those who are gifted with brute memory but little else, who are neither sufficiently adroit mathematically to succeed in the "real sciences" (chemistry, physics, astronomy), nor flat-out creative enough to pursue the arts or humanities. Alternatively, others turn away from the life sciences with awe for those who persevere, combined with a conviction of their own inferiority, concluding that the problem is themselves, that they simply aren't up to understanding material that is so sophisticated, so elaborate, so overwhelming.

It doesn't have to be this way. The study of life—human life not least—is genuinely lively, just as it is important. And it is conceptual more than rote. Nonetheless, you don't have to be a rocket scientist or even a biological scientist to enjoy a basic familiarity with today's biology. There is no reason why anyone capable of reading these words cannot understand what life is all about.

The philosopher Immanuel Kant once proposed this motto for the Enlightenment: "Sapere Aude" (Dare to Know). We offer it now. Dare to know that you are a biological animal, through and through. Dare to know, as well, what some of the major recent developments in life science are all about, and how they relate to that particular form of life known as *Homo sapiens*.

Please don't misunderstand, however. We will not transform you overnight into a top-notch molecular biologist, an insightful ecologist, or a pioneering neuroscientist. What we propose is to offer enough information, keeping it accurate and yet accessible, to enable every reader—regardless of background—to become bioliterate, able to understand issues likely to arise in the next few years, to follow attentively, to understand (not everything, but enough) so as to take part in the debates to come.

We will approach our subject from three different levels: close-up, at midrange, and from a distance, focusing in each case on three different components. First, the close-up view: we'll examine the crucial small stuff that underpins all biology (human and otherwise), notably genes and how they work, as well as how people have begun to work on their own DNA. Then, for the rest of our close-up perspective, we'll look at viruses and other tiny troublemakers, concluding with a tour through the building blocks of every body, the world of cells.

Next, we'll step back and examine human biology at a medium distance, the level of three different "systems," exploring the brain and behavior, sex and reproduction, and the ins and outs of food and energy. And finally, we'll get a wider perspective yet, from the revealing vantage points of ecology, evolution, and sociobiology.

In the interest of full disclosure, we must acknowledge some limitations. Although we have tried to be accurate, ours—like any mirror—has some distortion. Like a fun-house mirror, in some cases we magnify certain topics, while cutting off others. Our primer of bioliteracy, for example, does not reflect every part of the human creature. We have tried to identify areas of current mystification, as well as to anticipate issues that will be exciting and possibly troublesome in the years ahead. But no one can predict the future, especially not in science, which by definition is a way of looking into the unknown—far more than into a mirror.

When it comes to surveying the living world, most of us are too busy holding our own lives together to enter into a detailed study of all the panoramas open to us. To be sure, a little knowledge can be a dangerous thing. Better, then, to get a lot of knowledge. So if our abbreviated account whets your appetite, and you want more, we suggest additional readings at the back of the book, keyed to each chapter. It would be a great reward—to us, as well as to yourselves—if you outgrow the material in this book. The sooner the better.

Like Ray Bradbury's Martians, we can all profit by looking carefully at our own mammalian reflections. Like Immanuel Kant, we can all dare to know.

Acknowledgments

Isaac Newton once wrote that if he saw far, it is because he had stood on the shoulders of giants. We are not Isaac Newton, and moreover, we've been more concerned with seeing broadly than far. Nonetheless, our greatest debts are to those giants (and even some normal-sized folks) of biology and medical research whose findings and insights we have attempted to relate and explain. Fortunately, their shoulders are broad!

Ilona especially wants to thank her teachers at the University of Washington and at the University of California, San Diego, and David is particularly grateful to his students at the University of Washington. We also want to congratulate and thank . . . each other! Thus, despite the reality of "parent-offspring conflict" (described in the text), we confirmed what we always knew: that parent and offspring can indeed work together beautifully, creating a whole that is greater than the sum of its parts. Together, we are deeply grateful to our multifaceted editor, John Michel, who consistently displayed a necessary yet rare combination of humor, insight, and firmness, giving us a "Goldilocks" ideal of authorial freedom: not too much, not too little, but just right! Also project editor, Jane O'Neill, who did something remarkable: she made the production process so smooth as to be downright fun.

PART 1

UP CLOSE

In the summer of 1997 the *Martian Sojourner* wandered around the red planet, bumped into some rocks, and sent back a stream of information. Sadly, there were no sightings of live Martians. Let's imagine, however, that the tables were turned and intelligent Martians sent their own sojourner to planet Earth, to report back on its life-forms. Such an interplanetary gossip might be relatively unimpressed by the biological attributes of *Homo sapiens*, except, perhaps, for that species' oddly oversized cerebrum. Other species are bigger, stronger, faster. On the other hand, human beings would almost certainly attract more than their share of attention, if only because they occur in huge numbers (for a large mammal), occupy a remarkable range of environments, and, with such extrabiological achievements as culture and technology, have succeeded in making an immense impact, quite out of proportion to their biological uniqueness. The report back to Mars about life on planet Earth would likely have a lot to say about *Homo sapiens*.

We owe ourselves at least as much attention. We all share one of the world's great experiences: being human. And yet, we are generally strangers to ourselves. This is probably most true in the context of the very small, those aspects of human biology that function beyond the ability of the human eye to resolve. So, let's start our self-examination with some of the microscopic and submicroscopic aspects of that perfectly good mammal, ourselves.

KEY TO LIFE:
The ABCs of DNA

What do a patch of lichen struggling to survive on a dry boulder, a giant squid swimming two miles undersea, a grizzly bear munching blueberries in the Canadian Rockies, and you, a book-reading mammal, all have in common? The answer, of course, is life. And that crucial quality of aliveness, that shared something that makes the lichen, the squid, and you different from a rock, a button, or an automobile, is DNA. In a very real sense, DNA is the key to life. And also, to each particular life.

In the classic movie version of *The Wizard of Oz*, Bert Lahr asks "What makes the Hottentot so hot? What puts the ape in apricot? What have they got that I ain't got?" The answer, for Lahr's Cowardly Lion, is "courage." But for modern biologists, the key to specialness is more likely to be "DNA."

What do all mammals have in common? Most of us will answer: hair, warm-bloodedness, and lactation. But that's not all. All mammals share mammalian DNA, the genetic wherewithal that distinguishes us from birds, reptiles, or fish. And what makes you and me human is—you guessed it—human DNA. Human beings share about 90 percent of their DNA with the rest of the living world, and approximately 99 percent with our closest nonhuman relatives, the chimpanzees. Whatever it is that makes us uniquely human thus resides in that less than 1 percent of our DNA that we share only with other human beings. And what makes us *individually* unique—our personal claim to biological specialness—must therefore exist by virtue of even smaller snippets that we can call our own. Our DNA, our selves.

"If the property of complexity could somehow be transformed into visible brightness," wrote molecular biologist John Platt in *The Step to Man*, "the biological world would become a walking field of light compared to the physical world . . . an earthworm would be a beacon . . . human beings would stand out like blazing suns of complexity, flashing bursts of meaning to each other through the dull night of the physical world between."

Although it is possible to imagine alternative worlds in which life's complexity has developed around other classes of molecules, at this point, DNA and its allies have cornered the market on aliveness. It is DNA that determines why roses are red and violets blue, what it is that makes humans human, and hummingbirds hum. (Whether it also puts any ape in apricot is subject to debate.)

DNA is important, however, not only for its basic role in defining and creating life, but also as a star player in current and future events. If there is a single molecule that stands for all that is new and exciting—as well as confusing and troublesome—in modern biology, it is DNA. It is the closest we have to a master molecule, one that must be understood if we are to understand ourselves and the life around us, as well as some of the most dynamic frontiers of science.

In this chapter, we begin our close-up scrutiny of human biology by seeking to answer a few basic questions about DNA: What is it? What does it do? How does it do it? What have biologists been doing with it? And what are they likely to be doing in the future?

What Is DNA?

The Transforming Principle

It doesn't take a geneticist to know that like begets like. This is one of the basic rules of living things, so obvious that it can easily go unnoticed—and unappreciated. Salamanders give rise to other salamanders, bald eagles to other bald eagles, and human beings to other human beings. When a chicken egg hatches, out comes another chicken, not a condor. Moreover, different strains of chickens breed "true," just as tall parents are likely to have tall children, and blue-eyed parents, blue-eyed children. Why?

Something, it seems clear, is transmitted from parent to offspring, something that carries the basic information that says, in the case of chickens, "Make a bird, with weak wings, sharp nails, and a tendency to cluck." This something, which makes each generation a chip off the old block, is what is loosely referred to as "genes," and more precisely, as DNA. For a long time people thought that the hereditary material was carried in the blood: hence such expressions as "full-blooded," "blood relatives," "blood lines." Now we know better.

A major clue came from some simple yet perplexing experiments with bacteria that cause pneumonia. There are two different forms of *Streptococcus pneumoniae:* one makes smooth-looking colonies when grown in the laboratory, while the other produces a rough appearance. The two forms breed true, with smooth forms giving rise only to new, smooth colonies, and rough, only to rough. In 1928 Frederick Griffith made some puzzling observations. When he injected the smooth form of *Streptococcus pneumoniae* into rats, the animals died of pneumonia. When he injected the rough form, they did not. When he killed the bacteria by heating them before injecting them, neither form caused their hosts to die. So far, so good.

But here was the surprise: when Griffith injected rats with living rough bacteria (the kind that doesn't cause disease) along with *dead* smooth bacteria (the lethal kind), the rats died. Moreover, when Griffith took samples from these dead rats and "cultured" them in his laboratory (that is, grew the cells under artificial conditions), lo and behold, there were plenty of smooth colonies, as though they had risen from the dead. His conclusion? Something had been given off by the dead smooth bacteria that transformed the innocuous, rough form into killer smoothies. Griffith called it the "transforming principle."

Nearly twenty years later, three scientists, O. T. Avery, C. M. Macleod, and M. McCarty, were finally able to purify enough of this transforming principle to identify it as a long, chainlike chemical known as deoxyribonucleic acid, or DNA.

Still, some scientists were unconvinced that DNA was the molecule of heredity. DNA, after all, was a newcomer on the biologists' block; many believed that genetic information was carried by proteins, molecules that were relatively well known and, in a way, well loved.

But in 1952 a crucial experiment by Alfred Hershey and Martha Chase supplied strong additional evidence that DNA was indeed the crucial transforming principle. Hershey and Chase worked with a kind of virus known as bacteriophages ("bacteria eaters"). In particular, they worked with a strain that attacks the common intestinal bacteria *Escherichia coli*, taking over the metabolic machinery of each infected bacterium and turning it into a microminiaturized virus factory that produces hundreds more phage particles. Bacteriophages consist of just two kind of chemicals: protein and DNA. So when bacteria were infected, one of these chemicals was passed from phage to bacteria. Which one was the culprit, responsible for transforming an *E. coli* cell into lots of little phages?

To answer this question, Hershey and Chase took advantage of the fact that protein contains sulfur, which is not present in DNA, whereas DNA contains phosphorus, which is present only in very small quantities in protein. They incubated one batch of phage particles in a solution labeled with radioactive sulfur, and another with radioactive phosphorus. Then they exposed *E. coli* to each type. When the resulting mixtures were mashed up in a kitchen blender and then spun around in a centrifuge, the heavier parts (containing the *E. coli* cells plus whatever parts of the phage had entered them) were found at the bottom of the centrifuge test tubes while the lighter parts (containing those components of the phage that did not enter the cells) floated on the top.

The results? When *E. coli* cells were infected with phage whose *proteins* had been radioactively labeled, radioactivity was not found in the cells but only in the lightweight fluid; this showed that the protein had not entered the cells. On the other hand, when *E. coli* cells were infected with phage whose DNA had been radioactively labeled, radioactivity was found inside the cells. DNA had been caught red-handed, responsible for transforming normal *E. coli* into infected *E. coli* and ultimately into more bacteriophages.

Subsequent research has confirmed that DNA is the ultimate transformer: it makes the difference between life and nonlife.

The Structure of DNA

Let's examine the structure of DNA, not just because figuring it out was one of the great scientific detective stories of the twentieth century, but because its physical makeup is the key to how this molecule of heredity does its job.

When DNA is analyzed chemically, it is found to consist of three building blocks: a phosphate group, a type of sugar (deoxyribose), and a variety of so-called nitrogenous bases. These bases are the most interesting part. They come in four kinds: adenine, thymine, cytosine, and guanine (referred to as A, T, C, and G, respectively.)

Every cell of every animal and plant contains DNA. These DNAs differ somewhat from one species to the next, although in all cases this complicated chemical has been found in the nucleus of cells, never in the surrounding cytoplasm. In 1950, Erwin Chargaff made two important discoveries. First, he showed that the DNA of different species is different; specifically, it differs in the proportion of those nitrogenous bases, A, T, C, and G. This was crucial because for DNA to be the hereditary material, it could not very well be the same from one species to another. Second, Chargaff purified DNA from a number of sources and found that regardless of its origin, the amount of adenine (A) equaled that of thymine (T), and the amount of cytosine (C) always equaled that of guanine (G). This turned out to be portentous for revealing the structure of the DNA molecule.

Three years later, the now-famous double-helix structure of DNA was announced. Rosalind Franklin and M. H. F. Wilkins had bombarded DNA crystals with X rays, and by studying the scattering pattern that resulted, they obtained actual measurements of the spatial arrangement of the atoms comprising each molecule. It remained for two young researchers, American James Watson and Englishman Francis Crick, to put this information together and propose a structure for the enigmatic molecule. DNA, they showed, is like a ladder, whose structural sides are composed of the phosphate and sugar groups, and whose rungs are made up of nitrogenous bases, weakly joined together by so-called hydrogen bonds. These bases, furthermore, are organized so that A and T or C and G always pair up to make each rung. That's why the amounts of A and T are always equal, as with C and G. These base pairs are complementary; it is only when A is paired with T and C with G that a given rung of the DNA molecule is completed, and made structurally whole.

Finally, Watson and Crick suggested that the whole business is twisted, so that instead of a ladder, each DNA molecule is more like a spiral staircase.

Organization and Genes

Imagine a human DNA molecule, containing about 3 billion "nucleotides." (A nucleotide consists of a nitrogenous base—A, T, G, or C—combined with a sugar and a phosphate group; each nucleotide is a link in the spiral DNA chain.) If such a chain was arranged simply as a spiral staircase, it would be over a meter long; yet all those "rungs" manage to fit into a cell that is measured in microns (millionths of a meter). How does it do this? The answer lies in the way DNA is organized: wound around itself and some proteins in an exquisitely complicated structure.

First, DNA molecules do not normally stand naked to the world. Even within a cell, they are integrated into parts of physical structures inside the nucleus: the chromosomes. A typical bacterium such as *E. coli* has "only" about 5 million nucleotides and therefore has them organized as a single chromosome. Human beings, on the other hand, have a grand total of nucleotides that is about 700 times greater. These are distributed onto 46 chromosomes, each carrying about 50 to 250 million nucleotides (the notable exception is the Y chromosome, which confers maleness and has substantially less genetic information than its "sister" X chromosome). In any event, each of the other chromosomes would still be at least 1 centimeter long if it wasn't further organized into a substructure that is yet more complex and dense than the DNA molecule itself.

The enormous compaction that occurs, allowing all genetic information to be contained in a tiny package, occurs through the use of pebblelike helper proteins called histones. The long, stringy DNA molecules wrap themselves around nodules of these hugely helpful histones, rather like thread tightly wound on many tiny spools. The result is that a strand of DNA looks like a number of beads (the histones) on a string (the DNA between and around the histones). Beads and string are then further wound into a dense, complicated pattern, twisted around itself in various ways; this DNA packing must eventually be undone in order for it to exert its crucial biological role. Sometimes part of this unpacking doesn't occur, in which case that part remains biochemically and genetically inactive.

For all these twistings and turnings, however, the true importance of DNA is contained in its internal, linear sequence, the ordering of the nucleotide bases. Here lies all the information needed to create a functional being, provided the DNA gets a little help from its friends, as we will see. In general, more complex creatures need more DNA than their simpler counterparts. Mammals in general have more DNA than any fungus or bird and more than most insects, fish, and reptiles. It is easy, nonetheless, to wax indignant over the claim that a person is only 700 times more complicated than a bacterium.

However, there are some amphibians and plants that have more than 30 times the amount of DNA that a human has. One of the main reasons for this seeming discrepancy is that much of the DNA in a cell is junk, useless batches of meaningless jabberwocky (as in Lewis Carroll's "'Twas brillig and the slithey toves did gyre and gimble in the wabe. . . . "). It appears that the only useful parts of a DNA strand are those carrying the instructions needed to form a functional molecule called RNA, described below. Every such useful section of DNA is a gene. A gene, therefore, is not just a unit of heredity but also a packaging unit, by which we identify useful DNA as opposed to "slithey toves." To be "useful," the information contained in the nucleotide sequences of DNA must eventually direct the development and manufacture of something. Add up all those somethings, and you get a cell. Add up all those cells, and you get a porcupine, a porpoise, or a person.

What Does DNA Do?

Replication

What does a molecule of heredity have to do? First, it must be able to make accurate copies of itself, or replicate. This is why Gertrude Stein's observation "A rose is a rose is a rose" rings true, not only poetically but genetically, from generation to generation. For a rose to make a rose, which in turn makes yet more roses, rose DNA must create rose DNA, which in turn makes more rose DNA, every time each rose reproduces. In the biological world, DNA is unique in that it has the ability to direct its own replication.

But mere replication isn't enough. As the logicians might say, it is necessary but not sufficient. Living things, after all, are made up of pretty much the same atoms: carbon, hydrogen, nitrogen, phosphorus, sulfur, calcium, and so on. The immense difference between, say, roses and rhinoceroses is due to the way these atoms are put together to comprise the cells, tissues, and organs of individuals. And it is the job of heredity—of DNA—to organize this putting together and to keep each individual running smoothly. It is because of this organizational and maintenance role of DNA that a rose by any other name smells as sweet as it does. It is also why roses have sharp thorns and lovely petals. In short, it is what accounts for a rose being a rose, for its ineffable rosiness. And for a chicken's chickenness, and a human being's humanness.

This, then, is the other great task of DNA: to organize and maintain cells and bodies, making them what they are. This is done by
conveying information for the production of proteins. As we have
seen, proteins are not the hereditary material; but they are the initial
stuff upon which heredity acts. Proteins comprise the basic structural building blocks of living things; in addition (in their enzyme
form), they direct the various chemical reactions, such as digestion,
respiration, et cetera, by which life sustains itself.

First things first, however. How does DNA replicate itself?

Recall that the rungs of the spiral DNA ladder are made up of
paired halves: when A appears in one strand, it is always matched
with T in the other, and similarly T with A, G with C, and C with G.
The copying process relies on the fact that these complementary
pairings are highly specific: a chunk of DNA carrying G will match
up *only* with another chunk containing the complementary molecule, C. Once you know one half of a pair of bases (for example, A),
you know the other half (it must be T).

The first step in replication, then, is for DNA to uncoil and then
unzip, with the various A-T, T-A, G-C, and C-G bonds breaking
apart, right at the hyphen, which represents comparatively weak
hydrogen bonds. Each half of the old DNA molecule remains intact.
It then proceeds to reassemble itself because each of the nitrogenous
bases will pair up with only one type of partner; again, G with C, A
with T, and so forth. Extra supplies of all four bases are floating
about, needing only to be activated by particular enzymes, notably
one known as DNA polymerase. A polymer is any long molecule,
such as DNA, that is made up of many smaller component molecules,
all lined up in repeated sequence. The suffix *ase* indicates an enzyme,
which is a molecule (almost always a protein) that serves to speed up a
chemical process that is biologically important. DNA polymerase,
then, is an enzyme that helps DNA replicate by stringing unattached
nucleotides together to re-create a new, full-fledged DNA molecule,
using each half of the original DNA molecule as a template.

Through this process, a DNA molecule becomes two molecules,
with each half identically copied, based on the structural information (the sequence of the A, T, G, and C bases) present within itself.
Each strand—more precisely, each of the nitrogenous bases associated with each strand—provides the information needed for the manufacture of its complementary half.

So far so good. But simply making copies of itself is not good
enough. To be useful, and to count as a gene and a relevant unit of

life and of heredity, DNA must in a sense step outside of itself, get beyond its own selfish replication, and be pertinent to the affairs of the cell of which it is part. It does this by directing the production of proteins.

DNA's nitrogenous bases in their linear sequence do double duty: they carry not only the information needed for DNA to replicate itself but also precise instructions about how various proteins are to be put together. (These proteins, in turn, are the cellular handymen that go on to perform many of the functions of being alive.)

Here's how it's done.

Transcription

Each DNA molecule is a vast source of information, a kind of encyclopedic cookbook, chock full of recipes and instructions for the making of proteins. But these hereditary volumes, comprising the genetic library of each individual, do not circulate. Rather, they remain inside the central repository of each cell: the nucleus. (This is not surprising, since each one is precious and could be lost or damaged if it circulated freely; reference books that are especially valuable usually cannot be taken out from public libraries, either.) Somehow, then, the information contained within the DNA—the linear sequence of its nitrogenous bases—must be made available. In technical terms, it must be "transcribed" from its master repository, the DNA molecule, into a form that allows it to be conveyed to the cell's protein-building machinery and then deciphered. Enter DNA's henchman: ribonucleic acid, or RNA.

There are several kinds of RNA. Of special interest for the process of transcription is messenger RNA, known as mRNA. As its name suggests, mRNA's role is to carry a message: namely, the order in which the various bases are arranged to form the rungs of the DNA ladder. (This, as we shall see, is then "translated" into the order of amino acids making up each protein.) For the sequence of nitrogenous bases to be transcribed from DNA to mRNA, the DNA molecule must once again unzip, exposing those bases lined up along each chain. Recall that when DNA replicates itself, these chemicals serve as templates for each side of the molecule to reproduce its complementary half. During transcription, however, the DNA bases line up instead with complementary chemical chunks, which, when put together, comprise a slightly different molecule. This chemical, mRNA, is similar to DNA in that it consists of a phosphate group and a sugar (ribose instead of deoxyribose) along its backbone, as well as three of the four bases found in DNA: A, G, and C. However, a slightly different one, uracil (U), is present in place of DNA's thymine (T). When the DNA message is transcribed onto mRNA, the base pairings line up just as when DNA replicates, except that U, instead of T, bonds with A.

Under the influence of the enzyme RNA polymerase, messenger RNA is now assembled from the unzipped linear spines of the DNA molecule. For a brief time, therefore, a hybrid molecule is produced, one chain composed of DNA and the other of the newly formed mRNA. In a sense, each molecule of mRNA is the photographic negative of one strand of DNA: if part of that DNA consisted, for example, of A followed in turn by T, G, C, T, and another A, then the complementary strand of mRNA would be U followed by A, C, G, A, and then another U. Given a strand of mRNA, it is thus possible to piece together the DNA that gave rise to it; the information present in the parent DNA has been transcribed, via its complementary bases, onto mRNA. The DNA then zips itself back together, and the mRNA—a single-stranded molecule (unlike DNA)—moves out of the nucleus, carrying crucial chemical information in its very structure. As Marshall McLuhan might have said, the molecule *is* the message.

Now this message must be "translated" from the language of nitrogenous-base sequences into proteins; that is, from genetic, nucleic-acid code into structural, proteinaceous reality.

Translation

Faithful messenger that it is, mRNA makes its way out of the nucleus where it is clasped by two tiny halves of an odd little cell structure known as a *ribosome*.

To understand how the information encoded in mRNA is then translated into molecules of protein, we must first understand something about proteins. They are made up of smaller molecules, amino acids, strung together just as DNA is made up of smaller molecules, the nitrogenous bases, strung together. The proteins that result are not merely simple chains, however; rather, they are curled and twisted in ways that are characteristic of each protein (and which also recall the way DNA itself is curled and twisted to comprise a chromosome). Changing this aspect of their structure results in "denaturing" the protein, as when egg white is whipped or cooked. For our purpose, however, the linear or "primary" structure of the protein is what counts. Most often, once the linear sequence of its amino acids is in place, the curling and twisting that are necessary for the protein to do its job occur pretty much automatically.

Proteins are composed of only twenty different amino acids; since there can be hundreds of amino acids per protein, however, and they can be arranged in virtually any sequence, the number of possible proteins is almost infinite. In any event, specifying a protein boils down to specifying a particular sequence of amino acids, relatively simple building-block chemicals with names such as leucine, valine, lysine, alanine, et cetera.

Next, a little bit of arithmetic. The simplest way for DNA to specify a protein would be for each of its nitrogenous bases, transcribed into its "negative image" as mRNA, to stand for a different amino acid. But since there are only four different bases, such a singlet code would only allow for a total of four different amino acids, instead of the necessary twenty. What about a doublet code: two bases per amino acid? In that case, AA might specify, say, leucine; AT, valine; and so forth. But even then there are only sixteen possible combinations. A triplet code fits the bill, potentially specifying sixty-four different possibilities (forty-four more than are needed). Accordingly, it takes three nitrogenous bases, strung in a line, to specify one amino acid.

After much painstaking work, the entire code has in fact been deciphered. We now know, for example, that the triplet AAA (three adenines in a row) results eventually in a protein containing the amino acid phenylalanine; similarly, AAT codes for leucine, CCG for glycine, among others. Thus, if a sequence of bases along a DNA chain reads AAAAATCCG, then after it is transcribed onto mRNA, it will code for phenylalanine, leucine, and glycine, in that order. It is as though the instructions for protein synthesis are written in a string of three-letter words, for example, THE FAT CAT ATE THE BIG RED RAT, in which every word stands for a different amino acid (although it would appear more like THEFAT-CATATETHEBIGREDRAT). Each triplet of nucleotides is a "codon," just as each triplet of letters can make up a word.

Here is a simple yet weighty fact: the code is universal; that is, AAA specifies phenylalanine, and AAT, leucine, whether in pineapples, parrots, or people. The difference between different kinds of living things lies in which amino acids are called upon and in what order. This universality of the genetic code is doubly significant. First, it speaks to the common evolutionary origin of all forms of life. Second—as we will see later—it provides the opportunity for molecular biologists to modify genes from one species and then introduce them into another. And, more important, make them function within their new hosts.

The code is the key. But how is the code made real? Once the DNA's long string of base sequences has been transcribed onto mirror-image mRNA, how does a cell's machinery translate these three-letter "words," or codons, into a protein?

Recall that mRNA isn't an exact duplicate of its template DNA. Rather, it is made up of precisely structured, complementary bases: for every T in the original DNA, there is an A in the mRNA; for every C in the original DNA, there is a G in the mRNA. For every triplet in the original DNA, there is a particular triplet codon in the mRNA.

We turn next to another form of RNA, known as transfer RNA, or tRNA. The role of tRNA is to transfer the various amino acids floating about near the ribosomes onto the mRNA, whereupon the amino acids link up, forming a protein. The trick, of course, is to link up in the correct sequence: "correct" in each case means the sequence specified by the triplets constituting the mRNA, which in turn got *its* sequence from the DNA. (In the early twentieth century, it was said that among upper-crust, so-called Boston Brahmins, the Cabots spoke only to the Lodges, and the Lodges spoke only to God. Transfer RNA speaks only to mRNA, and mRNA speaks only to DNA.)

There are many different kinds of tRNA, each having a free end that binds to a particular amino acid, and also a particular sequence made up of three nitrogenous bases (AGA, GAU, etc.). The free end of each tRNA molecule first grabs its specified amino acid. The tRNA's triplet pattern—called an anticodon—matches up with its complementary three-letter sequence on the mRNA codon. This places the amino acids in the specified order. For example, let's say that a particular sequence of DNA is GCT. This would eventually be transcribed onto a section of mRNA as CGA. This section of mRNA, in turn, would bond to a chunk of tRNA that has the complementary sequence GCU, and that also has a particular affinity for only one amino acid, in this case, arginine. The end result is that DNA's original GCT would be *translated* into arginine. That arginine will then be ready to hook up with the next amino acid in line, which will have been obligingly transferred into position by its own particular tRNA, depending on the next three bases, which in turn depend on what was specified by the next triplet in the original DNA master copy.

You might want to think of tRNA as an interpreter. An interpreter's job is to recognize the words of one language and find their equivalent in another. In molecular genetics, the task is to translate the triplet code of nucleic acids into a different language, that of amino acids. One triplet code, or codon, "means" one amino acid. Just as the words of a translator must be strung together in linear sequence to make an understandable sentence, the amino acids brought to the mRNA and their associated ribosomes and by tRNA must also be connected.

Each amino acid is held by its tRNA only via comparatively weak hydrogen bonds, like those joining the nitrogenous bases in DNA. By contrast, because of its atomic structure, each amino acid is able to form a relatively powerful *peptide bond* with the next amino acid in line. The amino acids therefore abandon their temporary tRNA transfer vehicle and link up with the next amino acid in line, following the sequence specified by mRNA (and originally, lest you forget, by DNA), to form a protein. Which is the point of the whole affair.

There is a lot more to it, of course. Another form of RNA, for example, occurs in the ribosomes, and is involved in the actual construction of protein chains. There are codons that serve as punctuation marks, indicating when a particular sequence of DNA should be read, and when terminated. There are chemical codes that effectively inhibit these various processes, depending on what else is going on at the time, and whether other chemicals are present. There is an elaborate array of enzymes that control the fine tuning of replication, transcription, and translation, including an entire microminimolecular universe of gene regulation, whereby certain genes enhance or reduce the transcription rate of other genes, depending on, among other things, the kind of DNA packing that is encountered.

So-called gene expression is influenced, as well, by the presence—or absence—of many things, including hormones. For example, both male and female cells have the genetic information necessary to produce hair. When enough testosterone is present, cells in the human chin erupt in dense, hirsute sprouts; if not, then they don't. Provide enough testosterone and you can get bearded ladies, not because their DNA has been changed but because the genes in question, present in the smoothest-skinned young debutante, have been "turned on," chemically persuaded to express themselves.

Things get still more complex: in the cells of all higher organisms, messenger RNA undergoes what is called "processing" before it leaves the nucleus. Interspersed among the various meaningful base sequences of DNA are meaningless nonsense syllables, lots of them. So before the mRNA is translated into proteins, its nonsensical parts are enzymatically snipped out and the meaningful parts spliced together.

This is just a hint of what goes on in the wonderful world of DNA: *your* world, for you are enmeshed, enabled, or exalted (take your pick) in the immortal coils of those nucleic acids. DNA research is currently proceeding at an extraordinary pace. It's almost impossible to keep up with all the new developments in molecular biology. Nonetheless, to help you sort out some of the more recent advances and how they will (and already do) impact your life, let us sample a few of the latest and greatest stories in molecular biotechnology.

Biotechnology

A Basic Tool Kit

A typical cell in the human body is very tiny indeed, approximately .0025 centimeters across. Considering that only an extremely small fraction of that space is taken up by DNA, and moreover, that less than 1 percent of that DNA appears to contain information useful for the human body, it is amazing that human beings, regardless of their ingenuity, are able to manipulate individual genes at all. Nonetheless, many of the techniques used by molecular biologists do just that, with highly visible results.

One of the most important tools used by a molecular biologist is a *gel*, a thin layer of gooey substance used to separate strands of DNA by size. Large amounts of DNA are injected into little depressions in the gel, which keep the different samples in place. The gel and its DNA are then subjected to an electric field. DNA carries a distinct negative charge, so it tries to flow toward the positive side of the field. It is stopped, however, by the gel.

However, as their name suggests, gels are gelatinous. On the microscopic level, they contain millions of tiny holes, which act like a maze or obstacle course that the DNA must somehow navigate. Small pieces of DNA find it easy to snake through this maze, moving faster and further through the gel than the large, gangly chunks, which get hung up and wound about as they bumble along.

Researchers then remove the gel from its electric field and either expose it to X ray or dye the DNA and examine it under ultraviolet light. Remember that not all DNA is created equal or the same: the outcome of gel separation is that a glob of DNA, actually a complex gumbo of many different lengths and compositions, can be separated into its various subtypes. Depending on the makeup and density of the gel, this technique can distinguish between DNA strands differing in length by thousands of nucleotides or by a single nucleotide (just a few atoms!).

Molecular technology most often relies upon cutting and pasting together different DNA strands, visualizing the results via gels, and then, in many cases, inserting the newly created DNA into cells. To do this cutting and pasting, scientists use different types of naturally occurring enzymes as their scissors and glue. The scissors are called restriction enzymes; the glue is known as ligase.

Restriction enzymes were originally isolated from bacteria, where they serve as defense mechanisms that cut up unwanted, foreign DNA. There are thousands of restriction enzymes. Each cuts at a different, specific "recognition sequence," ranging from about four to eight base pairs (nucleotides) in length. It is a sophisticated bit of scissoring. Not only is the double-stranded DNA snipped in two, perpendicular to its long axis, but in most cases the cut is also staggered a bit, so that a few raggedy, unpaired bases are left at each cut end. The resulting DNA fragments have not only split ends, but "sticky ends" as well, so called because they are eager to pair with their complements: other sticky ends, produced by the cutting action of the same restriction enzyme on other DNA molecules.

This enzymatically induced Velcro effect is very useful. Start with DNA from two different sources, one containing a gene that is of particular interest, the other containing, perhaps, a bit of virus DNA or a small section of bacterial DNA known as a plasmid. (Both viruses and plasmids are possible "vectors" that can introduce bits of DNA from one cell into another, just as mosquitoes serve as vectors, introducing the malaria parasite from a victim to someone not yet infected.) Add a dollop of restriction enzyme, and stir. DNA from both the donors and the vectors are sliced and diced. Sometimes sticky ends from the target DNA simply rejoin; sometimes vector sticky ends do the same. But it also happens that a sticky end from a piece of target DNA meets up with a bit of complementary sticky end from the vector. The result is a pasted-together custom collage of DNA segments, called recombinant DNA because it is produced by recombining DNA from different sources (in this case, a target and a vector).

When a pair of sticky ends first meet, it is not quite love at first sight. Like a new marriage, the reconstituted chemical bonds are a bit shaky. This is where the ligases come in. They do the opposite of restriction enzymes, acting as a kind of nucleic-acid superglue and fusing the newly combined DNA fragments.

The resulting recombinant DNA is now ready to be ferried into another creature, where, if all goes well, it will take and produce a *transgenic* organism, carrying one or more of the donor genes (and also some vector genes). Molecular biologists often add other genes to the vector or use vectors that already carry certain genes, such as "markers" that convey antibiotic resistance. This is a convenient way of identifying those cells carrying the original, target gene. Once this target gene is combined with resistance to antibiotic, the cells can be grown in a medium containing this antibiotic; nonresistant cells die, leaving those that are resistant and that also carry the target gene.

Here is another nifty trick employed by genetic engineers. Let's say you want to obtain the gene responsible for manufacturing a particular liver enzyme. First of all, in this case you wouldn't concern yourself with skin or muscle cells, because although all human cells have a complete set of human DNA, only a very small number of the genes on board are turned on and functioning in any given cell type: only red blood cells make hemoglobin, only certain pancreatic cells make insulin, and so forth. So, you would grab a small chunk of liver cells, which are busily making the liver enzyme you want.

But how to isolate the special segment of DNA that is responsible? (After all, even liver cells have all sorts of other tasks; they don't only make liver enzymes.) The trick is to obtain some of the messenger RNA that is responsible for translating the enzymes in question—this is relatively easy to do—and then add an obliging chemical known as reverse transcriptase. Recall that transcription is the process whereby DNA transcribes its base-pair code onto messenger RNA. Reverse transcriptase has the useful ability to reverse this process: given mRNA, it directs the assembly of DNA! As a result, by intervening chemically and forcing the process to run backward, it is possible to synthesize a replica of the original, liver enzyme–specifying DNA. A major benefit is that this freshly synthesized DNA is composed entirely of functional material, free of the large amount of nonsense DNA that would otherwise stymie the biochemical machinery of the new cells into which the target genes are to be transplanted. (In the next chapter, we will encounter a considerably more malign side of reverse transcriptase: it is a key component of the AIDS virus's molecular machinery.)

For any of the above procedures to be successful, it is necessary to start with enough DNA, certainly more than one copy of each strand. There must be enough strands on hand so that when transferring material from one container to another, for example, there is plenty left over to begin cutting and pasting. It is not yet possible to put DNA in a copying machine and make immediate duplicates. So, it is usually necessary to simulate, in the laboratory, one of those things that DNA does so well inside a cell: making copies of itself. Such DNA amplification is most effectively done via an important little trick known as the polymerase chain reaction, or PCR.

Here's how PCR works. First, the DNA is heated to separate the two halves of each double-stranded molecule. Then it is attached to primers, which are little single-stranded pieces of DNA that match the sequence surrounding the region or piece of DNA to be collected. These primers are then extended so that they replicate the strand in question, in a procedure very much like normal DNA replication. This is the end of the first cycle: the DNA has been doubled. The newly formed double-stranded DNA molecules are then heated again, so that the strands separate, and new primers attach to the DNA, allowing it to replicate once again. Then the whole business starts over. With every cycle, the amount of DNA doubles, adding up to a huge amount in a short time. Starting with a single DNA molecule, for example, it is possible to obtain 100 billion copies in just a few hours.

This technique has already been used to amplify bits of DNA obtained from forty-thousand-year-old woolly mammoths frozen into the Siberian ice, bits of brain tissue from a five-thousand-year-old Egyptian mummy, and to identify, with certainty, the disputed remains of Czar Nicholas II, killed by the Bolsheviks following the Russian Revolution. It enabled the Defense Department to identify the remains of the Vietnam War's "Unknown Soldier" as Michael Blassie, and presumably provided key evidence as to Bill Clinton's dalliance with Monica Lewinsky, identifying—or threatening to identify—the source of the fabled stain on Ms. Lewinsky's infamous blue dress. If some mad scientist were to attempt in reality what the movie *Jurassic Park* portrayed in fiction, reconstituting living dinosaurs from tiny specks of seventy-million-year-old *Tyrannosaurus rex* blood, he would also likely have recourse to PCR (which might then stand for "plenty of cause for regret").

In the fairy tale Rumpelstiltskin, a miller's daughter is required to spin straw into gold. Of course, she cannot do so, except through the help of the magical but evil dwarf, whose assistance is purchased at a fearsome cost: the young woman's promise to surrender her future child. By virtue of the polymerase chain reaction, molecular biologists are able to spin a few strands of DNA into a mountain of golden possibilities . . . or a heap of trouble.

DNA Sequencing and the Human Genome Project

The uniqueness of the individual, although a cherished concept in Western morality, is not notably evident in most respects. When you come right down to it, there isn't really very much that each of us can, with absolute confidence, claim as our own. But every person has a genotype, a private, personalized claim to biological specialness. (The only exceptions are identical twins.) Just as this book is unique, composed of a particular combination of words, each of which is made up of individual letters, organized in particular ways so as to convey particular meanings, so each person is composed of a combination of genes, made up of nucleotide bases organized in a particular way. In theory, as expressed in the famous metaphor of the infinite number of monkeys banging away on an infinite number of typewriters, it is possible that another book, identical to this one, has already been published somewhere or will be in the future. In practice, however, we can say that the probability of this happening is essentially zero. Similarly, there is essentially no chance that two genetically identical people have ever or will ever be created by chance alone.

In creating this book—whatever its uniqueness—we have drawn from a limited number of building blocks, the mere twenty-six letters of the alphabet, preorganized into the tens of thousands of words constituting the English language. If the words of this book and the particular way they are strung together are analogous to its genotype, then the much larger well of verbal possibilities of which this book is just one example would be analogous to the human genome. It is, in a sense, the *Oxford English Dictionary* of human DNA.

The "human genome" is the full set of DNA present in every cell of every human body. It is unique not to an individual but to the species *Homo sapiens*. As we have seen, it is huge and immensely information-rich. Inside this enormous data bank can be found the blueprint, or recipe, for every structure and every function in the human body. Each organ, hormone, and cellular feature, in fact every molecule and the way in which that molecule is organized in respect to the other molecules comprising the body, is directed by the human genome. Decoding the details of this monolith has the been the focus of an enterprise that is nearly as immense as its subject: the Human Genome Project. Its goal is to catalog every base pair of human DNA (as well as those of a few other scientifically useful organisms such as bacteria and mice), so that researchers will be able to use this information for in-depth study of our innermost workings. The result will be a "consensus" human genome, revealing the most common DNA patterns, rather than the precise details of each person's genetic individuality.

Individual uniqueness—at the genetic level—is expressed by small base substitutions captured in the acronym SNP (pronounced "snip"), single nucleotide polymorphism. *Polymorphism*, from the Greek for "many bodies," refers to diversity, and each DNA base (adenine, thymine, et cetera) is a nucleotide. SNPs, therefore, are the peculiar and personalized variations that make you genetically distinctive from everyone else. Much time and effort is being expended in the search for SNPs and in providing low-cost techniques by which each individual's SNPs can be identified. One consequence could be the spawning of a new industry, pharmaceutical genomics, which might eventually produce personalized drugs, tailor-made to each person's genetic peculiarities and thus less likely to generate harmful side effects.

The basic technique used by researchers in the Human Genome Project is sequencing, which combines PCR with gel-based separation techniques. After the cells are collected, a complementary DNA (cDNA) library is created, generally using reverse transcriptase as described above. This library contains only DNA that would otherwise result in the production of useful proteins; that is, with junk DNA cut out. It is then chopped up by restriction enzymes into smaller pieces and attached to a longer segment of DNA—a vector—that can be kept active inside a bacterial or yeast cell. When it is time to sequence this DNA, the fragments can be isolated again from the vector.

Sequencing is achieved by setting up a PCR reaction with a small amount of the target DNA. In order for PCR to take place, however, the DNA strands must be able to make themselves longer once the primers have attached. Each nucleotide has attachment points on either side, enabling it to form a long chain, similar to people holding hands. A certain number of nucleotides with only one "hand" are added to the DNA mixture. When DNA polymerase tries to make a long strand of DNA, therefore, occasionally it will incorporate some of these "one-handed" nucleotides into the sequence, causing the chain to stop. Without an attachment point for the next nucleotide, the chain cannot grow.

In an actual sequencing routine, the DNA mixture is split into four groups; a different one-handed nucleotide is added to each group, so that in one there is some one-handed adenosine, in another, one-handed guanine, and so forth. In a long sequence and a reaction containing lots of different copies of the DNA, this "chain termination" happens randomly, at each of the different kinds of nucleotides, creating different-sized fragments in each group. When the reaction is finished and analyzed, these fragments can be visualized and the order of the nucleotides determined by the different sizes of the DNA strands that develop.

This procedure can now be automated, so that a computer spits out the DNA sequences automatically. There is such a huge amount to be sequenced, however, that the Human Genome Project is taking a long time, as well as a lot of money. As of fall 1997, scientists had finished sequencing 2 percent of the human genome, in addition to the entire genome of yeast, the molecular biologist's workhorse bacteria *E. coli*, and some other microbes. Then, in 1998, the world's largest manufacturer of automated DNA-sequencing machines announced that it was forming a new company, a for-profit venture that would, they claim, complete 99 percent of the task within three years and at a cost of $300 million, instead of the originally estimated $3 billion. The effort will rely on a new generation of machinery, more automated and capable of working twenty-four hours a day. It is a controversial effort, both scientifically and ethically; the ethical questions revolve primarily around whether private enterprise ought to patent and control genetic information, rather than whether such information ought to be ferreted out at all.

In any event, once the project is finished, researchers will have an enormous database of all the genes in a human body, from which it should be possible to uncover more of our own best-kept secrets—so long as they do not automatically become closely guarded industrial secrets.

DNA Fingerprinting

Let's hear it for spin-offs. They are unintended side effects of something else, and not uncommonly, they wind up being better than whatever they spun off from. One of the more socially useful and immediate spin-offs of the Human Genome Project is DNA fingerprinting. Because each person is different, unless he or she is an identical twin, it follows that each person's DNA is different. Given that humans share about 99 percent of their DNA with chimpanzees, however, any two people are guaranteed to be very close, genetically, to each other. But "close" only counts in horseshoes and hand grenades. When dealing with 100,000 or so genes and literally billions of nucleotides, even a 1 percent difference becomes a large number indeed. This provides an accurate way to tell people apart. Each person's DNA is unique, just like a fingerprint, only more so.

Surprisingly, perhaps, the most useful arena in which to search for one's distinctive DNA fingerprint is in the junk DNA, not in actual genes themselves. This is because genetic junk is really a kind of innocuous nucleic parasite, not related to the vital machinery that makes a person tick, so there is little biological pressure to eliminate changes that arise. Analogously, consider a population of race cars, each fine-tuned and carefully engineered to handle precisely and accelerate rapidly. It is likely that in their shared striving for high performance, each of these autos would have an efficient carburetor, a well-lubricated chassis, excellent tires, and so forth. In fact, there probably wouldn't be very much difference among the vehicles.

By contrast, think about those cars that are used by the rest of us. One may have a rabbit's foot hanging from the rearview mirror, another a photo affixed to the dashboard, another a little streamer tied to the radio antenna. Since such idiosyncratic modifications don't really affect a normal car's performance, and since it wouldn't really matter even if it did, there is much more opportunity for variation to arise, and to persist once it does.

In the case of junk DNA, a common pattern is for a small sequence to repeat itself within a functional stretch of DNA. The number of repeats varies from person to person and is for the most part inherited. Such repeating sequences are a kind of genetic baggage, but they don't seem to have any effect on health or personal attributes. As a result of these different-sized repeats, when the DNA from different people is cut with restriction enzymes, different patterns will be revealed on a gel, depending on how many repeats are present. The more repeats, the longer the junk fragments. Analysis of these differences is called restriction fragment length polymorphism, or RFLP (pronounced "RIF-lip"; *polymorphism* means "many shapes").

For a sample RFLP analysis, lets say that part of X's DNA is GCATGAATTCATATATATGAATTCTCGA, and Y's is GCATGAATTCATATGAATTCTCGA. The only difference between these two strands is that X contains two extra AT repeats. But of course, you, clever riflipping genetic sleuth that you are, don't know this yet. To perform the analysis, you cut both segments of DNA using a restriction enzyme that recognizes the sequence GAATTC and cuts between the G and the A. After being cut, person X's fragment would be AATTCATATATATG (fourteen base pairs), while Y's fragment would be AATTCATATG (ten base pairs). The difference in length between these two strands would show up easily when run on a gel, providing a way to distinguish the two people.

In a real case, numerous genetic fragments would be tested; the more that are run, the more accurate the outcome. The result is a molecular mug shot that looks remarkably like a supermarket bar code, with the number of lines equaling the number of DNA fragments, each one indicating the presence of pieces of a given length, and the thickness of these lines reflecting the abundance of each fragment.

DNA fingerprinting has many uses, most famously in criminal cases. Let us imagine that a crime is committed, and the criminal leaves behind a drop of blood, a hair, or a bit of semen following a rape. Investigators can collect this evidence even months after the crime was committed. Using PCR, they multiply the existing DNA. When this is cut by a restriction enzyme and the results analyzed, the DNA fingerprint can be compared with that of the suspect's. In a real case, the length between the fragments would probably be much longer than that illustrated above, and there would be many other fragments formed as well. A person's DNA fingerprint, therefore, can be an extraordinarily accurate method for pointing to a culprit. For forensics workers, some of the most satisfying cases are so-called cold hits, in which a perpetrator is collared solely on the basis of DNA evidence.

There are some drawbacks, however. The number of repeats in a DNA strand is usually inherited, meaning that a person's relatives may have similar DNA fingerprints, so there is at least the possibility that close relatives may be mistaken for each other. On the other hand, detailed DNA fingerprinting—involving a sufficiently large number of fragments—can reliably identify individuals to a confidence level of greater than one in a hundred million, far surpassing the reliability of most visual identifications. DNA fingerprinting is not perfect; it is, however, the most precise means of identification yet devised.

But this optimistic assessment assumes the impossible: that human error has been eliminated. The most common problem in using RFLP analysis in criminal cases is that the samples can be contaminated. PCR is especially sensitive to this problem because it can make such a large amount of DNA from a very small starting sample. This is why during the O. J. Simpson trial so much attention was focused on the evidence-gathering procedures of the Los Angeles Police Department and the care and cleanliness of the forensics lab where the DNA was analyzed. Since DNA evidence can be so damning, the danger exists that a jury may be persuaded by evidence that is fundamentally flawed, not in its science but via human error.

Done correctly, however, there is no question that DNA fingerprinting is wonderfully accurate. DNA fingerprinting also has another bow in its quiver: because DNA is not only a personalized biochemical identity card but also the molecule of heredity, it can be called upon to help determine paternity. (Maternity, too, although for obvious reasons, this is less likely to be in doubt.) If RFLP analysis is done on the mother, the child, and the presumed father, the child's DNA fingerprint should in various ways be a combination of the mother's and the father's. Such testing is a distinct advance over blood-type analysis, which is a much blunter instrument. Using the blood types A, B, AB, and O, even factoring in the options of Rh positive or negative, the outcome is far less precise, simply because there are too few categories to establish certainty: in such cases it can be possible to prove the negative—that a given man could *not* have been the father of an identified child—but it is impossible to demonstrate positively whether he actually *is* the father.

In at least one famous lawsuit, even this evidence was disregarded by a scientifically uninformed jury: Charlie Chaplin was convicted in a paternity suit, despite the fact that his blood type proved that he could not have been the father! With the precise identification offered by DNA fingerprinting, we can not only finger a culprit, but also prove innocence: for example, when a man accused of rape is shown to have DNA that does not match the semen found in a victim.

The inheritance of DNA also allows RFLP analysis to be used for diagnosis of some genetic diseases. If a disease is found to be consistently accompanied by a certain length of DNA fragment, people can be tested before they show any symptoms of the disease, so as to identify those who might be at risk. But this, too, can be a two-edged sword. Aside from the possibility of error, there arises the fascinating question of whether certain things are best left unknown, if not for all time, at least until adequate ethical guidelines have been established. For example, what about the morality of insurance companies denying coverage to someone whose genotype suggests that he is at 50 percent risk of developing a serious disease in the near future? And what about the personal consequences of having such information? As a general rule, we believe that more knowledge is better than less, and that in the long run the dilemmas caused by newly acquired genetic information are best resolved by obtaining yet more genetic information, not by ignoring it or sweeping it under the rug. (As has been suggested in a different context, only rarely does this solve a problem; more often, it just makes for a lumpy rug.)

Genetic Engineering

In the early 1980s Richard Palmiter and R. L. Brinster injected some raw, marked DNA into a fertilized mouse egg. Eventually the embryo developed, was born, and produced its own offspring. The first "transgenic" mouse had been created, opening a can of worms and flowers for the rest of us. Genetically engineered plants, animals, and insects are almost commonplace now, causing a great deal of consternation among some, who worry that "playing God" could get out of hand, and excitement among others, who see the possible good that might come of it. Before we briefly explore these alternatives, let us examine what actually can be done with current technology, and what might be done in the future.

A transgenic organism is any animal or plant containing one or more genetic stowaways, genes that have been artificially inserted into its personal freight of DNA. As we will discuss further when we turn to sex and reproduction, a single cell—the fertilized egg, or zygote—contains all the information necessary to make a fully functional being. When new genes are injected into such an egg, at least some of the time they are incorporated into the genotype and become part of the new, developing organism.

One way to create transgenic animals, therefore, is to inject genes directly into fertilized eggs. The resulting animal is not quite a Frankenstein's monster, who was, after all, a medley of bits and pieces scavenged from the local graveyard. The transgenic "creature" generally looks very much like its nonaltered fellows, because only a small amount of its genetic background has been artificially substituted. Moreover, unlike the Victor Frankenstein of Mary Shelley's imagination, genetic engineers are not creating life out of ghastly parts of criminal corpses. But by introducing genes where they would not otherwise occur, the molecular biologists of today and tomorrow are indeed fulfilling the prophetic subtitle of Shelley's famous novel: "The Modern Prometheus."

In Greek mythology, Prometheus brought fire to mankind. Genetic engineers bring new DNA. Prometheus was punished by the gods for his transgressions. Genetic engineers are feted, albeit warily, for their accomplishments.

The gene that is injected into an egg can be of many different types. Most often it is a "gain-of-function" gene, which provides some sort of enhancement to the developing organism. It may augment a trait already present, for example, adding growth hormone so that the animal gets larger, quicker. Or it may introduce something completely different, such as fluorescence from a firefly, so that the animal (or plant), originally no more luminous than the rest of us, glows. In either case, just one gene is typically altered, sometimes in a way that severely affects the organism, perhaps even causing its death. More often, however, the added gene makes a very subtle or perhaps not even noticeable change in the structure or physiology of the recipient.

On the other hand, the "trans genes" (artificial genes added by a molecular biologist) can also be "loss of function" constructs. When a modern biologist speaks admiringly about a real "knockout," don't expect the object of such attention to be Hollywood material. In today's genetics labs, a knockout is an animal or plant that has had a particular gene removed, or knocked out.

Knockouts are created by first manipulating genes outside the cell so a gene is replaced by another DNA sequence that lacks the specific action of the original (typically, failing to specify a particular enzyme necessary for some physiological process). When this new set of genes is then injected into cells and allowed to come into contact with the DNA already present, in a very small number of cases these knockout genes will replace the functional genes already there. The next step is to identify cells in which this substitution has occurred. These cells can then be injected, for example, into an early-stage mouse embryo.

The new knockout cells will eventually divide and form part of the developing organism in the same way that other, nonmanipulated cells normally function. When the unsuspecting host grows up, some of its cells will be normal, while some will contain the newly introduced knockout genes. If these genes happen to have lodged in the reproductive system of the animal, some of its offspring, in turn, will completely lack whatever gene was knocked out. Knockout technology takes time, patience, and luck. Only a small percentage of knockout genes enter into the target cells; only a small percentage of *them* succeed in becoming part of the new host; only a small percentage of *them* end up in the host's gonads; and only some of *them* find their way to the next generation. Nonetheless, knockout technology, when it works, is a knockout indeed: a powerful procedure enabling researchers to see what happens when an organism develops without a specific gene.

Much of biological science is concerned with finding out what specific, identified genes actually do. Knockout creatures often serve as the final test of a gene's proposed role, because if such a gene is vitally important for something, then that "something" should be impaired when the gene is removed. Thus, knockout mice are currently available as models for the study of such genetic diseases as cystic fibrosis, sickle-cell anemia, and certain forms of atherosclerosis. Interestingly, however, things are not always so simple: the body has marvelous ways of compensating for missing genes, often by harnessing others into a new role.

But what is it, specifically, that genetic engineers seek to accomplish? For some, it is science for its own sake, pushing things to their limits just to find those limits, and to better understand how DNA works—like Icarus flying not so much to reach the sun as to see how high he can get. In most cases, however, the goals are more immediate and practical. They can be divided in two: on the one hand, it might be a gene's product, some protein or enzyme that can contribute to human welfare if available in adequate quantities. In such cases, the idea could be to "cut out" target genes from one organism and implant them into another, generally a bacterium or yeast, which then reproduces rapidly and which, because of the new gene it carries, proceeds to make large amounts of the desired chemical.

For example, bovine growth hormone (BGH), produced by genetically engineered bacteria, boosts milk production in cows up to 25 percent. Human beings, too, can be the direct recipients of biologically active chemicals mass-produced via genetic engineering.

Examples include interferon and hepatitis-B vaccine, both produced by genetically engineered yeast, and interleukin for cancer treatment, as well as insulin for diabetics, both derived from genetically engineered bacteria. Until recently, insulin was derived from pig and cattle pancreases, collected from slaughterhouses. But neither pig nor cattle insulin is identical to the human variety, and occasionally serious allergic reactions ensued, as well as religious objections by orthodox Jews and Hindus. Thanks to genetic engineering, genuine *human* insulin is now available . . . made by bacteria!

The same is true of human growth hormone, needed by people whose bodies don't make enough and who would otherwise end up dwarfs. In the old days (a mere decade ago), this could only be obtained by literally grinding up thousands of pituitary glands, obtained from cadavers. Now there are swarms of genetically engineered bacteria into which the relevant human gene has been inserted; they crank out human growth hormone, at much less cost, by the vatful.

On the other hand, we might want the gene itself rather than industrial-strength quantities of its product. Here the procedure is to stick the desired gene into the needy organism, where it then acts directly. In some cases, the combinations have been downright weird. For example, flounders have a gene that is responsible for a kind of natural antifreeze, as befits a fish inhabiting very cold water. This gene has been inserted into tomatoes, thereby improving their resistance to early frost, the traditional bane of tomato growers.

Transgenic plants have also been created that manufacture their own natural insecticide; the result is reduced chemical pollution, since in such cases, there is less reason to broadcast insecticide onto growing fields. By a kind of inverse reasoning, other plants have been engineered to be genetically resistant to a common herbicide; then, when that herbicide is applied, it kills only the weeds.

A long-term goal is to engineer nitrogen-fixing bacteria, which make natural fertilizer directly from nitrogen in the atmosphere, and attach them to the roots of common crop plants such as corn or wheat. If successful, this could greatly reduce the need for chemical fertilizers, which are not only expensive but also pollute nearby fresh water. Another possibility is to increase the available pool of potential organ donors by genetically engineering animals—pigs are especially promising—whose internal organs resemble those of human beings but which are otherwise rejected by their hosts. The idea would be to knock out those genes responsible for labeling, say, a pig's kidneys as pig, so they might masquerade as human.

In addition to modifying the genes of other living things, the very real prospect exists of modifying human genes themselves, although gene therapy in human beings is still in its infancy. A young girl suffering from an inherited immune deficiency has already been treated (successfully, it appears) by introducing the healthy gene into her white blood cells. Other genetic diseases are being researched; for example, attempts have been made, although with only limited success thus far, to cure the lethal respiratory disease cystic fibrosis by using the common cold virus to ferry healthy genes into the lungs of victims. (In a homey but appropriate touch, the genetically engineered viruses in this case were administered by a standard nasal spray.) Gene therapies for a variety of cancers are also being actively pursued. The prospect of treating inherited diseases and even curing them by substituting healthy genes for defective disease-causers is very exciting.

There are two kinds of gene therapy, differing not so much in technology but in ethical implications. The first, which we have been discussing so far, is often called *somatic-cell therapy*, in that it involves correcting errors in the somatic or body cells of a disease victim. Successful somatic-cell therapy will cure a sufferer but will not affect the sufferer's descendants; thus it does not entail monkeying around with the human gene pool or the future of human evolution any more than does any other successful medical intervention, such as administering insulin for diabetes or penicillin for an ear infection. Accordingly, it is not especially controversial.

Germ-line therapy, on the other hand, involves modifying genes within the testes or ovaries of selected individuals, thereby altering the genes of their descendants. This is a whole new ball game.

There is real hesitancy about germ-line genetic engineering, not so much based on its technical difficulty, costliness, or other practical problems (all of which are genuine enough) but on philosophical and ethical grounds: Who are we, as perfectly good but not divinely wise mammals, to change nature and the way things are and perhaps should be? How can we be sure that the transgenic creatures we produce today won't come back to haunt, bite, infect, or otherwise plague us tomorrow? There are many images that reflect such worry, most of them antedating genetic engineering but nonetheless indicating a common concern that a little bit of knowledge can be a dangerous thing, especially when combined with a large dose of all-too-human hubris.

Icarus flew too close to the sun; the heat melted the wax holding together his artificial wings, whereupon Icarus plunged to his death.

Faust allegedly sold his soul in exchange for knowledge and power. Ibsen's Master Builder built higher than he could climb and died when he fell. And of course, there is the now-ubiquitous nuclear genie, released from its bottle by the too-smart physicists of the Manhattan Project. (Shortly afterward, Dr. Robert Oppenheimer, scientific leader of the project, recalled the words of the *Bhagavad-Gita*, "I have become Death, destroyer of worlds," and later noted, "We physicists have known sin.") Not to mention the Sorcerer's Apprentice, emblematic of magical forces run wild.

How do we know that changes made in an organism's genetics won't get out of hand, creating a modern-day Frankenstein's monster, which may eventually ravage our crops, our domestic animals, ourselves? Human mechanical engineering sometimes bears a discomfiting similarity to Icarus: bridges collapse, buildings crumble, airplanes crash. What if a piece of genetic sorcery initially thought beneficial were to backfire, creating a kind of superorganism that somehow exceeds critical mass and becomes difficult, if not impossible, for us—mere apprentices—to control? For example, what if domestic plants, engineered to resist insect and fungal attacks, became noxious weeds, as unkillable as they are unwanted? What to do with such innovations as so-called BT corn, genetically engineered to produce the naturally occurring bacterial toxin BT (and thereby enabling farmers to use less insecticide), but whose pollen appears to be lethal to monarch butterflies?

And what about intentional evil, particularly the engineering of vaccine-resistant strains of disease organisms for military purposes? Two Russian biochemists, working at the State Research Center for Applied Microbiology, reported in the December 1997 issue of the British journal *Vaccine* that they had successfully implanted two nonanthrax genes into anthrax bacteria. Such nefarious activities might well produce bacterial strains that are able to defeat vaccines, setting the stage for a whole new arena of arms races.

Beyond all this, there lingers yet another troublesome question: What remains of us, when and if we start inserting someone else's genes into our most private, secret places?

In the face of all these concerns, it is important to remember the good aspects of genetic engineering that have already been explored: using genetically engineered bacteria to create treatments and vaccines, adding growth hormone to pigs and cattle and thereby allowing them to grow faster and leaner and with less cholesterol at a lower cost, viral resistance engineered into crop plants so they stand up to devastating infestations. And who would object to a genetically engineered cure for sickle-cell disease, leukemia, or cystic fibrosis? The list could go on, and in fact is growing even as we write. The benefits may well outweigh the risks . . . or maybe not.

It is worth pointing out that for literally thousands of years, human beings have been interfering with the genetic makeup of the living things around them. It is called domestication, and plant- and animal-breeding. It is what gave us wheat, rice, and corn instead of inedible grasses, dogs from wolves and coyotes, cattle from water buffalo, chickens from wild jungle fowl, and so forth. And yet, just as most parents aren't especially swayed by the arguments of their teen-aged children that "everyone is doing it," we probably shouldn't let down our guard about possible dangers and abuses of genetic engineering. Because in the brave new world of DNA, "it"—although not yet done by everyone—is not only more promising but also more rapid and scary than ever before.

Genetic engineering, especially in the form of knockouts and transgenics, provides a full plate of potential worry fodder to go with its evident hope. But if that weren't enough, there is something even more exciting, promising, and worrisome.

Cloning

Cloning: a buzzword if ever there was one, conjuring images of totalitarian control, like that depicted in George Orwell's *1984*, or a future society of replicated genetic "perfection" as in sci-fi movies such as *Gattaca* (the title of which, incidentally, derives from DNA's nucleotide bases). In the public mind, cloning often seems to mean the ability to manufacture superbeings to particular specifications, allowing people to designate, perhaps, what type of baby they will have: how tall, how smart, what color hair, even such things as the baby's life span or sexual orientation. In scientific circles, however, much of this speculation is considered ridiculous and impossible.

This is not to say, however, that cloning is ridiculous or impossible.

A clone is a genetically identical copy of something, be it a human, sheep, mouse, bacterium, or even just a sequence of DNA. The cloning that scientists most often discuss—and do—is a much tamer variety than that depicted in sci-fi movies and books, or the overheated imagination of certain wealthy neurotics desperate to "Xerox" themselves, or headline-hungry politicians, equally desperate to prevent them. In fact, most of the time when biologists talk about cloning, they mean the cloning of genes, not animals. A gene is cloned when it is isolated, recombined, and copied, typically by the process of PCR, described earlier. For convenience, a cloned gene is usually attached to a vector, often a virus or bacterium, that can then be isolated and manipulated for further study. Cloning a gene has become relatively commonplace. It does not, by itself, have any serious biological consequences.

It can have legal and economic implications, however. For example, can a cloned gene be patented? Researchers who clone a gene are often considered—at least by themselves or their financial backers—to be its "inventors," who then "own" the gene. The meaning of this is not entirely clear, however. It brings to mind the claim that Newton invented gravity, which is sometimes made by schoolchildren and was never, as far as we know, advanced by the great Sir Isaac himself. (Just imagine the possible royalties!)

In reality, of course, things fell even before Newton told the world why. The scientist who clones a gene similarly did not *invent* it; it was there long before anyone even knew that genes existed. The scientist's accomplishment is to isolate and probably identify that gene for the first time, and to make it available in a biologically useful way for further study and development. Or to be inserted into bacteria and "pharmed" by the rapidly growing genetic engineering/pharmaceutical industry. (All the same, such "making available" is not simple at all; rather, each case is a significant accomplishment, one deserving of recognition, if not ownership per se.)

In 1997, however, scientists did something that approached what the public has long perceived as cloning: namely, they created the world's most famous sheep. Hello, Dolly!

There is nothing very special about making a new individual out of the genetic material of an early embryo. It happens all by itself, for example, when identical twins are formed: a two-celled embryo, called a zygote, splits apart, leaving two separate but genetically identical cells, each of which develops into an individual. Experimental embryologists have long been able to remove the nucleus of a fertilized egg and transplant it into one that has had its nucleus removed (this is called enucleation), whereupon this egg develops into an individual carrying the donor's genes.

What distinguished Dolly from earlier feats of genetic engineering and the transgenic animals described above was that she was a genetic replica of an *adult* mammal, created by using the adult's genetic material to develop a new organism. The procedure used to create Dolly was as follows: cells were taken from the udder of a white-faced ewe and put in a solution that deprived them of nutrients, thereby halting their development and making each fully adult cell act more like a sperm or an egg. Then an egg was taken from a black-faced ewe, and its nucleus, which contains the genetic information, was removed. The two cells—one an egg without a nucleus and the other an adult udder cell halted in development—were then subjected to a brief electric current, which fused them together. Next, a second electrical pulse caused the fused cells to begin dividing and acting like a normal fertilized egg. After a few divisions, this cell was reimplanted into another black-faced ewe and allowed to develop normally within this surrogate mother. It became a normal white-faced ewe, like its genetic (indeed, genetically identical) donor, and readily distinguished from its surrogate mother as well as the enucleated cell into which it had been transplanted, since both were black-faced sheep.

It sounds easy, but wasn't. Out of 277 different attempted egg/udder cell fusions, only 29 began to develop, and only one (Dolly) was ultimately successful.

Dolly nonetheless proved that it is possible to create a new being, genetically identical to an existing adult. In the process, Dolly also disproved what had been a standard article of faith among geneticists and developmental biologists. Thus, it has long been clear that the DNA within a fertilized egg is capable of giving rise, eventually, to all the highly specialized cells that make up a body: nerve cells, skin cells, muscle and bone cells, gland and hair cells, et cetera. It has also never been seriously questioned that these various specialized cells, found in an adult, are genetically identical to one another, even though somehow, somewhere along the line—from fertilized egg to multitrillion-celled adult—they become differentiated so that, for example, once they have taken form as liver cells they can only produce new liver cells, and not, say, bone.

Dolly changed all that. The only genetic instructions present in the single egg cell that eventually became a seemingly normal sheep were those derived from a cell that had already differentiated into glandular tissue as part of her "mother's" udder. (Hence the name Dolly, by the way: a tribute to the mammary endowment of singer Dolly Parton.) This degree of "multipotency" had already been shown in other animals, such as frogs and salamanders, but it had not been found in mammals. Ever since Dolly, it has been necessary to look with new respect at the DNA of every cell, no matter how specialized. It may or may not be possible to teach an old dog new tricks; it certainly *is* possible to get an old cell, fully differentiated, to make a complete new body.

Dolly also raises some difficult and worrisome questions for scientists, politicians, religious leaders, and ordinary citizens. Does Dolly's existence mean that it might soon be possible to clone other animals? Certainly. Rest assured, there *will* be—indeed, there already is—another ewe. What about other people? Probably. Perhaps another you. Is it desirable? Possibly. Such cloning could permit childless couples to reproduce, although it would raise a host of personal dilemmas, such as which parent should be cloned. It could also allow parents to produce a new, identical twin of a child who died of disease or accident, or provide a source of genetically identical organs that might eventually be transplanted into the "parent"—or into another child—if needed. (The ethical firestorms here can be easily imagined but not so readily doused.) Also, what about the soul of a child cloned from an adult? If, as Catholic doctrine currently holds, the human soul comes into being at the moment of fertilization, when sperm meets egg, is a cloned child simply a spiritual extension of its "parent," or does it have its own soul?

These conundrums aside, we are still a long way from having most of the capabilities so beloved of science fiction. For example, we still cannot manipulate any of the characteristics of a person's body via his or her genes. We don't even know, for instance, which genes act most effectively on intelligence or artistic ability, or even where they are; the likelihood is that no single gene exists for such characteristics. Even something so relatively simple as skin color, for example, is caused by a number of different genes acting in concert, resulting in immense variety that would be extremely difficult to specify or control.

In addition, all human traits—actually, all traits of all living things—develop out of the interaction of genes and environment, rather than either genes *or* environment acting alone. It will likely be *possible* in the near future to clone people who already exist, perhaps even tweaking their genetics slightly to make them more disease-resistant, say, or taller, or shorter, or of a specified blood type. It will be up to society, maybe to some of you, to determine how far cloning is allowed to go in terms of the manipulation of people.

As with other genetic technology, the prospect of cloning offers many possible benefits, even if the procedure is restricted to nonhuman animals, as Dolly's creators intend. Among the conceivable agricultural benefits is the exact replication of individual animals who, through accident of birth (actually, of conception), are especially good milk or meat producers. Cloning also could increase the power of transgenic animals, such as those rare individuals that have successfully passed through the various bottlenecks and have been genetically engineered to produce a medicine or drug of some sort. After producing one such individual via genetic engineering—a laborious process, as we have seen—cloning can make many. Cloners would reverse America's motto, to *E unum pluribus*, "out of one, many."

Dolly is a beginning, not the end. Eighteen months after she was unveiled, a team of reproductive biologists at the University of Hawaii announced the production of more than fifty mouse clones, some of which were clones of earlier clones. Since laboratory mice develop rapidly, and their biology is well known, the capacity to clone mice promises significant advances, and at a rapid pace.

Instead of using mammary cells as Dolly's designers did, the Hawaiian mouse makers obtained their donor nucleus from cumulus cells, which normally surround and nourish a developing egg. And instead of fusing the host and donor cells, they first removed nuclear material from the host egg, then, using a micropipette, injected it with donor DNA. After waiting six hours for the egg to reprogram* the newly arrived cumulus-cell genetic material so that it would behave like the DNA of a newly fertilized egg, they stimulated development using a chemical prod instead of an electrical jolt.

The Hawaii mouse demonstration was especially neat because of this visual touch: the donor mouse, whose cumulus-cell DNA was cloned, was coffee-colored; the recipient mouse, whose egg was enucleated and injected with donor DNA, was black; and the surrogate-mother mouse, into whose uterus the newly cloned cell was implanted and which gave birth to the cloned infants, was albino. Sure enough, the clones all turned out to be coffee-colored, like their genetic antecedent and unlike the black egg cell that received this DNA or the albino surrogate mother.

The future of genetic engineering is unforeseeable and almost breathtaking. Already, some transgenic cows have been cloned that contain a gene that will produce certain antibiotics in their milk. This could mean that instead of taking pills to obtain medication,

---

* *Exactly what happens during this "reprogramming" remains a particularly interesting mystery.*

people might soon drink milk produced by a clone of a genetically engineered cow.

It might also mean that people won't simply have to play the genetic hand they are dealt. "Know when to hold 'em, know when to fold 'em," exhorts the chorus of Kenny Rogers's song "The Gambler." "Know when to walk away and . . . know when to get yourself some new DNA"?

TINY AND TROUBLESOME:

Virus Pirates and Other Nasty Little Guys

Viruses are not a man's best friend. Or a woman's. The word comes from the Latin *viru*, which means "poison" (as in *virulent*), with a nod to the Old English *wose*, which has given rise to *ooze*. Not a pretty derivation, but appropriate for the agents that cause the common cold, encephalitis, hepatitis, influenza, measles, mumps, polio, rabies, shingles, smallpox, viral pneumonia, and yellow fever, as well as at least some cancers. And this list names only a few of the more than five hundred different viral diseases already identified and doesn't even include the most spectacular recent examples of widespread fear and viral violence: Ebola, Hanta, and, of course, acquired immune deficiency syndrome or AIDS.

Clearly, there is plenty of reason to take viruses seriously, and to get to know them—if only from a safe distance.

Most people already understand at least a few things about viruses. They are very small, very primitive (compared with elephants, for example, or even fleas), and they can make you sick. Since they are so tiny and so simple, it would seem reasonable that viruses would be among the first creatures to be found all by themselves, perhaps occupying a difficult environment such as the rocky reaches of Mars or the ocean floor. But in fact, viruses would probably be among the last things to live in such places. This is because viruses can't make it by themselves; they can only survive inside other life-forms.

Possession is a frightening and pervasive image: whether we imagine being possessed by a foreign and malign creature, like in the movie *Alien*, or demonic possession, reflecting an anxiety that pervades most human cultures and precedes any knowledge of viruses or other "germs." But possession is what viruses are up to, the work not of the devil but of themselves, and it has long bedeviled the human animal.

Most people think of free-living organisms, whether redwood trees or rhinoceroses, as somehow more legitimate than parasites or other disease-causing creatures; we also think of them as dominant and thus more abundant. But the opposite is true. There are far more of them than of us, since each free-living creature—including every mammal ever autopsied—carries a vast number of parasites and pathogens. (There is no real difference between parasites and pathogens, except size.) We are all inhabited, occupied by other life-forms, the tiniest of which are often the most troublesome.

What Is a Virus?

Many of us have a hard time believing in anything too small for the naked eye. Bacteria, for example, or protozoa. Viruses are even more challenging. At least bacteria can be seen under a standard light microscope, with a magnifying power of perhaps one hundred to five hundred times. Viruses, on the other hand, are so small they pass right through the finest porcelain filters. They were not seen until the advent of the electron microscope, offering magnifications of 100,000 power and more. Even today, it is easier to cast a miasma of blame ("it must be some virus") than to identify a culprit with any confidence. But there is nothing mythical about viruses; they are altogether real.

They come in different sizes, ranging from very small to unimaginably small, and various (relatively simple) shapes: cylindrical, spherical, polyhedral. Some viruses even look like small spaceships. All, however, are made of an outer protein coat and an inner package consisting of the virus's own genetic material, viral nucleic acid.

The protein coat of a virus is not mere packaging, however; it isn't inert. Rather, it is usually specialized to allow the virus to penetrate particular cells of its victim. This is probably why viruses tend to be highly specific: influenza, for example, is most likely to attack cells of the respiratory tract; polio and rabies go for the nervous system; the AIDS virus targets the immune system; and so forth. These attacks are not matters of brute force, of viral boarding parties simply ramming their way in. Rather, the outer protein surface of a virus often consists of materials derived from the membrane of a previously victimized cell, and it contains specific chemical sites that permit it to dock with its cell-victims and introduce its own payload of viral nucleic acid inside.

This nucleic acid is not always DNA; some viruses (including those responsible for polio, rabies, and AIDS) contain RNA instead. And a virus's nucleic acid can be either single-stranded or double-stranded. So, there are single-stranded DNA viruses, double-stranded DNA viruses, single-stranded RNA viruses, and double-stranded RNA viruses. Some viruses do their dirty work in a cell's cytoplasm. This is the case, for example, with all viruses that attack bacteria, since bacteria don't have a nucleus. Others migrate quickly to the nucleus of their victims and do their evil deeds there.

Viruses are not cells. They are more like old-time pirates of the Spanish Main: outside, a proteinaceous sailing ship, and inside, an aggressive boarding party consisting of either DNA or RNA. Once aboard its victim—fat, vulnerable merchant vessels such as bacteria or the cells of a plant or animal—the virus's nucleic acid masterminds a genetic takeover, murders the captain and crew, and proceeds to convert everyone and everything on board into more piratical viruses, which then jump ship and proceed to launch more attacks.

Viruses are known as "obligatory intracellular parasites." Parasites because they cannot reproduce on their own. Intracellular because they are inert and essentially lifeless unless they are within a living cell. And obligatory because this "lifestyle" is not optional; for viruses, it is the only game in town. They cannot even get by on a food-rich but nonliving laboratory-culture medium. According to some biologists, viruses are not even alive, since they have no internal raw materials, none of their own metabolic machinery, and—like salts, which indubitably are not alive—they can be crystallized and then activated when redissolved in water and offered a suitable victim. Instead of pirates, perhaps we should speak of viral vampires, the "undead," which require not their victim's blood but rather access to the energy-producing enzymes and other life-maintaining structures of living cells.

What Do Viruses Do?

The Lytic Cycle

We have already encountered the best-known, "classic" viruses, the bacteriophages—*phages* for short—that prey upon *E. coli* bacteria and are often used by genetic engineers to transfer host DNA from one cell to another. Like most viruses, bacteriophages do not really "eat" their victims; rather, they eventually cause them to burst, or "lyse." (Lysol disinfectant is named for its presumed ability to emulate phage viruses.)

The basic life cycle of a phage is as follows. First, the bacteriophage, more externally complicated than most viruses and looking a bit like one of NASA's Martian landers, attaches itself to the outside of a host cell. This attachment takes place because of a chemical match between "receptor sites" on the bacterial wall and the tail fibers of the phage virus. The phage's sheath then contracts, like a hypodermic syringe, and the viral DNA is injected into the victim, while the structural protein shell remains outside, inert and now abandoned.

The next ten to fifteen minutes is the "eclipse period," when, to a human observer peering through an electron microscope, nothing much seems to happen. But inside the phage-infected bacterium, all hell is breaking loose: within minutes, the viral DNA begins its unfriendly takeover. As in any well-orchestrated attack, the invading phage first decapitates its enemy. Of the approximately one hundred genes in the virus's DNA, the first to unzip and become active are translated into proteins by the bacterium's cellular machinery, which, naively, does not distinguish between the phage DNA and its own. This gene product is an especially deadly Trojan horse, an enzyme that chops up the host's own DNA, making it unable to replicate itself or to synthesize defensive proteins. Adding insult to injury, the victim's DNA is even used as raw material to be reconstituted by the invading viral legions into more copies of the virus's DNA.

Virus particles then begin to take form inside the doomed cell. Recall from the previous chapter that DNA has the crucial ability to replicate itself. This is what the viral DNA does inside its victim. The host's DNA no longer exists; it has been chopped up. Instead, the virus's nucleic-acid chain serves as a template to direct the production of more copies of itself, as genetic components of the hapless *E. coli* are salvaged and recycled into the DNA of new phage particles. But what about that outer protein coating, without which a virus is just a naked coil of nucleic acid? New viruses will require new proteins.

No problem. The manufacture of proteins, of course, is what DNA's other function, translation, is all about. Within the occupied cell, new messenger RNA is produced by the same basic mechanism that the doomed bacterium would have employed if it were healthy and intact. And this mRNA goes on to do its thing, that is, provide a template for the eventual specification of proteins, only instead of *E. coli* proteins, these are the building blocks for the characteristic outer core of the next generation of phage viruses. The newly minted phages assemble themselves under the influence of appropriate enzymes and chemical correspondences, so that within a half hour, the bacterium is phage-filled and ready to burst. All that remains is the getaway: one set of phage genes directs the manufacture of an enzyme known as lysozyme, which begins digesting the inside of the bacterial cell wall. Water leaks in, and within minutes the cell pops open, liberating one hundred or more viruses, each of them ready to repeat the whole nasty enterprise.

The Lysogenic Cycle

The lytic cycle is nature's version of a hostile takeover: fast, violent, and nearly irresistible. Many viruses, however, have another way of doing business. Known as the lysogenic cycle, it is kinder and gentler than the lytic cycle; it doesn't destroy its host—at least not right away.

Here's how the lysogenic cycle works. A phage-virus boarding party has just entered a bacterial cell. Instead of hacking up the host's DNA and hijacking its innards, viral nucleic acid simply incorporates itself into the bacterial DNA at a specific receptor site and then, in a sense, goes to sleep, like an unwanted houseguest who settles in and refuses to leave. Cozily ensconced within the host DNA, this sleepy newcomer is now called a prophage. It stays drowsy because of a repressor protein that is coded by one of the phage genes and that for the most part keeps the prophage quiet. So quiet, in fact, that when the host bacterium reproduces, the prophage simply goes along for the ride. As a result, it gets reproduced too, in this way giving rise to a huge population of daughter cells, each carrying a dormant lysogenic virus in prophage form. It is interesting to ponder whether some of the normal makeup of bacterial chromosomes may have originally arrived as prophage viruses that "came to dinner" as part of a lysogenic cycle, never left, and were somehow tamed over time.

In any event, the typical prophage is not so cozily domestic. The sleepy guest eventually wakes up and transforms from the benign, lysogenic Dr. Jekyll to the murderous, lytic Mr. Hyde. First, it expels the host chromosome. Then it goes lytic altogether, destroying the host DNA, replicating active phages, and breaking down the host cell. Certain phage viruses commonly alternate between lysogenic and lytic cycles, waiting like a viral version of Sleeping Beauty for an energizing kiss by some sort of environmental trigger—often radiation or chemical—that causes the prophage to switch from lysogenic to lytic.

Actually, even during a lysogenic cycle, some viral genes remain active. And when they do, they are generally up to no good. Diphtheria, scarlet fever, and botulism, for example, are caused by bacteria that in themselves are harmless. The human disease in each case results from a toxin that the bacteria produces under prodding from prophage genes, which infect these bacteria and apparently do a bit of sleepwalking, even when lysogenic. (The fact that bacteria, small and troublesome as they often are, find themselves invaded by even smaller and more troublesome creatures brings to mind Oliver Wendell Holmes's observation that "even fleas have smaller fleas, upon their backs to bite 'em. And so it goes, *ad infinitum*." It might even make some of us feel more empathic toward bacteria.)

Some "regular" viruses also undergo a lysogenic phase. This is what happens when someone has an "inactive" form of a disease. Just as an inactive, incorporated bacteriophage is called a prophage, an inactive virus incorporated into otherwise normal cells is known as a provirus. For example, once someone is infected with the herpes virus, which in one form produces canker sores and in another, genital sores, it is a permanent resident. It alternates, however, between lytic episodes, when the disease is active and infectious, and lysogenic phases, when it is a provirus: quiescent but latent.

Vaccines

Considering the abundance and rapacity of viruses, it seems almost miraculous that any uninfected bacteria have survived at all, or any plants or animals (including people) for that matter. But where there are moves, there are countermoves. Another way of looking at it: any cells that have been able, somehow, to resist the incursions of virus pirates enjoy a substantial advantage over their more vulnerable counterparts and were doubtless favored by natural selection.

Not surprisingly, therefore, not all of those cellular merchant ships are sitting ducks; many are armed. Some have powerful armor in the form of receptor sites to which viral proteins cannot attach. Others produce restriction enzymes, the same ones that molecular biologists have appropriated for DNA technology, which do battle with foreign nucleic acids.

But in general, viruses are frustrating and difficult opponents for medical science as well as for cells under attack. The problem is reminiscent of the leaky-roof dilemma: no one wants to fix a leaky roof while it's raining, and when the sun is shining, you don't need repairs. Similarly, when a virus is not inside a cell, it is a kind of spore, inert at best, and thus invulnerable to traditional antibiotics. (You can almost hear the virus taunting, "You can't kill me, I'm already dead!") On the other hand, once it has invaded a cell, a virus becomes so fully incorporated into its innards that the only way to kill it, it seems, is to kill the cell, too. Which brings to mind the infamous Vietnam War observation, "We had to destroy the village in order to save it."

But happily, all is not lost. Some antiviral drugs have proven at least mildly effective, notably a few, such as acyclovir, that interfere with viral nucleic-acid synthesis. Others are in development.

At present, however, the most powerful of the available antiviral agents make use of the body's natural defense, its immune system. For medium-sized animals such as human beings, the world is a dangerous place, filled not only with large threats—hungry beasts, volcanoes, and nuclear weapons—but also with small ones, such as bacteria and viruses. The immune system is our most important way of countering these physically small but potentially lethal threats. To appreciate how the immune system works against viruses, as well as how medical researchers can manipulate immunity for our benefit, let's take a quick trip back in time.

The year is 1796. The place, Gloucestershire, in not-so-merry Olde England. The problem is smallpox, long one of the most feared diseases. It is not uncommon for smallpox to kill one child in three and for the survivors to be left either blind or permanently scarred, their skin pitted with unsightly pockmarks. "A pox on you and your house" is an oft-heard curse, and smallpox was a curse indeed. The Spanish conquistadors carried smallpox with them to the New World, where it wiped out half the Aztec nation, far more than ever died in battle. Colonial Americans used to employ smallpox-infected blankets in germ warfare against Native Americans; the horrors of smallpox were seemingly endless.

But among the dairy farmers of Gloucestershire, it was widely known that anyone who contracted a similar but much milder disease, cowpox, was safe from its dreaded cousin, smallpox. Although folk wisdom is not always wise, sometimes it contains more than a grain of truth. It is said that for generations Romanian peasants used to keep moldy bread in the rafters of their cottages, to eat when they were threatened with infections (can you say *penicillin*?).

In any event, Edward Jenner, an English physician, decided to test the widespread Gloucestershire myth that cowpox protects against smallpox. On May 14, 1796, he removed a small drop of pus from the wrist of a milkmaid, Sarah Nelms, who had cowpox, and placed it on the skin of a young boy (an unwitting hero, largely unsung even to this day, named James Phipps). As expected, Phipps developed the characteristic cowpox pustule. Then, six weeks later, Jenner intentionally exposed him to pus from a smallpox victim. Young Master Phipps did not contract the disease. Over the ensuing years, this human guinea pig was repeatedly tested with live, virulent smallpox fluid, in a way that would have spread the disease to anyone else. Phipps never developed smallpox. He was "immune."

Jenner had no idea what was the actual agent causing cowpox, or smallpox . . . or immunity, for that matter. The French mocked his procedure as "vaccination," derived from *vache*, the French word for "cow." (In translation, *vaccination* would mean something like "being made into a cow," or "encowment.") Nonetheless, Jenner had in fact developed the first vaccine, and in time, both the word and the procedure became eminently respectable. To this day, smallpox vaccine is made by scratching a small amount of cowpox pus into the skin of a calf. Fluid is then collected from the pustules that develop, and this, in turn, is scratched into the skin of a person to be inoculated.

How does smallpox vaccine work? Both cowpox and smallpox are caused by viruses that resemble each other, especially in the makeup of their protein coats. The cowpox virus has only a minor effect on human beings (typically it produces a single pustule at the point of exposure), but it stimulates the body's immune system to produce antibodies, chemicals that neutralize the outer protein coat of cowpox virus. Since the molecular architecture of cowpox virus is similar to that of smallpox, antibodies to the former work against the latter as well. They "cross-react," and so, if and when an immunized person is exposed to smallpox virus, the body "remembers" (falsely, it turns out!) that this has happened before, and it quickly mobilizes a barrage of antibodies that stop the invaders in their tracks.

As with any vaccine, infection may still occur; it is *disease* that is prevented. Such prevention has been so successful, however, that smallpox has been eliminated worldwide. This is easy to say, perhaps too easy, its quickness and simplicity tending to obscure the enormity of the triumph. In an age of misery—much of it human-caused—the triumph over smallpox is a matter for rejoicing.

Procedures similar to Jenner's have been followed ever since in the development of vaccines against other viruses. Consider another grand success story: polio. Between 1950 and 1954 there were about 20,000 cases of polio per year in the United States alone. Summer was the "polio season," when families lived in great fear, especially for their children. In many communities, municipal swimming pools were virtually abandoned as parents kept their kids home to minimize the chance of infection. Polio, like smallpox, is caused by a virus, but there is no harmless pseudo-polio virus, no cowpox equivalent. So Dr. Jonas Salk exposed regular polio virus to chemicals that essentially knocked out its disease-causing ability but left its protein shell intact. When injected into people, the Salk vaccine stimulated them to make antibodies that protected against full-fledged polio viruses. By 1961, there were only sixty-one cases of polio contracted in the entire United States. Later, scientists isolated a mutant strain of polio virus, which, after being grown in the cells of monkey kidneys and then taken orally, stimulated antibody production without producing the disease. The Sabin vaccine was even more reliable than its predecessor.

Measles, chicken pox, mumps, diphtheria, whooping cough, yellow fever: all have been successfully fought—more accurately, prevented—through vaccines. In such cases, an ounce of prevention is worth megatons of cure, especially when, as we have seen, viral diseases generally cannot be "cured." The best that modern medicine can do in most cases is simply to treat the symptoms and wait for the victim's natural immune system to overcome the invading virus, or vice versa. Not surprisingly, there is a lot of hope riding on the possibility of developing vaccines against the remaining virus pirates that plague mankind, including, of course, the one that produces AIDS.

Edward Jenner and the dairy farmers of Gloucestershire—not to mention Sarah Nelms and James Phipps—would be proud.

But we cannot end this part of our account, hopeful as it is, without pointing out that viruses and their cousins remain formidable adversaries: abundant, resourceful, and likely to dog humanity forever. Not that people should give up. If anything, the challenges posed by viruses make it all the more important that they and their kin be confronted, even though they constantly mutate to new forms and thus will probably never be totally overcome. We are reminded of the closing lines of Albert Camus's great novel *The Plague*, in which the novelist/philosopher observed that one learns in time of pestilence that "there are more things to admire in men than to despise." A plague had just been beaten back, and the populace was rejoicing. Nonetheless, the hero, Dr. Rieux,

> knew that the tale he had to tell could not be one of final victory. It could be only the record of what had to be done, and what assuredly would have to be done again in the never ending fight against terror and its relentless onslaughts, despite their personal afflictions, by all who, while unable to be saints but refusing to bow down to pestilences, strive their utmost to be healers.
>
> And, instead, as he listened to the cries of joy rising from the town, Rieux remembered that such joy is always imperiled. He knew what those jubilant crowds did not know but could have learned from books: that the plague bacillus *[Note: and most viruses as well]* never dies or disappears for good; that it can lie dormant for years and years in furniture and linen-chests; that it bides its time in bedrooms, cellars, trunks, and bookshelves; and that perhaps the day would come when, for the bane and enlightening of men, it would rouse up its rats again and send them forth to die in a happy city.

Like the rats of Camus's plague, human beings are also mammals, vulnerable to many of the ills that trouble mammalian flesh. In the next few sections, therefore, we present a rogue's gallery of some of the more dramatic and/or prevalent viruses as well as a near-virus or two, plus some other tiny and troublesome little creatures.

Types of Viruses

HIV/AIDS

Human immunodeficiency virus, or HIV, has become perhaps the most notorious of all viruses that afflict people in the modern world. It is certainly the one that Americans hear the most about, since it causes acquired immune deficiency syndrome, or AIDS. The United Nations estimates that HIV infects 16,000 new individuals every day, or nearly 6 million per year. By the year 2000, 40 million people will have died from AIDS, most of them in the developing world. Because HIV is a virus, there is no cure. Because AIDS attacks the body's defense system, it is extremely deadly—although people die not "of AIDS" but of illnesses that they cannot fight off because of AIDS. Since HIV mutates so quickly, it is a very elusive, moving target, and accordingly, there is not (yet) an effective vaccine against it.

AIDS is not an especially infectious disease, in that it cannot spread through the air (unlike the most common forms of influenza, for example). In many ways, HIV is an oddly delicate little creature. It can only be transmitted from person to person via direct transmission of body fluids, notably blood or semen; saliva, for instance, does not appear to suffice. But since HIV is easily transmitted via sexual contact, AIDS is thought to be the ultimate in modern ailments, without a cure and spreading fast.

Recently, however, there have been some advances in drug "cocktails," or combinations that control the virus once it is lodged inside a victim. To understand how these new drugs work, however, it is important to understand how HIV itself works.

As mentioned above, viruses can be roughly divided into two classes: those whose genetic material is DNA and those that employ RNA. HIV belongs to the latter group; it consists of RNA inside a protein coat. But host cells are DNA-based. Therefore, in order for HIV to perform its genetic takeover of the infected host, its RNA must be remade into DNA. This is accomplished with an enzyme that we encountered in our discussion of genetic engineering: reverse transcriptase. In fact, viruses of this sort—so-called retroviruses—were the original source of this enzyme, creatively hijacked by molecular biologists for their own purposes. (It is somewhat reassuring to note that viruses aren't the only organisms capable of hijacking nucleic acids and their associated enzymes for their benefit.)

In any event, it is because of this reverse process—going from RNA to DNA instead of vice versa—that HIV is called a retrovirus. It processes genetic information backward.

An HIV particle contains a minimal but effective tool kit, consisting of two identical copies of an RNA strand, plus a few molecules of reverse transcriptase. The outer shell of the virus is made of molecular signals that allow it to fuse with the outer membrane of special cells called CD4 or helper T-cells, which are vital for an active human immune system. By attacking the immune system first, HIV is essentially following the same tactics as opponents in the early stages of a protracted air war: first, destroy the other side's radar and antiaircraft capability so the victim cannot shoot down the attackers. After suppressing these cells, HIV can then sit back and "bomb" its target at will (see the next chapter for details). Once inside a cell, the virus's RNA is changed into DNA by reverse transcriptase and then integrated into the cell's DNA as a provirus, whereupon it begins a lysogenic life cycle.

While hidden inside host cells in this condition, HIV is especially resistant to attack. Even when the virus is actively replicating and thus potentially vulnerable, some of the particles apparently remain as nearly untouchable proviruses, making it especially difficult to rid infected people of *all* their HIV. After years of drug treatment, during which it may appear that viral levels have essentially reached zero, it takes only a small number of proviruses—no one knows exactly how many—to have weathered the storm in their seemingly "benign" state, to reemerge and resume their ravages. (HIV has yet another dirty trick up its sleeve: it sometimes manages to insert itself directly from one infected cell into another; when this happens, the virus is continually "under cover," never directly exposed to attack by antibodies or other chemicals.)

Even during its relatively quiescent period, however, the HIV provirus is active. Having gotten itself incorporated into a host cell, as DNA, it gets itself re-created when the host's DNA (some of which actually comes from the virus) goes about doing what DNA does so well: making RNA. Except that the RNA spun off is not messenger RNA in the service of the host cell, but the RNA that "is" HIV. Not only that, but some of the messenger RNA is obediently translated into the reverse transcriptase needed to make more HIV, along with the necessary protein coat.

The newly formed virus particles eventually leak out and infect other helper T-cells. (On the way, they surround themselves with some of their host's membrane, which provides them with a biochemical match that makes it easier for them to gain entry into other victims whose cell membranes are similar.) Typically, there is a protracted lag time, usually about ten years, before a person infected with HIV develops full-blown AIDS. From this point, the disease generally progresses very quickly because during its inactive phase it established an arsenal of invaded cells ready to serve as a kind of immunologic fifth column. Upon command, these HIV-laden cells become adept at sending out vast numbers of HIV particles that destroy the immune system, after which other diseases take advantage of the AIDS victim's weakened state.

Current treatments for AIDS rely upon some of the virus's specific needs. Many drugs—with such names as AZT, 3TC, ddC, and ddI—throw a monkey wrench into the workings of reverse transcriptase, necessary for HIV to proliferate. These drugs are chemically similar to some of the components of human DNA and are thus mistakenly incorporated into the newly produced viral DNA. But they are not identical, and they glue up the works, preventing the viral RNA from being fully transcribed into DNA. They only succeed, however, against actively replicating viruses.

Newer drugs have also been developed, such as the protease inhibitors. An essential step in the life cycle of HIV is the cleaving of some important proteins, which is accomplished by an enzyme known as protease. A protease inhibitor, therefore, limits the amount of cleaved protein available for use by the virus. No drugs developed thus far are 100 percent effective, however, and people infected with HIV generally do best by taking a combination of many different drugs. The most effective combination to date is 3TC, AZT, and a protease inhibitor.

Although retro fashions may enjoy occasional popularity, the retro aspect of HIV is unlikely to win it friends within the biomedical research community. Indeed, it contributes to making HIV such a frustrating opponent. With each switch from RNA to DNA, a certain number of mistakes are made by reverse transcriptase; it may even be that reverse transcriptase is especially error-prone. These mistakes—essentially, mutations—sometimes result in DNA that does not form a workable virus, but occasionally it results in the virus simply evolving much more quickly than would a normal, "forward-going" virus. Every little mistake provides the opportunity for a new and more effective AIDS virus to develop, perhaps one that is not recognized by the antibodies produced by the latest vaccine. These "mistakes" also present immense difficulties to researchers and help explain why an HIV vaccine has not yet been developed. They also explain why multiple-drug cocktails work better than singletons: to survive an attack from two or three different medications, HIV would need two or three resistant mutations, independently and simultaneously.

HIV is such a quick-change artist, however, that even this is not beyond its capacities. In fact, it mutates so rapidly that a person suffering from AIDS will often have different kinds of HIV at different times and in some cases may be home to a variety of mutated strains at the same time!

As to vaccines, HIV's capacity to mutate introduces an insidious danger: if living but attenuated (inactivated) virus is eventually used to generate an immune response, the vaccine itself could possibly mutate into full-fledged HIV, thereby switching from cure to culprit. One exciting possibility, bypassing this risk, is to make a "DNA vaccine." This would involve taking a chunk of the viral gene after it has been reverse-translated into DNA and tucked away into the nucleus of a host cell. The idea would be to choose some of the nucleic acid that, when activated, codes for producing the protein found on HIV's surface, and inject that DNA into a healthy person. If all goes well, this viral-generated DNA might then induce an immune response, with the production of antibodies against the virus's outside and no potential dangers, because the rest of the viral nucleic acid—the inside—is never introduced.

Ebola

Ebola is another virus of recent concern, especially because of the dreadful nature of the disease. If HIV attacks the human body like a besieging medieval army settling in for a protracted war of attrition, Ebola is more like a nuclear attack. It is not known precisely how Ebola kills its victims; internal organs are basically liquefied, resulting in massive hemorrhage from every orifice, even the eyes.

There have been a number of isolated outbreaks of Ebola, the two worst occurring in Zaire in 1976 and 1995. Most diseases that are highly lethal in human beings employ other living things as their primary host. This makes sense, because if a virus consistently kills its victims, then for all its horror, it is unlikely to persist for long; having run out of habitat, it too will generally die away. Accordingly, such diseases as sleeping sickness, malaria, yellow fever, and others reside in various animal "reservoirs," from which they may be transmitted to human beings by a vector such as the tsetse fly, in the case of sleeping sickness, the *Anopheles* mosquito, for malaria, or a tick that bites someone after having first dined on an infected deer, in the case of Lyme disease. For bubonic plague, rats are the most common reservoir and fleas, the vector.

No one knows the original reservoir of HIV, although it is widely believed to have arisen as a mutated virus that normally afflicts one of several different African monkey species or, possibly, chimpanzees. The identity of Ebola's original host species is also unknown, although presumably, between its periodic jumps to *Homo sapiens*, this deadly virus resides in some other animal species. Also mysterious is what causes sudden, terrifying Ebola outbreaks to occur. These unknowns as well as its unpredictability make Ebola especially scary.

To make matters worse, Ebola has an extremely high death rate: nearly 90 percent for some strains, probably the highest of any known infectious agent. Once the symptoms start, death is just around the corner, and Ebola is highly contagious. All these threats have led to Ebola's getting quite a bit of press since its 1995 outbreak, which also coincided with publication of *The Hot Zone* by Richard Preston, a riveting, best-selling account of the illness and its "near-miss" eruption in a monkey colony near Washington, D.C.

Paradoxically, however, because it kills its victims so quickly (in just a few days), Ebola is unlikely to become a truly devastating worldwide threat because people who are infected are likely to succumb before they can infect others. It burns itself out so intensely and so quickly that it seems unable to persist for long. Preston suggests that whereas HIV is like a simmering fire in a coal mine—slow, persistent, and very difficult to extinguish—Ebola is like a flash fire in a pile of dry tinder: extremely hot but localized and unlikely to spread very far. The Ebola virus thus appears to be one of those creatures that is too effective—in this case, too lethal—for its own good.

On the other hand, Ebola has also begun to alert people to a unique danger of modern times, something that otherwise seems rather benign: namely, worldwide transportation. It is unclear whether viruses are busier these days, literally evolving more rapidly, or if they are simply more likely to infect people hundreds or even thousands of miles from their place of origin . . . or perhaps people worldwide are simply more aware of their existence. In any event, it seems clear that outbreaks which in previous centuries would have remained isolated now have the potential of becoming global. Every year approximately 400 million people travel internationally, an unprecedented number. Someone unknowingly carrying, say, Ebola virus can walk onto an airplane in Nairobi and within a day or two be anywhere in the world, along with virus particles in the hundreds of millions.

In addition, only about 30 percent of human deaths worldwide are medically documented. What we don't know—about the other 70 percent—can indeed hurt us.

One of the downsides of the earth becoming a global village is that we now share the liabilities as well as the benefits of being involved, literally, in everyone else's lives. In addition to the worry that a virus long restricted to occasional local outbreaks may suddenly spread worldwide, there is another fear: that viruses may jump from one host to another. Thus, it is virtually certain that Ebola itself, as well as HIV, evolved originally in some other mammal species, then began multiplying in human beings, perhaps with an assist from a mutation or two. During the 1950s, a virus known as monkeypox (similar to smallpox, although less severe) jumped successfully from a laboratory population of African monkeys to an Asian species. More recently, monkeypox has made the leap to human beings in central Africa, raising the specter of a renewal of smallpox or something equivalent just when that disease had finally been eradicated. (Actual smallpox virus survives today only in tightly controlled laboratories in the United States and Russia.)

Ebola belongs to a group of viruses called filoviruses or *thread viruses* (the Latin *fil* means "thread"). They resemble spaghetti. Like HIV, filoviruses are made of RNA instead of DNA. Unlike HIV, however, the Ebola virus does not have to be transferred to DNA in order to continue infecting cells. Ebola reproduces as RNA instead. As a result, the Ebola viruses do not have a method for proofreading themselves, and their rate of mutation is even faster than that of HIV.

It appears that Ebola works its lethal spell by producing two different versions of a glycoprotein, a protein combined with some sugar molecules. One of these glycoproteins is secreted by the virus, whereupon it acts as a toxin that blocks the body's early-warning inflammation response. This, in turn, helps the virus get a toehold on its quarry. The other glycoprotein, present on the virus's coat, preferentially binds to endothelial cells, which comprise the lining of blood vessels. In this way, Ebola is able to gain entry to these cells, causing them to rupture and giving rise to the hemmorhages that eventually lead to death.

There is a second paradox associated with Ebola (aside from its self-limiting lethality). Now that the endothelial-binding glycoprotein has been identified, it might well be possible to use it as a prefabricated key and attach it to some harmless viral vehicle, which could then deliver genetically engineered genes to blood vessels that might be under siege from coronary artery disease, cancer, sickle-cell anemia, or other diseases. Considering all the misery that Ebola has caused, such an outcome would be poetic justice indeed.

Hanta

Hantavirus is another headline maker. It was first identified in the United States in 1993, when scores of people in the Four Corners area (the junction of Utah, Colorado, Arizona, and New Mexico) developed respiratory infections that became progressively more severe, resulting in more than a dozen deaths. The infectious agent was quickly recognized as a virus that had been known to Western science since the Korean War (and named for a river in Korea) but not previously reported in the United States.

It erupted after the deer-mouse population exploded during a stretch of wet weather unusual in the arid Southwest. Hantavirus is spreading throughout the United States, rendered especially dangerous because it is difficult to identify. By the time it is accurately diagnosed, Hanta has typically gotten such a head start that it is hard to treat effectively.

It belongs to a family called Bunyaviruses, which have single-stranded RNA that does not need to be changed into DNA before replication, much like the filoviruses such as Ebola. Hanta is transmitted by contact with mouse and rat feces as well as rodent saliva. As a result, it is relatively easy to avoid, so long as one is not dealing with a large rodent infestation or cleaning up an area recently inhabited by mice or rats. It is also possible to contract Hanta through rodent-contaminated water or food, although this has not yet proved to be a major problem. Faced with an unwanted mouse metropolis, it is wise to take precautions in the cleanup, by wearing latex gloves, taking care not to sweep or vacuum the material—as this is likely to stir up virus-laden dust—wetting the area with a disinfectant and wiping with a damp towel, double-bagging all material removed, and carefully washing hands, gloves, and all clothing when finished. (As a general rule, by the way, when dealing with any germs, viral or otherwise, the most effective single preventive technique is also the oldest and simplest: wash your hands with soap and hot water. Although many different brands of germicidal soaps are available, none of them are especially worthwhile. What makes hand washing so effective is not that it kills disease organisms but that it flushes them off your body!)

Hantavirus is generally a respiratory and cardiovascular assailant. It starts out acting like a normal cold, with fevers, congestion, body aches, and so forth, but then progresses to pulmonary edema (filling of the lungs with fluid) and associated lack of oxygen and hypotension (low blood pressure). By this time, patients are usually in the hospital on a ventilator.

Hantavirus has likely been around for many decades, if not centuries, only now spreading to new areas. But it is also possible that Hanta is simply being diagnosed more often, because improved communication and better-quality rural medical treatment have made physicians more aware of the danger.

How did Hanta get to the United States? Did it come from Korea, perhaps via a returning member of the armed forces? No one knows. We do know, however, that whereas old-fashioned pirates are virtually extinct, virus pirates are popping up in areas where they had not previously been reported. To get a sense of just how troublesome this has become, consider an everyday stubborn problem, often comical in itself but now with lethal implications. If you have ever tried to get water out of an old automobile tire, you know how ridiculously difficult this can be. This is not just a metaphor: it seems to be precisely the cause of the rapid spread of a lethal viral disease, dengue fever, which is carried by *Aedes* mosquitoes. These insects breed in stagnant water and have spread from the Old World tropics to the Caribbean, traveling in barges full of tires, which were themselves carrying infected mosquitoes, happily breeding in their small, stubborn reservoirs of tire water.

The identification of Hantavirus as the culprit in the 1993 outbreak was a triumph of scientific sleuthing on the part of physicians, virologists, gene analyzers (who used PCR and gene-sequencing techniques), wildlife biologists, and government officials, who achieved positive identification of the perpetrator in a matter of weeks. Results in other cases have been more delayed, and the prospects uncertain. For example, the widespread but frustratingly diffuse disease known as chronic fatigue syndrome has long baffled researchers. Maybe it is caused by some as yet unknown virus. Or maybe not.

Influenza

Influenza can be extremely dangerous, although it does not usually get the attention it deserves because everyone knows about it—or thinks so. People often say they are sick with the flu, which has become a catchall term for a general malaise that is probably closer to the common cold (which in turn is caused by a single-stranded RNA virus called a rhinovirus). Why can't we find a cure for something so commonplace as the common cold? First, because it is virus-caused, and there are no simple cures for viruses. And second, because "the" common cold doesn't exist! There are more than 150 different common cold–causing viruses. Each local region has its own distinctive lineup of rhinoviruses, so that over time, people develop natural immunity to their neighborhood array. Travel somewhere else, however, or encounter someone who is a newcomer in your neck of the woods, and coming down with a "cold" is all too likely.

Throughout history, local immunity combined with sensitivity to geographically distant infectious agents has caused disease, explaining, for example, the rapid genocide by which Westerners managed to wipe out huge numbers of indigenous New World inhabitants, from the Great Plains of North America to the Andes to Hawaii. H. G. Wells, who was an accomplished biologist as well as novelist, used this theme in *The War of the Worlds*. In that science fiction classic, creatures from Mars invaded Earth and used their ruthlessness and superior technology to devastate much of Europe. But just when it looked like the end for *Homo sapiens*, the aliens died: despite their ability to command immense destructive power on a grand scale, their bodies succumbed to the humblest of earthly life-forms, lowly cold-causing "germs" to which they had never been previously exposed and to which they were lethally vulnerable.

For human beings, symptoms of rhinovirus—the common cold—are unpleasant but far from lethal. They include overall malaise, headaches, stuffy/runny nose, cough, and maybe a fever. Fortunately, with a little bit of time, chicken soup, and rest, colds usually go away. When viruses attack the nervous system, as with polio, the damage is likely to be permanent. By contrast, people generally recover completely from a rhinovirus "flu," because the injured cells—most of them within the respiratory system—are readily replaced.

The *real* flu, however, can be quite dangerous. During the Middle Ages, European astrologers thought that influenza—and just about everything else—was caused by heavenly bodies; in fact, *influenza* is Italian for "influence," implying astrologic powers. There have been a number of deadly influenza epidemics, the most dramatic being the pandemic of 1918 that killed about 20 million people worldwide, including nearly 700,000 in the United States. Fears of a new influenza pandemic surface regularly, and in fact, lethal worldwide influenza outbreaks appear to occur every few decades. After 1918 the next major episode was the "Hong Kong flu" of 1968. We may be due for another.

The most recent scare appears to have been a false alarm . . . or else a potential killer that was aborted by prompt action. This was in 1997, when it appeared that "chicken flu," long endemic to Hong Kong poultry, was beginning to spread to people. As a preventative (or a palliative, to inhibit panic and give at least the impression that something was being done), about 1.5 million chickens, ducks, and geese were slaughtered.

But what makes the flu so dangerous?

"There is nothing new under the sun," claimed Ecclesiastes, but that was before the age of virology (which is novel in itself). Now we know better. There *is* something new; it was new last year, is new this year, and will be new, yet again, next year. Like so many other viruses, influenza mutates rapidly, not as rapidly as HIV, but quickly nonetheless. One year's flu is different from the next: sometimes very different, sometimes only slightly. Change is the one great constant in life, and nowhere is this more clearly demonstrated than in the external structure of flu viruses. As a result, the human immune system is not likely to recognize its latest manifestation from one year to the next. These changes can occur via a gradual series of accumulated mutations or by an abrupt modification in the proteins on the surface of the virus.

It is probably no coincidence that the most troublesome new strains of influenza arise in southern China. Agricultural practices there tend to be unusually integrated, involving close associations among people, poultry, pigs, and fish. The result is close physical contact between human beings and the animals they eat, in rearing conditions, at farmer's markets, and even through the common practice of slaughtering poultry, fish, and even pigs by the consumer, at home. Animal manure is used to fertilize both fields and fish ponds, with the crops and the fish consumed by people. Many opportunities thus exist for viral genes to hop back and forth, regularly confronting the world with new permutations that go beyond mutation alone. Thus, wholly new strains appear to be generated from time to time when viral components from pigs, poultry, and people swap genes, eventually coming up with the ability to replicate inside human beings.

Whatever its origin, a new strain of influenza can spread rapidly, if two conditions apply. First, if it presents the immune system with something that is not biochemically recognized as a previous invader;

that is, if prior immunity does not exist. And second, if it is contagious; that is, if it travels easily from person to person. (The Latin *contagio* means "touch" or "contact.") Unlike HIV or Ebola, influenza is readily transmitted through the air: whenever someone carrying the flu sneezes or coughs, millions of flu particles are released in tiny droplets.

This in turn leads to an intriguing possibility. Traditionally, medical science looks at disease symptoms as the nasty effects of the invading organism, whether via toxins or invasion of body cells. Added to this is the recognition that sometimes the consequences of a disease represent the body's response to being invaded, as with fever, which apparently serves, at least in part, to help combat certain infections. But here is the new wrinkle: What if some disease symptoms represent instances in which the disease organisms are actually manipulating the victim's body, for their own benefit?

After all, disease organisms have no conscience. What they have is an evolutionary past, which insures that only those that have been successful so far are likely to be encountered. Part of that success, of course, takes place at the expense of its victims; if there were no costs to housing an invader, we wouldn't call it a disease. When subjected to disease, the body responds to protect itself, for example, by mobilizing its immune system. Not surprisingly, therefore, some disease organisms (notably HIV) go after the immune system. In such cases, the interests of the disease and that of the infected individual are diametrically opposed.

In other situations, however, a sick person and the creatures that cause the sickness may actually be similarly inclined. Influenza viruses inside a flu victim, for example, would like to get out and into other potential victims (since, through evolutionary time, those that managed to do so left more descendants). Part of the flu virus's strategy may well be, therefore, to induce its host to cough and sneeze, thereby spreading copies of itself to other potential hosts. In this case, the body may be strangely complicit: after all, it would just as soon get rid of some of those troublesome invaders, too! Maybe this is why colds and flu produce a runny nose, coughing, and sneezing. The viruses make us do it, and we gladly oblige.

All this would be consistent with influenza being an older, better-adapted disease than AIDS, for instance, which is less adroit at jumping from person to person. It also suggests a new way of looking at other diseases such as rabies, for example. If you were *Formido inexoribilis*, the virus that causes rabies, what better way to get yourself spread than to lodge in the saliva of dogs, raccoons, and other victims, and then induce them to go mad, foaming at the mouth and biting anyone nearby? It might also be possible to carry this interpretation too far. For example, does Ebola virus cause liquefaction of the internal organs and connective tissue *in order to* spread itself via the massive hemorrhages that result? Maybe. To take another case, there is no evidence that infection with any of the various sexually transmitted diseases (AIDS, gonorrhea, syphilis, chlamydia) actually causes enhanced libido and greater promiscuity, although that would seem to be in the interest of the infecting organisms.

In any event, winter is the "flu season," not so much because people are likely to get chilled outdoors at this time, but for precisely the opposite reason: they spend more time indoors and are therefore more exposed to contagion. Yearly flu vaccinations are widely recommended, especially for anyone whose resistance is likely to be low or who comes into contact with many potential flu carriers, which would include teachers or health care workers. Nonetheless, it is not possible to vaccinate for all strains of flu (especially those that do not even exist at the time the vaccines are prepared). By February or so each year, scientists and physicians in Geneva make their best guess as to the *next* flu season's likely villains, then put together an injectable cocktail made of those anticipated virus elements. The immune system prepares antibodies against these elements but not against others.

Once someone gets influenza, the normal symptoms are fever and respiratory ailments, headaches, all-over body ache, and fatigue—much like the common cold, although more serious and prolonged. Even the real influenza virus usually causes illness for only one to two weeks, although some people develop life-threatening complications. The very young, the very old, or the immune-compromised can get pneumonia and other serious respiratory illnesses. Each year, tens of thousands of people in the United States die from influenza. (Note, by the way, that *pneumonia* is a general term for inflammation of the lungs. It does not refer to a specific disease process. Although classic pneumonia is caused by a particular *Pneumococcus* bacterium, people can also develop, for example, viral pneumonia from several different sources, or even a range of chemically induced pneumonias from inhaling things like glue or gasoline.)

Although influenza is caused by a single-stranded RNA virus, it doesn't bother to get itself reverse-transcribed into DNA, as HIV does. Instead, influenza RNA serves as a template from which mRNA is made. As a result, it must be packaged along with an enzyme called replicase, not normally found in animal cells. As its name suggests, replicase serves to replicate the viral RNA. Some research groups are focusing on this enzyme as a potential target for antiviral therapy, although the most promising anti-influenza drugs, including amantidine, seem to be inhibitors of neuramidinase, a protein on the outside surface of the virus. These drugs and others, plus the development of a new test that will be able to diagnose influenza within fifteen minutes at a doctor's office, should aid the treatment of influenza and might eventually help prevent future pandemics.

Another fascinating prospect is to reconstruct the genetic make-up of the killer flu of 1918, to find out why it was so lethal and perhaps to get a head start developing effective vaccines before a similar strain repeats itself. Toward this end, tissue samples from victims of that outbreak are currently being analyzed and subjected to genetic sequencing.

Herpes

Herpes simplex viruses (HSV, or herpes, pronounced "HER-peez") are extremely widespread and contagious. Herpes infection is not usually life-threatening, except sometimes indirectly, in that the sores it creates make the transmission of other diseases, such as AIDS, more likely. But once infected with herpes, you have it for the rest of your life, as do more than 30 million people in the United States alone. (If herpes were lethal, it is unlikely that so many people would carry it; as it is, anyone with herpes has a full lifetime to infect others.)

Herpes viruses cause painful blisters around the mouth and genitals. HSV is divided into two types, I and II. (There are other forms of herpes virus, including Epstein-Barr, which appears to exist harmlessly in the majority of otherwise healthy people, and which, under certain circumstances, also causes infectious mononucleosis.) Either can infect both body areas, although it is more common to have HSV-I in the mouth and HSV-II in the genitals. Either can also be transmitted through contact with an open herpes sore, which also allows for possible AIDS transmission by anyone carrying both herpes and HIV. People infected with herpes often are without symptoms for long periods. Outbreaks are a result of something (unknown) causing the viral genes to be activated—awakened from their lysogenic slumber—and transcribed. It almost certainly breaks out at some point, however, with recurrences triggered by sexual intercourse, stress, or another illness.

The prolonged inactive stages of herpes occur because the viruses are made of double-stranded DNA that must be within a nucleus in order to be replicated and transcribed. The herpes virus is a classic lysogenic lurker, becoming integrated into the host cell's genome as a provirus. At this time, it is also impossible to remove.

The most devastating effects of herpes occur when it is transmitted to newborns, often during childbirth. This can result in mental retardation or death, but more often causes blindness when the sores occur in the eye. Contact-lens users with herpes also need to be aware of this problem.

Although there is no cure for herpes, there are drugs that seem to inhibit outbreaks. Acyclovir, the most common, works by inhibiting certain viral genes necessary for the virus's replication. Other drugs are being developed, although cure seems a long way off.

Prions

Viruses are pretty strange and dangerous creatures, especially since they are so tiny and difficult to fight, not to mention that they inhabit a gray area between the living (when inside a host cell) and the dead (when between hosts). But even then, there is something fundamentally "normal" about them: they have DNA or RNA as their genetic material. Sure, they require a host to replicate. Sure, they look more like interplanetary probes, Leggos, or the leftovers of a microscopic food fight than an elephant or a chimpanzee. But their basic defining attribute is still their nucleic acid, which dictates the cells they infect and pretty much every move they make after doing so.

If viruses seem to have carried minimalism to an extreme, consisting of merely a protein and nucleic acid, prions (pronounced "PREE-ons") have gone further yet. This is one reason why they are so foreign, so little understood, and so controversial among biologists and laypersons alike: they contain *no* genetic material! Yet prions appear to cause some extraordinarily harmful diseases and may well be involved in many other biological phenomena of which we are not yet even aware.

Stanley Prusiner, a neurologist at the University of California at San Francisco, had long been intrigued by certain degenerative diseases of the mammalian nervous system, ailments for which no straightforward pathogen had been discovered. In 1982 Prusiner isolated some odd protein particles from hamsters that had been infected with a sheep disease known as scrapie. Scrapie, in turn, resembles a lethal brain disease known as *kuru*, found among certain human inhabitants of New Guinea.

Ever since, Prusiner has single-mindedly pursued the possibility that the scrapie substance is fundamentally the same as that causing *kuru*, believing the perpetrator to be found among those oddball protein fragments from infected hamsters, which he named prions. Prusiner's work was rewarded with a Nobel Prize in 1997, and although no one in the biomedical research community doubts that prions are real, many remain unconvinced that they cause disease.

Prions are proteins, long chains of amino acids strung together and then contorted in interesting ways. Common shapes for proteins include alpha-helixes (similar to the shape of DNA, looking like a spiral staircase) and beta-sheets (long strands of amino acids lying next to each other). The conformation that each protein assumes is responsible for its particular activity. Prions come in two forms: pN, normally found in every body, and pD, which can cause disease. Interestingly, the amino-acid components of both the pN and the pD forms are identical.

The difference lies in their secondary structure, their conformation, the way in which they fold up on each other, and the different amounts of alpha-helixes as compared to beta-sheets that make up that structure. A small amount of pD does not directly do harm, but in sufficient quantity the prion protein seems to have an unnerving ability to propagate by associating with normal protein and causing it to change its conformation and cause disease. Thus, prions may be self-catalyzing agents: once enough of them are introduced into a cell, they appear to catalyze the conversion of normal protein into disease prions, via a chain reaction. Like the proverbial bad apple that can spoil a whole barrel, bad prions can spoil a cell, and eventually a life, without themselves being alive!

Many ailments are currently labeled prion diseases, including scrapie in sheep, Creutzfeld-Jakob disease (CJD), *kuru*, and Gerstmann-Sträussler-Scheinker (GSS) in people, as well as bovine spongiform encephalopathy (BSE, or mad-cow disease), which might be transmissible to human beings. These illnesses are spread in various ways. They can be caused by direct contact such as an organ transplant from a patient suffering, for example, from CJD. They can also be caused by eating the brains of infected animals (or people), as was the case with the Fore tribe in Papua New Guinea who practiced ritual cannibalism as a way to honor the dead, and thereby contracted *kuru*. Another example is the cows in England who appear to have contracted BSE by ingesting contaminated food, namely, the ground-up remains of other, infected animals. (The resulting anxiety caused by mad-cow disease, and the possibility that it might spread to people, resulted in massive, preventive kill-offs of British beef cattle.) Some prion diseases, such as CJD and GSS, can even be inherited.

In all cases, disease progresses very slowly, as pN is gradually converted to pD. The symptoms of prion diseases usually involve degeneration of brain tissue, with dementia, confusion, and bumbling behavior. The brains of people and animals who have suffered from prion diseases are usually spongy, with holes where bits of brain used to be. The mechanism whereby this occurs is not known, although the normal prion protein appears to be involved with programmed cell death (apoptosis, described in the next chapter), so it is possible that disease-causing prions act by speeding up normal cell loss. Unfortunately, given current limited knowledge about prions, there are no cures for prion diseases; once infected, victims inexorably deteriorate.

Prions appear to be newcomers on the roster of harmful invaders of the human and animal body. More likely, however, they have actually been around for a long time and are only now being recognized. Prions or prionlike phenomena may be responsible for other neurodegenerative disorders and maybe even for some muscle degenerative diseases. Could Alzheimer's be caused or at least influenced by prion-type behavior among proteins? (Stanley Prusiner thinks so.) What about Parkinson's? These and other prion possibilities are being investigated by various research groups around the world.

Bacteria

Bacteria constitute a large portion of the little critters that inhabit our bodies, and whereas prions are new and poorly understood, bacteria have been known for decades, in some cases for centuries. There are good bacteria and bad bacteria, bacteria that we could not live without and others that are extremely deadly. In contrast with both prions and viruses, however, bacteria are unequivocally alive and therefore usually vulnerable to antibiotics. Given sufficient resources and nutrients, a bacterium will eat, metabolize, move, and reproduce, just like the rest of us. Instead of the conglomeration of billions of little cells that makes up a human body, however, each bacterium is a single cell, an independent (although simple) organism. Because they are so simple, bacteria serve as a good model from which people can learn about how living systems work. Because they tend to be parasitic, however, they can also be dangerous.

Bacteria cause more suffering than most people realize. Before 1998, the tuberculosis bacterium, for example, killed more people worldwide than any other disease.* Even such relatively common ailments as stomach and intestinal ulcers, once thought to be caused largely by stress, have been unmasked as the nefarious work of a bacterium, *Helicobacter pylori*. As happens with viruses, there can be sudden and unexplained bacterial outbreaks, occasionally of species previously unknown. In 1976, for example, a group of people attending an American Legion convention in Philadelphia came down with a mysterious illness, characterized by fever, cough, headache, and pneumonia. Several died. The illness was dubbed Legionnaires' disease, and frantic on-the-spot sleuthing attributed it to a new bacterial species, designated *Legionella pneumophila*, which had somehow gotten into the hotel's air-conditioning system.

Bacteria are probably similar to some of the first organisms to occupy the earth. Most of them are found in one of three basic shapes, often reflected in their scientific names: spherical (*coccus*), rodlike (*bacillus*), or spiral (*spirilla*). But regardless of shape, they always consist of a single cell, containing many of the same basic subcellular components—organelles—found in the animal cells we will discuss in the next chapter. Like plant cells, however, bacteria also sport a rigid external skeleton or cell wall. Most dramatically, bacteria don't have a nucleus, the membranous sac that, in animal cells, holds the DNA. A bacterium's DNA generally consists of a relatively long double-stranded chain, which is circular, like a snake holding its tail. In addition, most bacteria possess a number of smaller ringlets of DNA called plasmids, which may contain genes and may be transferred among different bacteria. Plasmids might have originally taken up residence as proviruses, which then became fully domesticated.

There are vast numbers of different bacteria, and even some organisms that are not quite bacteria. The human mind seeks order and structure, not just in its own creations (such as mathematics) but also in the world around us. We like to pigeonhole things, but some pigeons just won't quite fit into their designated slots. As we have

---

According to a 1999 report from the World Health Organization, AIDS currently leads in this dubious competition.

already seen, even the distinction between life and nonlife can be blurred, as with viruses. So it goes for other distinctions, such as between certain bacteria and viruses. There is, for example, a group of peculiar, in-between organisms known as rickettsias (named for American pathologist Howard T. Ricketts), which are responsible for typhus and Rocky Mountain spotted fever (which, incidentally, is more prevalent in the Southeast than in the Rocky Mountains). Rickettsias are tiny—about the size of viruses—but contain a cell wall, like bacteria. Go figure.

Another group of peculiar in-betweeners are the mycoplasmas, which are even smaller than rickettsias and lack cell walls, although they don't quite meet criteria for viruses, either. They are especially implicated in certain difficult-to-treat forms of pneumonia; they cannot be stopped by penicillin or its relatives, since these drugs destroy bacteria by interfering with the construction of their cell walls—which mycoplasma don't have.

One of the most fascinating, useful, and also dangerous aspects of bacteria and their near relatives is their ability to reproduce with extraordinary speed. Some bacteria are anaerobic—living happily in an environment without oxygen—while others are aerobic, requiring oxygen. Either way, and unlike viruses, which require the innards of a cell to reproduce, bacteria need only a suitable medium, essentially a supply of food. Then they can engage in binary fission (splitting in two) as rapidly as every twenty minutes. In this process, the bacterium's DNA is replicated, after which the cell splits into two identical "daughter" cells ("son" cells aren't spoken of, although as we shall see, there is a kind of maleness and femaleness in the bacterial world).

A single bacterium, if placed in optimum conditions for dividing, can become more than 10 million cells overnight. Most of these 10 million bacteria will be identical clones, but as with viruses, mistakes always happen. When such errors occur during DNA replication—for instance, when T matches up with G instead of A—we call the result a mutation. These mutations, although rare, can multiply quickly because of the rapidity of bacterial division. The bacterial population is therefore constantly changing, adapting to new situations and finding new ways to avoid people's immune system and their antibiotics.

Binary fission occurs quickly and completely, and it is asexual. As strange as it may seem, however, bacteria are capable of having sex of a sort: a "male" and a "female" bacteria can share genes in a process called conjugation. The male bacterium extends something called a sex pili, a kind of protopenis that contacts the female and eventually forms a bridge between the two bacteria's internal liquid, their cytoplasm. Over this bridge, DNA can be transferred between the two "lovers." The DNA that is transferred is often in the form of plasmids, although occasionally part or all of the genome can migrate, too (as the DNA is transferred, it is replicated so that the "male" retains all its genetic material). The ability to form a sex pili and therefore be "male" is itself carried on a plasmid, so that it is possible for a "female" bacterium to become "male," if conjugation transfers the sex plasmid to the "female." In bacteria, maleness is contagious!

It isn't only bacteria that can transfer plasmids; biologists have learned to do it, too. Because they contain many different genes, are relatively easy to manipulate, and have a kind of biochemical safe-conduct pass that allows them to enter and leave many bacterial cells with very little difficulty, plasmids have come into their own as primary "vehicles" whereby scientists accomplish much of the genetic engineering and molecular biology discussed in Chapter 1. As we have seen, bacteria grow quickly and easily in a laboratory setting. It is relatively simple to add and take out plasmids from a colony of bacteria, and to test whether genes are present inside the bacteria. Moreover, plasmids are treated inside a bacterial cell in much the same way as the bacterium's "normal" genome. So if a plasmid contains an introduced gene for the production of a certain product, perhaps a hormone or a protein, the bacterium will produce it.

Antibiotics

Bacteria, as we all know, are not all good. When they do harm, it is most commonly by producing a toxin. These toxins can do any number of things, from generating the muscle spasms associated with tetanus to causing the severe dehydration and diarrhea of cholera. Bacterial infections are extremely widespread but fairly easy to treat with antibiotics, which are, in a sense, toxins produced by bacteria or fungi for use against their competitors (and which people can use, too).

Because a bacterium is clearly alive, with structures and functions that are necessary for its survival and reproduction, it is possible to find or develop substances that interfere with these functions and thereby stop bacteria in their tracks. In this crucial sense, bacteria have more chinks in their armor than do the semi-alive viruses. The first challenge in developing antibiotics is to find a substance that will kill the bacteria but not harm the body's own cells. Many of the different processes necessary for the functioning of bacteria are also necessary for the functioning of a healthy body, so an ideal drug would be something of a biochemical cruise missile, programmed to strike its target while minimizing the collateral damage. As a result, most antibiotics focus on the differences between bacterial and animal cells in order to limit their attack to the former.

It is important, by the way, to distinguish between antibiotics and antiseptics. Antiseptics, also known as disinfectants, are strong poisons, effective at killing bacteria but too powerful to be used inside the human body. Chlorine, for example, is a powerful antiseptic, effective in keeping bacteria—as well as algae and just about anything—from growing in swimming pools, hot tubs, and (in suitably small amounts) drinking reservoirs. So are iodine, rubbing alcohol, and hydrogen peroxide. By contrast, antibiotics are much more precisely targeted, "designer drugs" created to attack bacteria or other infectious agents when they live inside human beings or domestic animals.

The most common place of attack for antibiotics is the cell wall. A bacterium's cell wall protects it from invaders and also gives the little creature its shape. Without a cell wall, bacteria absorb water and then burst like a bubble. Animal cells, however, do not have cell walls. The possibility therefore presents itself: find something that will breach or prevent the construction of cell walls and only cell walls, and you have found something that will kill bacteria while leaving animal cells, such as those comprising the human body, intact. This is how penicillin works.

There are also a number of different antibiotics that take advantage of the fact that protein synthesis occurs somewhat differently in bacteria than in animal cells. Common antibiotics of this sort include tetracycline, streptomycin, and erythromycin. They operate by preventing cell division, so that rather than killing bacteria as such, they prevent the next generation from appearing. As already explained, however, no known antibiotic has any effect on viruses.

There are some problems with antibiotics. One is that although they distinguish between animal and bacterial cells, once let loose among bacteria, they tend to be undiscriminating. They kill any vulnerable bacteria, even the good ones essential for digestion and found inside everyone's digestive tract. Large doses of antibiotics, therefore, commonly destroy helpful intestinal bacteria, causing diarrhea and stomachaches. If your doctor has ever recommended yogurt or buttermilk when taking an antibiotic, it is because the bacteria found in these foods can replace some of those that the antibiotic inadvertently kills.

The biggest problem with antibiotics, however, is that many bacteria have developed resistance to them, making them difficult to destroy. Bacteria don't have the wherewithal to become "immune" to antibiotics in the same way the human immune system, for example, responds to smallpox: by stimulating specialized immune cells to manufacture antibodies. After all, bacteria are just single cells, without an immune system in the human sense.

But don't forget those plasmids. Bacterial resistance to antibiotics typically develops in the form of genes on specialized R (for *resistance*) plasmids. Antibiotic-resistant genes often operate via enzymes that destroy the antibiotic. For example, certain strains of the bacteria that cause syphilis have developed an enzyme called penicillinase, which destroys penicillin. R plasmids can be transferred between bacteria during conjugation, so that a small number of resistant bacteria can share their newfound trait with others. (What else are friends for?)

The result is a seemingly unending race between the discovery and development of new antibiotics and the development and spread of antibiotic resistance. The rapid replacement of previous generations of antibiotics with the latest "miracle drugs" is not driven simply by the same consumerist mentality that leads to "new, improved" brands of laundry detergents or underarm deodorants. The issue is more somber: like chewing gum kept too long in the mouth, antibiotics quickly lose their flavor. Bacteria that wilted or simply split asunder when exposed to, say, penicillin, eventually gargle with the stuff. Some strains even find older-generation wonder drugs wonderful indeed: they metabolize them for food!

There is in fact considerable fear in the biomedical research community that it may become impossible to develop antibiotics fast enough to outpace the evolution of antibiotic resistance in bacteria. This explains why doctors are increasingly hesitant to prescribe certain antibiotics unless absolutely necessary. And also why it is very important that if antibiotics are used, they be taken in adequate quantity and persistence to kill off all the disease organisms. If a bacterial population is exposed to moderate or (worse yet) low doses of an antibiotic, it is more liable to develop resistance. So the idea is to practice a kind of scorched-earth policy: take no prisoners, and leave no survivors that might eventually retaliate by becoming drug-resistant.

For a time in the early 1950s, penicillin, the "wonder drug," was routinely added to mouthwash, toothpaste, and even chewing gum, in the hope of wiping out disease-causing microbes, until resistant bacteria began to appear and the connection was made. Even today, animal feeds are commonly laced with penicillin and tetracycline, a practice that must end if these drugs are to retain any effectiveness.

Not that the antibiotic actually *produces* resistance. Rather, by a straightforward process of very small-scale natural selection, any vulnerable bacteria are promptly killed. What is left? Those that are less vulnerable, perhaps because they have a mutated gene that conveys antibiotic resistance. Normally, these bacteria are no better off than their nonresistant fellows, but in a world flooded with antibiotics, any reduced susceptibility to these "toxins" conveys an immense benefit. Resistant bacteria thrive while their less fortunate neighbors are destroyed. Not only that, but the survivors may even find themselves in bacteria heaven, since they no longer have to compete with so many other bacteria. They proliferate rapidly and might well transfer their resistance to any other bacterial cells that happen to have survived (perhaps because the antibiotic was not continued long enough), making yet more resistant cells. These bacteria could then be transferred to other hosts, resulting in even more people being infected by drug-resistant bacterial strains.

This, incidentally, is why "germs" picked up in hospitals tend to be especially troublesome: above all, they are likely to be survivors, genetically toughened descendants of disease organisms that have already been exposed to a continuing array of the best and the brightest antibiotics. Only the resistant survived.

Already, gonorrhea, which used to be among the diseases most easily cured with antibiotics, has developed several drug-resistant strains, apparently because certain antibiotics have been chronically consumed by prostitutes, especially in Southeast Asia, where such drugs are available over-the-counter. Tuberculosis has proven even more worrisome, since it can be deadly. TB can be cured by antibiotics and appeared to be on the ropes, but since the 1980s it has been making a comeback, especially in two new human populations: the homeless and AIDS patients. (The former often lack the money for medication and, in crowded shelters, are especially likely to sleep in close proximity to other infected individuals, whereas AIDS patients, because they are susceptible to so many diseases, often receive chronic doses of antibiotics, which sets up the opportunity for drug resistance to develop.)

Tuberculosis is caused by a bacterium with an unusually thick cell wall. The lungs respond, in turn, by walling off infected cells, forming scarlike "tubercles," which, as they accumulate, reduce the lungs' ability to function. Because the protective cell wall of *Mycobacterium tuberculosis* is so difficult to penetrate, as are the tubercles themselves, it is necessary for antibiotic treatment to be continued for as long as six to twelve months, to make certain that all the culprits are killed. Since most TB symptoms remit long before the last bacteria bite the dust, it is tempting for patients to stop taking their drugs, especially if they also suffer from mental illness and confusion. As a result, a small number of disease-causing TB bacteria get a second chance to reproduce and, moreover, to do so in an environment containing intermediate to low amounts of antibiotic; it is a perfect setup for the evolution of drug resistance. And indeed, one in seven strains of TB are now antibiotic-resistant.

Finally, there is the periodic appearance of so-called flesh-eating bacteria. These are strains of streptococcus, a relatively common troublemaker, most often implicated in strep throat and in a cut or abrasion that we say has become "infected." Nearly always such infections can be treated by standard antibiotics. In 1994, however, a particularly virulent and drug-resistant strain of strep surfaced in Britain, killing eleven people and giving rise to the notable tabloid headline KILLER BUG ATE MY FACE! The "eating" was actually done by toxins produced by the bacteria, assisted perhaps by a malfunctioning immune response on the part of the victims. The result was a rampaging series of infections that could literally demolish a human limb within a day. Fortunately, even these strains eventually succumbed to new drug combinations, but not before contributing to some of medical science's worst nightmares.*

E. coli: The Good, the Bad, and the Ugly

Of the many different kinds of bacteria, *Escherichia coli* is perhaps the most widespread. It swarms by the billions in the intestinal system of every human being, indeed, every known terrestrial mammal. It is also the most prominent laboratory organism, because of its adapt-

---

Here's some limited good news on this front: some success has been reported in developing a vaccine for staphylococcus. Tests on human beings should be completed by 2001 or so.

ability and the ease with which it can be cultivated and manipulated. Without *E. coli*, everyone would suffer from chronic, lethal diarrhea, while many scientists would experience, in addition, a kind of professional constipation, unable to conduct many of their experiments that call for a convenient research subject.

Some *E. coli* strains, however, are very harmful to people and have caused numerous deaths. Because of the variety of situations in which *E. coli* is found, it is a good example of the complicated relationship between *Homo sapiens* and bacteria.

E. coli is one of a few essential microorganisms found within everybody's digestive system. Most *E. coli* inhabit the colon, or large intestine, where their usefulness is threefold: they produce enzymes that help digest different types of molecules; they produce vitamins such as vitamin K and many B vitamins that we cannot produce ourselves but are necessary for certain metabolic functions; and they help solidify the feces. Intestinal gas and the distinctive smell of feces is caused by the metabolism of bacteria such as *E. coli*. In addition, as much as 50 percent of human feces consists of dead bacterial cells. "Shit happens," as the saying goes; for better or worse, much of it is *E. coli*.

Because they are so common, these bacteria have developed many different strains, some of which secrete deadly toxins. This makes *E. coli* a dangerous foodborne pathogen. The most notorious of these is a strain known as 0157:H7, which is occasionally found in the gut of otherwise healthy cows and is therefore transmitted to people when they eat undercooked beef. It has also been found in unpasteurized drinks, such as milk and apple juice. In the latter case, contamination is acquired when fallen apples are bathed in rainwater that has passed through contaminated cow poop. Hamburger meat is particularly susceptible to *E. coli* contamination. Infection with *E. coli* 0157:H7 causes severe bloody diarrhea and stomach cramps, which in most cases resolves in about a week. But it can cause acute kidney failure in children, leading to lifelong complications and sometimes death.

What causes normally benign bacteria such as *E. coli* to become so dangerous? It appears that genes can jump around between species of bacteria, some of which—genes as well as bacteria—are harmful. The toxin secreted by *E. coli* 0157:H7 is the same as that secreted by the *shigella* bacterium, which causes dysentery. It is thought that the problem arises when the relevant gene somehow makes the leap between the two bacteria, perhaps with the help of bacteriophages that infect both types.

Other common harmful foodborne bacteria are campylobacter, which can lead to deadly infections and is found in undercooked chicken, and salmonella, which occurs in eggs and chicken and causes diarrhea and sometimes death. All of these can be avoided by cooking food thoroughly, and all are treatable by antibiotics—provided the strains are not resistant. In addition, salmonella has the odd distinction of being passed through reptile feces and is thus a little-known downside of keeping a pet turtle, snake, or lizard (also, for some unknown reason, at least one perfectly nonreptilian mammal, the African pygmy hedgehog).

Protozoans

Moving up in size and complexity among the various small and harmful invaders, after bacteria we run into protozoans. As we have seen, prions are "merely" proteins and probably cannot be considered alive. Viruses are also extremely small and unable to reproduce and grow without the metabolic machinery of a genuine, living cell. Bacteria, although unquestionably alive, are still relatively simple, and given their cell walls and lack of a nucleus, they are readily distinguished from animal cells. Protozoans, on the other hand, are much more similar to mammals and other animals. They are the smallest of the so-called eukaryotes, meaning that they have a nucleus and other membrane-bound organelles with various functions. Bacteria, which are prokaryotes ("before a nucleus"), do not have any of these structures. Protozoans and all eukaryotes, unlike bacteria, have no cell walls. These differences between eukaryotes and prokaryotes make protozoans more similar to animal cells than to bacteria, even though upon initial inspection a protozoan and a bacterium seem similar, at least in overall size.

All animal cells—therefore all human cells—are eukaryotic. In the not-too-distant past, bacteria were considered plants and protozoans, animals. No longer. Bacteria were unceremoniously evicted from the plant kingdom because they lack both chlorophyll and a nucleus. Protozoans were kicked out of the animal kingdom because of their stubborn individualism: their insistence on living as a distinct single-celled creature, a eukaryotic cell independent of the other cells surrounding it. An animal cell, on the other hand, is simply one out of millions that are working together to create an organism. This independent lifestyle of protozoans wouldn't be troublesome if they would only be consistent and truly live their own lives. In fact, many do, including the well-known amoebas that can be taken from many freshwater ponds and admired under a child's microscope.

But a number of protozoans insist on combining the trappings of independence with residence inside the human body, which makes them very effective parasites, often creating unique problems associated with harboring an unwanted, semi-independent lodger who takes advantage of his host.

There are many different types of protozoans. All of them, like miniature animals, feed off bacteria, other one-celled organisms, or dead material. Some are parasites, but others have important and necessary relationships with their landlord. These protozoans often live inside the digestive tract of a larger animal. In this relationship, called symbiotic, the protozoan needs the host animal in order to survive, and the host needs the protozoan to perform some vital task. One striking example is a type of protozoan called a flagellate, which inhabits the intestines of termites, where it has the job of digesting wood. The flagellate secretes cellulase, an enzyme needed to break down cellulose, without which termites would be no more able to live off wood than human beings are. (When termites molt, they lose their intestinal lining with its trove of flagellates; accordingly, a crucial ritual of polite termite society calls for a freshly molted termite to lick someone else's anus, thereby reinfecting itself—so it can continue digesting wood.)

Protozoans also live within the human digestive tract, much like bacteria such as *E. coli*. There is considerable variation among different types of protozoans, however, not all of it beneficial.

Probably the best-known protozoans are amoebas, characterized by their strange lack of structure and their odd way of extending bits of themselves outward to form pseudopods ("false feet"), which are used to move and to corral bits of food. The common pond-dwelling amoeba is technically named *Amoeba proteus*, after the Greek god notable for his ability to change shape at will (this same god also gave his name to proteins, those ubiquitous chemicals that come in a seemingly infinite variety of shapes). Although most amoebas are harmless and live freely, some cause dysentery when ingested by people.

Another well-known group of protozoans are the ciliates, such as paramecia, characterized by little "hairs" on their surface. In the case of paramecia, these help the creature to move and also to transfer food into its oral groove, which functions like a mouth. Ciliates do not generally pose much of a problem to people, except for the occasional freshman biology student who may have trouble recalling them on a test.

Other types of protozoans can be extremely dangerous, however. Malaria, a terrible disease transmitted by the *Anopheles* mosquito, is caused by *Plasmodium*, a type of protozoan called a sporozoite. Sporozoites have extremely complicated life cycles, making them very hard to track and contributing to their other name, *apicomplexa*. About half the life cycle of the malaria plasmodium is spent inside a mosquito, and half inside an unwitting human host. When inside people, it penetrates the liver and then the blood cells, somehow hiding from the body's immune system and, at the same time, regularly changing its outer structure in order to hide even more effectively. Periodically, the plasmodia emerge from the victim's red blood cells, causing the bouts of weakness and high fever that are characteristic of malaria.

Another type of protozoan, the zooflagellates, can also be very dangerous. The creatures that cause African sleeping sickness are zooflagellates known as *Trypanosoma*. They, too, are quite skilled at hiding from the body's immune system; like plasmodia, trypanosomes change their exterior coatings very quickly. And so, like a skillful "most wanted" criminal, they elude detection by constantly slipping into a new disguise. Zooflagellates are characterized by a long, thin, whiplike appendage issuing from the cell's surface. This flagellum, as it is called, is used primarily for movement.

Among the unicellular—or, as some experts like to say, acellular—eukaryotes that cause disease in people, the intestinal parasite giardia is yet another "in-between" creature. Giardia is believed to be partway between prokaryote and eukaryote, having some features common to both. For example, prokaryotes such as *E. coli* have only one set of DNA; they are "haploid" like human eggs or sperm. Eukaryotes have two sets of genes and chromosomes housed inside one nucleus; they are "diploid" like other human body cells. Giardia and other creatures of their peculiar ilk have adopted a strange system indeed: not one nucleus but two, each of which is haploid.

Seen under the microscope, giardia look eerily like a human face, with each primitive nucleus resembling an eye. Some have suggested that the group to which giardia belongs, the archezoans ("old animals"), warrant their own kindgom. Regardless, they may well be a sort of living fossil or "missing link" between prokaryotes and eukaryotes, as though some ancestral prokaryote doubled its nuclear material but the two haploid nuclei have not yet fused to form a diploid nucleus as in eukaryotes.

Giardia are the bane of campers, hikers, and travelers to developing countries because they are commonly found in contaminated drinking water. As recently as the 1970s, before giardiasis—unjustly called "beaver fever"—had spread throughout North America, it was possible to hike in the Rocky Mountains, the Cascades, and the Sierra Nevada and to drink without fear from nearly every stream. No longer. Giardia cysts are transmitted by fecal material and when consumed produce the disease giardiasis, which results in severe cramping and diarrhea. Effective treatments are available, although the most widely used medication, flagyl, causes side effects that are nearly as unpleasant as the disease it cures. Giardia spores can be killed by boiling unclean water or treating it with iodine; they are also large enough to be removed by standard purifying filters. Interestingly, many people carry inactive giardia and can transmit the virulent disease without actually suffering from it themselves.

CELL-SIZED:

Building Blocks of the Body

"Just another brick in the wall," repeats the chorus of "The Wall," by the British rock group Pink Floyd. They sing it mournfully but angrily as well, and no wonder: we all value our specialness, our individuality. No one wants to be just another brick in the wall. And biologically, at least, we're not. Each of us is unique, especially when it comes to our genetics and our brain. We aren't bricks in a wall; rather, each of us *is* a wall (of sorts), composed of a very large number of very tiny "bricks" known as cells.

It isn't immediately obvious why this is so. Based on principles of engineering, for example, there is no reason why a human being has to be cobbled together from innumerable little building blocks, instead of being fashioned in one piece using the kind of unibody construction so popular in automobile manufacturing. The most likely explanation is evolution. That is, human beings as a species were not "created" anew, from scratch, any more than is an individual human being. Rather, we are contingent creatures, having evolved gradually from living things that preceded us, which in turn evolved from other creatures, and so forth, back to the primordial ooze. Although life itself presumably began when molecules started to self-replicate, initially copying themselves and drawing energy from the nonliving nutrient broth that surrounded them, in all probability these early life-forms quickly discovered that there was a benefit to being surrounded by some sort of structural membrane, if only to keep themselves intact.

Among such early life-forms, those that just stayed put, happily ensconced within their private, fortresslike protective membranes, didn't leave any footprints upon the future. To do so, they would have to reproduce themselves so that their descendants *were* the future. Any stay-at-homes eventually died, leaving no trace. So, it wasn't the meek that inherited the earth; it was the reproductive. The early earth environment came to be dominated by primitive one-celled—or no-celled—organisms, notably those who got into the habit of dividing in two every once in a while.

But how did this lead to multicellular bodies as we know them today? No one knows, but it probably began when one or more of these unibody life-forms divided and then, instead of each offspring going its own way, they either stuck together or perhaps rejoined.

At this point, they (now "it"?) might well have stumbled upon the benefits of division of labor, and various subunits began to specialize—in food-getting, locomotion, digestion, coordinating the actions of other subunits, reproducing, and so on. It stands to reason that a cell specialized for, say, contracting (i.e., a muscle cell) wouldn't be nearly as good at filtering as a kidney cell, whereas if both muscle and kidney cells could be combined, the resulting "individual" would have the advantage of being both a good contractor and a good filterer, instead of being a mass of general-purpose cells that aren't terribly good at anything.

But why should these different cells agree to work together? And why should kidney cells, for example, cheerfully undertake to do such a nasty job as detoxifying the blood, leaving all the fun of reproducing to the gonads? The answer is not simply that these cells constitute a single body and thus have no choice, because it is quite possible to imagine systems in which cells—perhaps those dissatisfied with their lot—go elsewhere. But cells are unlikely to be cantankerous or otherwise jealous of one another's success for one crucial reason: since they all derived from the same dividing "mother cell"—these days, a fertilized egg—they are genetically identical. Accordingly, there was (and still is) good reason for each of these subunits to cooperate. After all, success for those cells doing the digesting or contracting resulted in identical success for those doing the thinking and the filtering, and vice versa. The success of one is not only achieved via the success of the others, it is exactly equivalent. Accordingly, when those cells that specialize in reproducing (either as testes or ovaries) succeed in their mission, cells in the kidneys or salivary glands are also hitting the evolutionary jackpot. The result? Bodies (walls) made of innumerable small cells (bricks). Moreover, cooperation among those cells.

Such cooperation is a key concept in this chapter and should not be taken for granted. It is worth appreciating, for example, that when a cell settles down to become, say, part of your biceps, it is undergoing a remarkable, indeed a truly historic, transition. For more than 4 billion years, its ancestors had reproduced and kept on reproducing. That cell in your biceps, after all, did not derive from a nonreproducing biceps cell in your mother or father but from actively dividing egg and sperm cells, which in turn owed their existence to successful reproduction of the preceding generation's egg and sperm cells, and so forth, back to the earliest living globs. For every one of the trillions of cells in your body—except your gonads—the "decision" to settle down and refrain from reproducing, or at least to specialize in being a muscle, gland, or nerve cell rather than pass its direct descendants into the future, had never been made before, since life began! That is, for 99.9999 percent of your cells, this body that you call yourself is the first and last dead end ever encountered.

In this chapter we shall take a look at the world of these cells, because each us is made up of many thems.

Parts of a Cell

Every mammal, bird, reptile, amphibian, and fish is a multiplex conglomerate of cells. Being cell-built is part of being distinctly mammalian, perfectly animal, and altogether alive. Cells, however, are not the most basic of the building blocks. We have already seen that DNA and its relatives are comparatively simple molecules that are necessary for life. And viruses, whose very aliveness is arguable, are a lot smaller then cells. It is better, perhaps, to think of cells as basic packages containing all that is necessary to be alive. But although they are the "bricks" of a body, they are not homogeneous. Every cell is a complex, highly organized corporation of different chemicals and mini-organs, so-called organelles, that collectively keep a body moving and functioning and in fact provide much of the structure that makes that body a body.

More than three hundred years ago, the great British physicist and chemist Robert Boyle wrote that it is "highly dishonorable for a Reasonable Soul to live in so Divinely built a Mansion as the Body she resides in, altogether unacquainted with the exquisite structure of it." As a Reasonable Soul, it is time to become better acquainted with the exquisite structure of your body.

To pursue such an acquaintance, it is necessary to have a good feeling for the basic parts of a cell and what they do. Interestingly, cellular organelles vary dramatically among living things. There are two main subdivisions: eukaryotic and prokaryotic. (*Eukaryotic* is from the Greek: *eu* means "true," and *karyon* means "kernel," referring to the nucleus.) Prokaryotic ("before the kernel") cells such as bacteria are the most primitive; they lack a nucleus and most of the other organelles that are found among those eukaryotic cells that make up any mammal, including *Homo sapiens*. There are two basic types of eukaryotic cells: animal and plant. The main differences are that plant cells have a cell wall, which animals lack, and also chloroplasts, which they use to harvest the energy of the sun directly, by photosynthesis. People, our focus, are made of eukaryotic animal cells. (*Eu*, incidentally, is pronounced "you." Therefore, if a *Homo sapiens* is interrogated by an extraterrestrial who asks "You karyotic?" the correct answer is "Yes, and eu?")

The human body contains approximately 10 trillion cells, of about 100 different types. And these various types are concentrated in particular parts of the human geography. If the body were a city, it would be a peculiarly segregated one, divided into numerous neighborhoods, each with its own characteristic makeup; the plumbers would live in one part of town, the lawyers in another. When we speak of lungs, heart, liver, brain, bone, et cetera, we are identifying not only an organ but also a kind of cell, each with a particular "profession," even though all are genetically identical.

Rather than proceed through a catalog of cell specialization, in this chapter we'll look at some of the things that all body cells have in common, first at the level of physical structure and then with regard to the universal cellular timetable. Specifically, we'll start with tiny bits of subcellular anatomy known as organelles, before turning to the basic pattern of life, division, and death, known as the cell cycle. Then we'll look at cell communication and examine one of the most common things that goes wrong with cells: cancer. We'll conclude with some of the ways the body uses its cells to preserve itself.

The first organelle that demands attention is the nucleus, the widely touted "control center," the bull's eye. The nucleus houses a creature's DNA, as well as an array of helper proteins. The inside of the nucleus is strictly controlled, a limited-access zone with small, selective pores that open to allow only molecules that must interact directly with the precious load of DNA. These molecules, such as steroid hormones like testosterone and estrogen, as well as other messengers, are essential for controlling which genes get turned on and off at any particular time. When turned on, a gene is transcribed into mRNA, which is then escorted out of the nucleus into a large system of membranes surrounding the nucleus called the endoplasmic reticulum, or ER. (In biomedical science, ER doesn't only stand for "emergency room.")

Reticulum is Latin for "net" or "network," and *endoplasmic* simply means that this particular network lies deep "in" (*endo*) the bowels of the cell's substance, or plasm. The ER is made of several different types of molecules that form a netlike barrier within a cell. This barrier is usually impermeable to most waterborne substances. But it has embedded proteins that serve like selective gates to keep certain substances in and others out. The ER looks like a collection of sheets that were folded by a two-year-old, with the corners not quite matching and the folds uneven. It comes in two varieties: rough and smooth. The smooth ER helps out in the synthesis of lipids, molecules such as fats, oils, and steroids that do not dissolve in water. The rough ER has stuck on its outside a dense speckling of ribosomes, bumpy little organelles that are visited by tRNA, whereupon they manufacture proteins using messenger RNA as a template.

Since there is no such thing as a typical cell, it can be very misleading to generalize about how prominent the ER is in "the" cell. For example, there is virtually no ER in red blood cells (which don't synthesize much in the way of proteins, and which have a very short life span and don't even divide), whereas liver cells have a huge ER, consistent with their job as biochemical factories.

Most proteins in an animal cell are made by the rough ER, although some ribosomes are also found floating loose in the cell's fluid. Those proteins made on the ribosomes of the rough ER are bound for further processing in our next organelle of interest, the *Golgi apparatus* (discovered by—no surprise here—Golgi).

Looked at through the electron microscope, the Golgi apparatus consists of numerous platelike structures, flattened in the center and expanded on the outside, arranged in a stack. After leaving the ER, newly minted proteins or lipids are bound up in a tiny pouch or vesicle made of a small amount of membrane. This membrane empties its load into the Golgi, where the contents become separated according to their destinations—inside the cell, outside the cell, to another organelle, and so forth—and then processed and repackaged for their journey. The Golgi is, in many respects, the central dispatching station that organizes and puts together the cell's products and sends them out to wherever they are needed. It is our own private delivery service, the UPS within.

All this manufacturing and processing takes energy, however, which is provided by structures known as mitochondria in animal cells, or chloroplasts in plant cells. We'll spend more time with mitochondria when we investigate energy in Chapter 6. For now, note that after food is eaten, it is eventually broken down into its component parts. These food molecules make their way to the mitochondria, which are membranous sacs with lots of internal cavities containing many different enzymes. In turn, these enzymes, in a carefully controlled manner, break down the molecules even further to carbon dioxide and water, releasing energy along the way.

To review: the nucleus houses the DNA as well as some RNA and proteins needed for the control of gene transcription. Ribosomes manufacture proteins and are thus the site of translation. The rough and smooth ER serve as factories to put together proteins and lipids, respectively. The Golgi apparatus sorts, modifies, and sends out different molecules from the ER on their way to their final destination. And finally the mitochondria (in animal cells) and chloroplasts (in plant cells) act as energy banks. In addition to these major organelles, most cells also have other, smaller pouches known as vacuoles (from the Latin for "empty"), which serve as the cell's disposal sites and storage dumps. They don't stay empty long.

There is even a particular type of structure known as a peroxisome whose sole purpose appears to be to store hydrogen peroxide, a by-product of certain metabolic reactions. Since peroxide is a potent antiseptic—which is why it is also used to treat cuts and abrasions—cells find it necessary to isolate it inside these tiny biohazard containers, within which the poisonous peroxide (H_2O_2) is gradually broken down into harmless water (H_2O) plus oxygen.

Another notable organelle is the lysosome, a sac filled with highly acidic digestive enzymes, formed in the Golgi. Lysosomes travel about, latching on to the tiny food-filled vesicles that are pouched off from the cell membrane when the cell "eats" or "drinks," and helping to break down complex molecules into simpler ones.

One of the best ways to clarify the role of any biological structure is to see what happens when it malfunctions. In the case of lysosomes, for example, there are a variety of so-called lysosome storage

diseases, genetic illnesses that occur when, because of certain mutations, lysosomes lack one or more appropriate enzymes. As a result,
molecules that would otherwise be broken down by the enzyme in
question simply accumulate in the lysosomes, which end up storing
things rather than digesting them. A dramatic and tragic example is
Tay-Sachs disease, a hereditary, congenital illness in which infants
appear normal at birth, but their ability to move deteriorates progressively, leading to paralysis and death by age three or four. Tay-
Sachs babies are unable to metabolize fats in their central nervous
system; the lysosomes become immense and distended with undigested fatty chemicals. There is as yet no cure for Tay-Sachs,
although it is possible to identify carriers of the culprit genes.

The image you might have now of the human cell is as a membrane sac filled with little pouches and membranous sheets, sloshing
about at random inside a pool of liquid glop. The reality is quite different. Any healthy cell is highly organized, with its own characteristic internal structure, called a cytoskeleton, giving it shape and substance and providing a means whereby different chemicals can
dutifully move around within the cell. There are, in turn, three different components of this cytoskeleton: microfilaments, intermediate filaments, and microtubules. Each type of skeletal structure has a
different, specialized job, and they all work together to keep things
moving correctly within the cell. The microfilaments are mostly
responsible for changes in cell shape, such as muscle-cell contraction
and division of the cytoplasm, which we shall discuss shortly. Intermediate filaments are mostly important for maintaining cell shape,
much like the poles of a tent. And finally, microtubules are essentially sluiceways that allow different organelles and structures inside a
cell to move around. All three are present in varying patterns and
amounts; they form a complex network that crisscrosses each cell,
giving order to the cytoplasm and giving the cell its own distinctive
structure, shape, and—to some extent—function.

Cell Division: Mitosis

Imagine for a moment that you are a single cell within a multicellular creature, perhaps a human being. You have a nucleus, a collection
of intracellular membranes such as the Golgi apparatus and your
various endoplasmic reticuli, a handful of mitochondria, and a huge
number of crisscrossing filaments and tubules making up your

cytoskeleton. Around you are millions of other cells, the nearest ones probably quite similar to yourself, all working together to make a living individual.

Many questions arise. For starters, how do you coordinate your actions with your fellow cells? After all, a body's motto could well be drawn from *The Three Musketeers*: "All for one and one for all." Easier said than done, however. And yet there is little room for loners or soloists here. Although cells are the bricks that make up a being, they are also active and changing every second, and to be good team players, they have to adjust to their fellows. In short, they must communicate with other cells, as we shall see later in this chapter. At the same time, and like the individual they constitute, individual cells grow, die, and reproduce. Some even move around. Each cell passes through its own personal life cycle, and this dictates what the cell is capable of doing at any instant.

Perhaps the most noticeable part of any cell's life cycle occurs when it divides, becoming two identical "daughter" cells, each containing the DNA and all the organelles of its parent. The process is called mitosis, and although a very complicated series of events, it can proceed with great accuracy and speed, leading to vast numbers of identical descendants. Mitosis is necessary for growth (not of the cell, but of the body). It is the way we get from fertilized egg to full-fledged adult, but it doesn't stop when growth stops. It occurs constantly in people, if only to replace cells that have died or are damaged.

In some cases, the turnover rate of body cells is very high indeed; red blood cells, for example, are produced in just a few hours. You doubtless recall your past, and think of yourself as having experienced continuity of existence as far back as your memories stretch. (This may be due, in no small part, to the fact that cells of the central nervous system are among those that do *not* undergo mitosis!) But for most of the body's cells, there is a great deal of truth to the old advertising jingle, "There's a new you coming every day"—thanks to mitosis.

The different stages of mitosis have all been named. (There is nothing like naming something to give the illusion of control or comprehension. On the other hand, only after things are named can they be focused on and eventually understood.) The mitotic stages are: prophase, prometaphase, metaphase, anaphase, telophase, and finally cytokinesis, which is not strictly part of mitosis but occurs at roughly the same time as telophase. Watched through a microscope, the various phases of mitosis are beautifully choreographed, literally a dance of life.

During prophase, DNA in the nucleus becomes tightly wound until each chromosome looks like the letter *X*. By this stage, the DNA has already replicated, at an earlier step, and every chromosome consists of two "sister chromatids," which are identical copies of the original chromosome. Each sister is two, adjacent legs of the *X*, and their meeting point is called the centromere (the Greek *mere* means "part"; the centromere is simply the central part of the replicated chromosome). Also during prophase, formations of microtubules emanating from something called the centriole move to either side of the cell body, trailing bundles of microtubules between them.

During the next step, prometaphase, the nuclear membrane breaks, allowing the microtubules, which had formed outside the nucleus, to come into contact with the sister chromatids. Some of the microtubules attach at the centromere and begin to jerk the sister chromatids around until, at metaphase, the microtubules have lined up the sisters in the center of the cell, on an imaginary line called the metaphase plate. The two sister chromatids orient entirely opposite each other, with microtubules pulling each to a different side.

With the onset of anaphase, the sister chromatids split apart. At the same time, the cell begins to lengthen until each end has a full set of DNA and the two ends begin to separate. During the next step, telophase, nuclear membranes begin to re-form around the two separated sets of chromosomes, and the DNA becomes slightly less condensed. At the same time, cytokinesis (literally "cell movement") is taking place, and the contents of the cytoplasm are split, just as the DNA had been. This is the final step in cell division, when a single cell has become two, and the two daughter cells pinch apart, because of a cleavage furrow made from a ring of contracting microfilaments. The resulting cells are fully functional, capable of undergoing the whole process again if conditions warrant. Each is identical to the other and to the parent cell, so long as no mistakes were made.

The Cell Cycle

Mitosis, for all its phases, is only a small part of the entire cell cycle. Even when a cell is not actively dividing, it is always doing something, either preparing to divide or fulfilling its function as an active member of that community we call a body. (Very few cells, if any, are simply along for the ride.) As a result, the cell cycle is usefully divided into two main segments: the mitotic phase and interphase. The mitotic phase, or M, consists of the mitotic dance plus cytokinesis, during which the nucleus and all the organelles are split into two separate (but identical!) daughter cells.

Interphase is the rest of the time. It is what cells do when they aren't busy on the mitotic dance floor; in short, when they are going about their usual lives. Interphase consists of two different "gaps," G_1 and G_2, separated by a "synthesis" phase, S, when the amount of DNA in the nucleus is doubled, preparatory to dividing once again. The life of a cell, therefore, proceeds as follows: G_1, when it synthesizes proteins and grows; S, when it replicates its DNA in preparation for division; G_2, when it grows some more and makes final preparations for cell division, including more protein and lipid synthesis; and M, when the cell divides into two daughters. These offspring may then repeat the cycle. In addition, there is a quiescent phase, called G_0, during which the cell is not en route toward active division.

Different cell types cycle through their lives at different rates. Those lining the intestine may take only a day or so to divide; skin cells require a bit longer; and nerve cells almost never do so. Interestingly, the duration of mitosis itself hardly varies at all among different cell types; what matters is how long a cell spends in G_1, S, and G_2. It is as though each cell type takes a different, circuitous route to the same waterfall. Some travel for hours, days, or even weeks along twisty backwaters. When—and if—they reach the falls, however, it takes about the same amount of time to go over. Continuing the metaphor: by the time they arrive at the bottom, they have been split in two, although the dividing is a gradual and rather elegant process, nothing traumatic like being cracked asunder at the bottom of Niagara!

Also, cells don't wander aimlessly before they divide. A number of crucial steps occur during these different stages, each of which is important for understanding how any cell works. In an adult human being, most cells opt out of the cycle and spend their time in G_0, which is a kind of limbo, somewhat removed from interphase. Once they are fully developed, for example, nerve and muscle cells are permanently in G_0 until they die. These cells have very specialized functions and shapes and do not divide. As a result, when you lift weights to "build up muscle mass," you are actually making individual cells stronger and bigger, not creating a greater number of muscle cells. Similarly, when adults learn something new, they are either recruiting already present but inactive nerve cells or simply setting up new connections in the brain among those nerve cells already there. This, at least, has long been the received wisdom among neurobiologists: new nerve cells, like matter according to the Newtonian view, are not created.* (Although unlike matter, they can be destroyed.) Others, such as liver cells, spend most of their time in G_0 but are able to re-enter the cell cycle to regenerate tissue if conditions warrant.

The decision for a cell to go into semiretirement in G_0 or to continue along the cell-cycle pathway is made at a point in G_1 just before S phase, or DNA synthesis, commences. It is called the restriction point and is controlled by different molecules that cycle through the cell's system.

Finally, cells die. They might rupture from too much water flowing in or shrivel from too much cytoplasm leaking out, they might be eaten from within by viruses or bacteria, dissolved by toxins, sliced, crushed, starved (of food or oxygen), irradiated, or chemically poisoned. As if these slings and arrows of fortune weren't enough, most cells face something inexorable, if not outrageous: programmed cell death, also known as *apoptosis* (pronounced "a-poe-TOE-sis" and derived from the Greek, referring to "a tree that sheds unneeded leaves"). During the development of an organism, the number of cells increases, but certain cells also have to die for most living things to reach normal adulthood. In a dramatic example of this, the worm *Caenorhabditis elegans* (a simple, tiny nematode often studied genetically) starts out with 1,090 larval cells. Before it reaches adulthood, exactly 131 of these cells die, in every animal! And it is the same particular cells each time.

People are made of many more cells, and indeed, a huge number must die for the body to survive. It appears, for example, that nearly half of all brain cells commit apoptotic suicide—or should we call it "cellicide"?—in the course of every life. This pruning creates a kind of neuronal topiary, with emphasis on function instead of beauty. Apoptosis is normal, natural, and necessary. Generally it is an orderly process, as opposed to the messy business of unplanned cell death, or *necrosis*. But the possibility exists that, at least on occasion, apoptosis gets out of hand, leading to the excessive nerve-cell die-off associated with Alzheimer's disease. In this case, drugs that block overenthusiastic apoptosis could contribute to Alzheimer's prevention or even cure.

---

*Because of new findings in the late 1990s, this view has been changing. Although it is still clear that brain cells do not divide in anything like the frequency of skin cells, or those of the intestinal lining, it is now clear that—at least in certain cases—nerve cells are born anew in a region known as the hippocampus (see Chapter 4).

Cancer's Causes

Cancer is perhaps the best-known and most deadly example of the cell cycle gone awry. About one in three Americans will develop some form of cancer eventually. Cancer is not one illness, however; rather, it is a catchall term that includes more than a hundred different diseases, usually named for the organ in which it first arises (liver cancer, breast cancer, skin cancer, etc.). All cancers have two things in common: first, they are characterized by indiscriminate growth on the part of cells, an out-of-control process that, unless stopped, can lead to death. Second, all cancers are in a sense genetic diseases in that they are caused by malfunctions in the genes that otherwise keep cell division in check. (This does not mean that all cancers are literally inherited, although in many cases a predisposition to cancer is hereditary. Rather, even when induced by some external agent such as exposure to radiation or certain chemicals, cancers are caused by disruption in the normal DNA of a cell.) For scientists, cancer is a valuable learning tool for understanding how the cell cycle is controlled. It provides a window into the basic control mechanisms of living cells, because of the dire consequences when these mechanisms fail.

To begin peeking through that window, let's step outside the human body for a moment and visit cells grown in vitro, that is, in a controlled laboratory situation. Scientists have become quite adept at growing cells, or "culturing" them, on small plates called petri dishes. Usually, these dishes are coated with different nutrients and food substances that enable a variety of cells to divide and grow. Certain things are necessary for cell division, such as the essential amino acids that we will discuss in our chapter on energy. Cells need other chemicals as well, in addition to basic "food." These essentials are often called growth factors. Recall that at a particular stage in a cell's life cycle, it reaches the so-called restriction point, a kind of crossroads at which every cell must travel down the path to G_0 (inactivity) or eventually to mitosis. Growth factors work by attaching to specific receptors on the outside of cells and initiating a cascade of events that push the cells past the restriction point between the G_1 and S phases. This occurs when cells are cultured in a laboratory as well as during natural growth inside a human being.

Another powerful controller of cell division appears to be their density. When cells are cultured with all the necessary ingredients, including growth factors and nutrients, they usually stop dividing when they cover the bottom of the petri dish. If some of these cells are then taken away with a small spatula, those remaining around the newly created hole will begin dividing again until the bottom is covered once more. Healthy cells will not crowd on top of one another, probably because they need direct contact with the growth medium on the bottom of the petri dish; deprived of contact, they are unable to make it past the restriction point and they stop growing. Cancerous cells, however, ignore whatever signals cause normal cells to avoid excessive crowding. They climb all over one another, piling up in multiple layers on a petri dish. This behavior extends beyond cells in culture: healthy cells in the body are generally able to grow only when they are attached in the correct place to the extracellular matrix (a set of molecules that forms a structure between cells in the body), and they stop growing when they have taken up the space "set aside" for them. Cancerous cells do not respect this boundary. They are pushy and obnoxious. They crowd beyond their rightful territory onto the turf of neighboring cells.

As if this weren't bad enough, cancer cells, in culture, also show abnormal life spans. A healthy cell, even if given plenty of nutrients and exposed to all the "right" conditions, usually is only capable of dividing five to sixty times. A cancer cell can last much longer, more than forty years in some cases and perhaps indefinitely, dividing all the while. In addition, even when and if these troublemakers stop their relentless reproduction, they don't always do so at the normal checkpoint. Cancer cells, as a result, are typically uncontrollable invaders that overstep their boundaries, take over other tissues, and keep on going.

The development of most cancers is as follows: first, a normal cell is somehow "transformed"—the official term—into a cancerous cell. (We explain how this happens below.) This cell begins to divide uncontrollably, but in some cases the body's immune system is able to recognize such aberrant behavior and destroy it before its descendants become too numerous. If the transformed cells evade detection, however, they and their progeny can form a tumor, which is essentially any abnormal growth. If these cells show at least a modicum of manners and remain in place, the tumor is benign and is in most cases easily removed by surgery. By this time, it usually consists of from 1 million to 100 million cells.

Tumors, especially benign ones, are relatively common. Warts and moles are actually small benign tumors. If, however, things are really awry in the cells' control system and the tumor continues to grow, perhaps getting in the way of an organ or system, the tumor is considered malignant, and the person or animal is said to "have cancer." These cells generally look weird: they frequently have ragged edges and large, oddly shaped nuclei; when looked at on a microscope slide, an unusual number are typically caught in flagrante delicto, actively dividing. Cancerous cells are also undifferentiated to varying degrees; in general, the farther along the malignancy, the more undifferentiated the cells. (Thus, a cancerous muscle cell, for example, ceases to look or function like a normal muscle cell. Just as it has become a reproductive freelancer, the cancer becomes less specialized the more it advances; it is no longer subordinated to the good of the body.)

Malignant cells can continue growing more or less in place, or they can move to other sites in the body. When this happens, the cancer is said to metastasize. This is really dangerous.

A metastasizing tumor can kill in a number of ways, depending on its type and location. Most of the time, cancer cells are derelict in their duty: they stop producing the correct substances and contributing to the body's well-being in the same way their healthy predecessors did. If nothing else, therefore, cancers are troublesome freeloaders, using up nutrients and giving nothing back in return. But this is the least of the problems they cause. In fact, cancerous cells often excrete toxins that actually poison other cells, in some cases killing the body as well. A tumor may also grow in a way that impairs the functions of an organ or organ system; more commonly, the metastasizing cells form new growths in detrimental places within the body. Metastasizing cells can get so out of hand that they basically take over the body, overwhelming the immune system and compromising other functions. Also, they are unusually efficient at transporting and using nutrients. So they are not only ill-mannered, pushy, and hurtful, they are also greedy, stealing foodstuff that would otherwise go to normal, essential body cells.

The best defense against cancer is to avoid getting it in the first place. But this is more easily said than done because many cancers are, in a sense, "natural." Most of them are less the result of infection by some exotic agent—like Ebola or a dangerous strain of *E. coli* bacteria, for example—than a result of normal processes, often speeded up by exposure to chemicals and energy that cause cell damage, especially mutations, or abnormal changes in the DNA sequence of a cell. These mutations typically affect the genes in charge of the cell cycle, most often those controlling the checkpoint, that stage at which cells normally cease dividing. Like most well-guarded checkpoints, those in the cell cycle are generally equipped with redundancies and multiple safety backup systems. Accordingly, in most cases it requires multiple mutations to overcome these fail-safe mechanisms.

Unfortunately, although individual mutations are rare—and, therefore, multiple mutations even rarer—there are enormous numbers of cells and, thus, immense numbers of cell divisions. As a result, there is an impressive possibility that even such a low-probability event as several malignancy-generating mutations can occur simultaneously in the same cell.

In a sense, we could say that given enough time and enough cells, mutations can be expected. This might seem surprising, given that the rate is only about 10^{-6} mutations per gene per cell division. (That is, one mutation in a million, per gene.) But consider the huge number of cell divisions that occur during a human lifetime. As a result, each gene has many opportunities to mutate, although it is important to remember that most mutations have no effect. In order for cancer to develop, somewhere between three and seven independent mutations of a certain type have to occur in the same cell.

Anything that generates mutations is likely to contribute to cancer. Such "mutagens" include ultraviolet radiation, high doses of X rays, and tobacco smoke. It appears that certain chemicals, known as free radicals, contribute to mutations by helping to oxidize and thus break down nucleic-acid molecules; this has led to increased interest in antioxidants such as beta-carotene and vitamin E as a way of countering the mutational drift toward cancer. In fact, a standard laboratory procedure known as the Ames test is used to evaluate all new drugs for whether they are mutagenic, therefore likely to increase the risk of cancer. In the Ames test, the substance in question is added to a strain of salmonella bacteria, which normally cannot grow without a particular nutrient. If the bacteria manage to grow, it must be because the added substance induced a mutation; this is enough to disqualify the substance for human consumption, since it is generally assumed that anything so mutagenic stands a good chance of being carcinogenic, too.

As we age, our bodies increasingly become the sum total of our experiences, notably the nicks and dings that we accumulate over time. Most important are those bodily insults that cannot be so readily seen. This doubtless helps explain why cancers are more common among the elderly: their cells have had more time to accumulate enough mutations to enable them to break free of the usual restraints imposed by the cell cycle, to the detriment of the rest of the body.

For cancer to ensue, mutations must occur in one of two types of genes: proto-oncogenes or tumor-suppressor genes. Oncology is the medical specialty concerned with cancer, and proto-oncogenes are any sort of gene involved in the control of cell division, usually promoting it. Proto-oncogenes are normally present within everyone's DNA. The problem arises when a proto-oncogene undergoes a mutation that makes it hyperactive, transforming it into a genuine oncogene, which increases cell proliferation beyond the usual levels.

Tumor-suppressor genes are normal genes as well, but instead of coding for substances that increase the amount of cell division as do proto-oncogenes, they *slow* the cell cycle and the amount of cell division. A cancer can develop if there is a mutation in a tumor-suppressor gene that impairs the gene's ability to repress unwanted cell growth. For example, retinoblastoma, a lethal disease in children, arises among individuals who have only a single copy of the relevant tumor-suppressor gene; one "hit" on the remaining gene, and the victim cannot suppress the uncontrollable development of eye tumors.

There are many different types of both proto-oncogenes and tumor-suppressor genes, and mutations usually have to occur in both types in order for a cancer to develop. It is as though proto-oncogenes represent the accelerator pedal and tumor-suppressor genes, the brake. To go faster (become cancerous), you must press down harder on the accelerator pedal and also ease up on the brake.

Recently, a number of genes have been identified for different types of cancers; each leads to an increased likelihood that someone carrying the gene will develop cancer. These genes are either mutated proto-oncogenes or tumor-suppressor genes, and they do not in themselves cause cancer. Rather, they give their carriers a push along the path toward developing the necessary array of mutations that lead to it. As a result, someone who carries a "breast cancer gene" already has at least one of these mutated genes in her cells. She (or rarely, he) is therefore more likely to develop breast cancer than someone whose genes are unmutated and who must develop this particular mutation independently, by chance, in addition to any others that might lead eventually to the disease. Cancer itself is very rarely inherited; what is hereditary is a predisposition to various forms of cancer, which simply means the presence of one or more contributing mutations.

These mutations can occur in a variety of places. For example, there are anti-oncogenes that normally function as self-destruct mechanisms, causing the membrane of an abberant cell to bunch up in peculiar ways, forming ugly little "blebs" that eventually kill the sick cell. This happens when a cell's DNA is severely damaged. If these anti-oncogenes themselves mutate, then they are no longer able to perform their job, "causing" cancer not so much by producing uncontrolled cell division as by failing to prevent it—an error of omission rather than of commission—in this case, failing to cause an out-of-control cell to bleb itself to death.

Viruses can also bring about cancer. Certain retroviruses, for example, incorporate their own proto-oncogene into their host's DNA, which then functions as though the resident proto-oncogene had undergone its own predisposing mutation. In addition, there are DNA-tumor viruses, which carry tumor genes and insert them directly into the DNA of an otherwise normal cell. The common Epstein-Barr virus seems to carry genes for Burkitt's lymphoma, a form of cancer that used to be relatively rare but is often suffered by AIDS victims; HIV apparently reduces immune resistance to Epstein-Barr virus and its associated genetic hangers-on.

Recall our earlier discussion about how pathogens might well have their own strategies for spreading, such as influenza viruses inducing their victims to cough and sneeze. The great majority of the body's cells are good citizens, unless they carry multiple mutations or have been hijacked to do the bidding of some foreign invader. In this sense, it is ghoulishly consistent that certain viruses could have evolved with their own cancer-causing genes, whose effect is to induce cells carrying these genes (and, of course, the rest of the virus as well) to proliferate wildly and to spread beyond the usual limits for normal, uninfected cells. If coughing and sneezing are effective ways for influenza viruses to get out into other bodies, then becoming part of a cancer cell may well be an effective way for certain other viruses to get themselves spread around inside a body.

Cancer Treatments

The best way to cure cancer is to get rid of the cancer cells. In the simplest case, they are directly cut out by a surgeon's scalpel. For this tactic to succeed, however, the cancerous tumor must be contained within itself, not yet metastasized. This is why early detection is so important. For example, in the early stages of tumor development, the transformed cells generally do not have their own blood supply and may be vulnerable to their own toxins. Under normal circumstances new blood vessels do not develop in the mammalian body, except following injury. But cancerous tumors typically secrete chemicals that induce the development of their own blood vessels, which not only enable the transformed cells to gain more nutrients, but also provide an avenue of spread to the rest of the body.

If cancerous cells have become incorporated into other organs so that they cannot be surgically removed, they must be stopped or killed in place. In itself, this is not difficult. The challenge is to knock out malignant cells selectively, without killing other, normal cells. Chemotherapy refers generally to the use of chemicals to destroy cancerous cells. In most cases, the weak point of cancer is precisely what makes it so troublesome: cancer cells' rapid reproduction. As a result, most chemotherapies focus on destroying rapidly dividing cells, since rapid division is one of the hallmarks of cancer. The hope, therefore, is that the rapidly growing cancer cells will be more affected than the body's healthy cells, which generally replicate more slowly.

Recall that in mitosis, cells avail themselves of microtubules, which provide the internal ligaments whereby the replicated chromosomes pull apart, forming two new cells. Some of the most effective chemotherapies interfere in various ways with microtubule development. For example, chemicals derived from two flowers—colchicine, from the crocus plant (genus *Colchicum*), and the vinca alkaloids, from a tropical periwinkle (genus *Vinca*)—inhibit the assembly of microtubules. When cancer cells can't divide, they die. Unfortunately, other actively dividing cells are also killed, notably the epithelial cells of the digestive tract as well as hair follicles. This leads to the nausea, bleeding gums, and temporary hair loss that is characteristic of persons undergoing cancer chemotherapy.

Another notable chemotherapeutic agent is taxol, derived from the Pacific yew tree (genus *Taxus*), which also homes in on microtubules. Instead of preventing microtubules from developing, however, taxol works by stabilizing them, thus putting a dividing cell in a kind of straitjacket, so that the two daughter cells cannot separate, and both die.

One distinguishing characteristic of cancer cells is that without medical intervention, they are nearly immortal. In tissue culture in a laboratory, as we have seen, they often seem capable of reproducing indefinitely, whereas healthy cells give up after several dozen divisions. How do cells keep count, and might this provide a way of attacking cancer cells? At the end of chromosomes are structures known as *telomeres* (*telos* in Greek means "end"), which, like the plastic tips on shoelaces, seem to protect the rest of the structure. These telomeres are produced by the enzyme telomerase, and most cells stop producing telomerase after the embryo is fully formed. Each time a cell divides, a bit of the telomere on each chromosome is shaved off, so that eventually the functional DNA of a chromosome is exposed and a little bit is sliced away with every cell division; once a cell runs out of telomere, it eventually runs out of time. It may be that unlike normal cells, whose telomerase production is turned off at a certain point, cancer cells have their telomerase-making genes turned back on. If so, then one long-term treatment for cancer might involve figuring out how to switch them back off again—so long as this doesn't result in premature aging on the part of normal cells.*

Radiation therapy is also used to combat cancer cells. Like chemotherapy, it homes in on rapidly dividing cells and has the unwelcome side effects of killing any cells that divide frequently. Ideally, radiation therapy works not so much by destroying cells as by injuring their DNA, after which the damaged cells can usually be counted on to commit suicide, undergoing self-destruction via the blebbing process described earlier. Faulty DNA itself, however, does not trigger this process. Rather, it is under the influence of genes, notably one called p53. If a cell's DNA is damaged, it is p53's job to halt cell division until the DNA is repaired; if the cell has been totaled, p53 then initiates cell death. But if, as part of the cancer process, p53 has also mutated, then this genetic guardian fails to bring about self-destruction (remember apoptosis?).

The result is that higher levels of radiation are then needed, not just to damage the DNA of targeted cells, but actually to destroy them. And the problem in such cases is that heavy doses of radiation kill normal cells too. Although it is quite possible to survive the death of hair follicles (all the more so since they eventually are

---

* Interest in telomerase genes also focuses in the other direction: stimulating them to keep performing, thereby perhaps retarding aging and prolonging life spans . . . so long as this doesn't precipitate cancer! Also noteworthy is a report in 1999 that Dolly, the famous cloned sheep, appears to have shorter telomeres than would a normal sheep of her age. This is not surprising, since she was initiated with chromosomes from an already adult (6-year-old) ewe. As one biologist noted, she was "a sheep in lamb's clothing." In any event, Dolly's troublesome telomeres may foretell a situation in which cloned animals are subject to what looks like premature aging, but is actually appropriate to the chromosomes from which they are derived.

replaced), far more troublesome is the loss of bone-marrow cells, which are important as a source of immunity and which—because they divide rapidly—are especially vulnerable to radiation. This is why heavy doses of radiation are often necessary for especially resistant cancers, and also why such radiation therapy is typically combined with bone-marrow transplants, in an effort to replace the patient's devastated immune system.

Since gene mutations are so important to the production of cancer, the search for cancer cures is focusing in part on possibilities for gene therapy. There are prospects aplenty, including the possibility of eventually inserting healthy tumor-suppressor genes and oncogenes into cancer cells. For example, reintroduction of healthy p53 genes could lead a sick cell to identify itself as a loose cannon, potentially lethal to the rest of the body, and to commit suicide—as it should and would if the original p53 had not mutated. Also under study are experimental treatments for leukemia that involve inserting into cancer cells DNA that interferes with the functioning of newly mutated oncogenes. Another promising gene therapy, being tested for prostate, ovarian, and brain cancers, uses viruses to insert a gene that causes tumor cells to secrete a particular enzyme, which in turn makes those cells sensitive to a certain drug. Since the drug is activated only in those cells producing the enzyme, it might eventually be possible to target just the tumor cells.

A different kind of targeting involves tamoxifen, recently approved as both a treatment and—uniquely—a drug that might in some circumstances prevent breast cancer. It has long been known that in some cases women with breast cancer do better after ovariectomy; ovaries make estrogen, and estrogen can speed up the growth of breast cancer (which is why breast cancer is generally more lethal in young, premenopausal women than when its victims are older). Some breast cancers contain estrogen receptors, which stimulate tumor growth. Originally studied as a possible birth control agent, tamoxifen is similar to estrogen, which it blocks or displaces from these receptors, thereby diminishing the hormone' s pro-cancerous effects. But tamoxifen has a downside: tamoxifen in the uterus mimics estrogen, increasing the risk of uterine cancer.

Yet another potential treatment might derive from interfering with a powerful, peculiar ability that many tumors possess: that of inducing their own blood supply. As we have seen, a fully formed mammalian body does not normally grow major new blood vessels; tumors are an exception, secreting angiogenic (blood vessel–producing) chemicals that enhance circulation, which in turn feeds the tumor. This, in turn, has stimulated a vigorous search for antiangiogenic drugs; two natural proteins, angiostatin and endostatin, have proved effective in treating cancerous tumors in mice. It will take several years, at least, to determine whether human tumors are similarly affected; that is, whether they can be starved by denial of blood supply. Such uncertainty, however, didn't prevent a brief media frenzy when these prospects were made public—and incautiously hyped by *The New York Times*—in 1998.

Cell Signaling

As we have repeatedly pointed out, our bodies are made of trillions of cells, all filled with organelles and undergoing some stage of the cell cycle. Simply being the structural units that make up our bodies and grow and change as we age, however, is not what really occupies the time and effort of a cell. Most cells have some other role, notably creating substances or doing work to allow the body to function effectively. In order to accomplish these goals, however, the cells need to communicate with one another. As you read this, the cells in your eyes are sending messages to different cells in your brain, which in turn are sending signals to one another as you make sense of the little squiggles on this sheet of paper. At the same time, your heart is beating and you are breathing, at (we hope) a fairly regular rhythm that is (probably) resulting in exactly the correct amount of oxygen getting to your brain and muscles according to your physical exertion. You may well be digesting food eaten several hours earlier. Hormones and other signals are also making their way through your body, perhaps telling you that you are getting hungry or thirsty, or that you have to go to the bathroom, or adding to your perception that the person sitting across from you is really sexy. All from different signals traveling among cells! This cell signaling takes place in many different ways, and without it, the immense number of cells known as "you" couldn't possibly work together.

All cell communication occurs because some kind of signal molecule comes into contact with some kind of receptor molecule. The signal molecule is called a ligand (from the Latin word for "tie" or "bind," which is also the root of *ligament*). Either it is secreted by a cell and sent downstream, or is actually bound to the cell's plasma membrane. The receptor molecule is most often attached directly to the outside of a "target" cell. The relationship between the two molecules—sender and receiver, ligand and receptor—is very specific: one ligand will fit in a certain way with one receptor. When they merge, there is usually a change in the shape of the receptor molecule, which in turn causes a cascade of events leading to an outcome such as the activation of genes or the secretion of more signal molecules. There are three main types of cell signaling, all based on ligands contacting receptors: *paracrine, endocrine,* and *synaptic.* Each is important.

Paracrine signaling occurs among cells that are very close to one another. The signaling cell releases the ligand, often called a local mediator because it only acts at short range before being destroyed. This local mediator diffuses to cells in the immediate vicinity, making contact with the cell receptors. A good example is the inflammatory response, which occurs shortly after a wound or a local infection. When tissue is injured, perhaps because of a cut, the cells surrounding the injury such as nerve endings, some of the blood cells, and the damaged tissue itself send out paracrine signals to the nearby tissue. These signals attract a large number of white blood cells, which are involved in healing, as well as destroying bacteria or other invading organisms that may have entered through the wound.

Another important example of a paracrine signal is nitric oxide (otherwise known as NO), which has an important role in the dilation of blood vessels. Take erections, for example. When a man is sexually aroused, nerves in the penis stimulate the production of NO. This, in turn, ultimately increases blood flow into the spongy tissue of the penis. But the route is circuitous. First, the nitric oxide diffuses into the smooth muscle cells of the penis. Here it activates an enzyme, which in turn causes blood vessels to dilate and an erection to occur, via another chemical. This erection-inducing chemical, however, is itself degraded by yet another enzyme! And this is where the new wonder drug, Viagra, comes in: Viagra enhances the effect of NO by inhibiting the enzyme that degrades the chemical that NO activates, in this roundabout way turning NO into YES. (This is why Viagra alone does not produce an erection. Sexual stimulation is still needed, because the necessary nitric oxide must first be produced.)

Nitric oxide also works on heart muscle, once again dilating the blood vessels. This explains the seemingly peculiar fact that nitroglycerine, better known as a powerful explosive, also helps relieve the painful symptoms of angina pectoris, in which atherosclerotic plaque restricts blood flow to the heart. Taken orally in small amounts, nitroglycerine does not explode or otherwise blast open the heart's blood vessels, except metaphorically. Instead, it is metabolized into nitric oxide, which, in its role as a paracrine transmitter, dilates the heart vessels, allowing for improved blood flow and thereby reducing chest pain.

Unlike the short-range distress signals involved in paracrine signaling, cells also send chemical messages over long distances, through the blood. This is endocrine signaling.

Compared with its paracrine cousin, endocrine signaling acts over an extended time frame as well as greater distances. Most often, these signals have broad effects on the behavior or general functioning of the body as a whole. Endocrine signals work in very small doses, and because the signal must first circulate through the blood, its effect is far from immediate. Endocrine signals are usually produced by so-called endocrine glands, and the signal released is called a hormone (derived from a Greek root meaning "to excite, stimulate, or set in motion").

Hormones, in turn, can be usefully divided into two major types: those that cross the membrane into the cell nucleus of the target cells and those that don't. Among the latter are peptide hormones, which are small proteins, as well as modified amino acids, the building blocks of proteins. Peptide hormones include insulin, which decreases blood glucose levels and is well known because of its role in diabetes (we discuss this in the chapter on energy). Also growth hormone, which encourages growth in animals, and luteinizing hormone, which encourages ovulation. There are many others. Modified amino acids include epinephrine and norepinephrine (formerly called adrenaline and noradrenaline), which are especially important in the so-called flight-or-fight response. This prepares the body for emergencies by raising the blood-sugar level, increasing the metabolic rate, and stimulating heart contractions. All of these hormones attach, in various ways, to specific receptors on the cell's surface and do their work from the outside.

The other main type of hormones are the steroids. These include glucocorticoids, such as cortisone, which increase blood sugar and have some anti-inflammatory effects; testosterone, which causes males to look and act more malelike and to produce sperm; and estrogen, which causes females to look and act more femalelike and to produce eggs. Steroids act by drifting through the membrane of target cells into the nucleus, where they bind to receptors that interact directly with RNA polymerase to control gene transcription. As a result, steroids are able to influence the activity of genes within target cells, typically causing them to create and then release new substances or to perform a different activity, such as turning themselves on or off.

Many different tissues release hormones. One of the most important is the pituitary gland, which is controlled by the hypothalamus, a small area in the brain that actually secretes peptide hormones of its own, with the pituitary as its target tissue. When stimulated, the pituitary releases many different hormones, ranging from antidiuretic hormone (ADH, which helps prevent dehydration) to growth hormone to luteinizing hormone. These different hormones have widely varying target tissues, ranging from very specific (ADH targets the kidneys) to very general (growth hormone affects most cells, stimulating growth in all of them). Some of the target tissues are even endocrine glands in their own right. Luteinizing hormone, for example, acts on the ovaries, which also secrete their own hormones such as estrogen. Other endocrine glands include the thyroid, the adrenal gland, the pancreas, the digestive tract, and the heart. Almost every type of cell is a target for some sort of hormone.

As we have seen, endocrine signaling covers a broad range of relatively slow-acting, long-term signals that diffuse through the bloodstream. By contrast, paracrine signals affect only those cells that are nearby. Synaptic signals are similar to their endocrine cousins in that they can act over long distances, but they resemble paracrine signals in that they involve close physical contact between parts of the signaling tissue and the target cells. In addition, synaptic signaling can be very rapid. It is also the most target-specific of the three ways cells "speak" to one another; one cell literally reaches out and touches—or very nearly touches—another. Most often seen in nerve cells, synaptic signaling requires the sending cell to have very long projections that snake their way through the body until they reach the target cell, which they bombard with small signaling molecules known as neurotransmitters. We'll examine synaptic signaling in Chapter 4, when we explore the connection between brains and behavior. We'll visit some of the endocrine glands and their work in Chapter 5 and again in Chapter 6.

With all three types of signals, it is important to remember that different types of molecules may have different effects on different tissues. For example, the signaling molecule acetylcholine will cause a skeletal-muscle cell to contract, a heart-muscle cell to relax, and a secretory cell to secrete. Different combinations of signals will cause different cells to do different things as well. Many cell types will die without any outside signals. If they receive specific signals, they will survive, and with even more, they grow. In addition, the response of a cell to a signal can change over time, as the cells get desensitized. If a cell is overloaded with a signal, its number of receptors will decrease over time, an adaptive response that makes it less sensitive and prevents it from being overstimulated. This sort of adaptation may explain why people who take certain drugs regularly may need to increase the dose in order to experience the same effect.

The Immune System

The immune system is one of the most fascinating parts of the mammalian body. Also one of the most difficult to understand. Consisting of an array of cells spread throughout the body yet closely interconnected via chemical signaling, the immune system provides protection against a host of different invading organisms, from cancerous cells to bacteria to viruses. These invaders can cause great damage, often in very different ways. They have one thing in common, however: they are typically small, too small to be dealt with by the body's more obvious defense mechanisms, such as hands or teeth. The immune system is based on the ability of cells to communicate with one another about specific problems they are encountering. It also has a "memory," so that if it meets a problem once, it will be better able to react next time.

But the precision of immunologic memory comes at a cost: something it doesn't know is likely to be ignored. Small mutations in the invading creature can render the immune system worthless, at least until it has time to react to the new version of the troublemaker by creating new cells specific to that mutation. The cells of the immune system provide excellent examples of the fussy specificity that cells can have.

The body's first line of defense against dangerous microscopic invaders consists of its outer barriers, which prevent some troublemakers from entering in the first place. The most important and often overlooked of these barriers is skin, actually the layer of material on the outside of skin called keratin. The skin barrier is waterproof, and for most disease- and infection-causing agents, it is impermeable. Skin breaks down easily, however; cuts and scrapes allow foreign objects to pass unencumbered into the body. Another important outer barrier consists of the mucous membranes inside the digestive tract and lungs.

These prevent many different invaders from actually entering the body, although obviously certain ones such as viruses are very good at getting through. Tears, which contain an enzyme (lysozyme) that digests bacterial cells, are a surprisingly effective barrier, as are saliva and the nasal passages themselves. Considering the number of different irritants that lurk in the outside world, these systems deserve more acknowledgment than they usually receive. But as soon as a virus or bacterium gets beyond these outer lines of defense, different mechanisms are called upon to protect the body against invaders.

The immune system is a potentially confusing, Byzantine world of numerous types of cells doing many different things. As a group, these cells are loosely called white blood cells, but in fact, there are many different kinds. For clarity, it is helpful to divide the immune system into two parts: one that is specific and one that isn't. The nonspecific part consists of those cells that ingest invading particles and destroy them with enzymes. Another important nonspecific activity is to destroy body cells that have already been invaded and mop up the mess. This is largely the job of "natural killer cells" (that's really what they're called!), especially a group known as neutrophils, drawn to a cut or scrape by paracrine signaling, which serves as a homing signal that attracts these benevolent killers. They destroy cells that are lethally infected by bacteria or viruses and also act against many types of cancer cells.

As for ingesting bacteria and viruses that have not yet entered a "host" cell, this is primarily the job of certain especially strange cells known as macrophages ("large eaters"). These begin to develop within the blood, after which they migrate to different tissues where they destroy bacteria and some viruses by reaching out long filaments that clasp the invaders and drag them into the macrophage, where they are destroyed. A popular British children's book titled *Mr. Tickle* features a cheerful globular fellow with immensely long and flexible arms that he uses to tickle people from far away; macrophages are a bit like Mr. Tickle, except that for bacteria, the tickle is terminal. Unfortunately, however, many bacteria have developed ways to avoid the searching arms of the macrophages, and so are tickleproof.

Other types of nonspecific immune cells include eosinophils, which mainly attack larger invaders such as parasites and worms. Eosinophils function like a platoon of mobile, chemical artillery, discharging corrosive enzymes at the outside of the invader. (*Phil* means "love" in Latin and is the root, among other things, of *philosophy*, "love of knowledge." In any event, the various cell types whose names end in *phil* are so designated because of how readily they absorb—"love"—certain chemical dyes that are applied before viewing these cells under a microscope.)

The inflammatory response, often considered a problem, is actually an important component of the body's nonspecific immune system. When there is an injury, cells called basophils and mast cells—small, mobile connective-tissue cells—release proteins known as histamines, which cause swelling and redness as the capillaries become more leaky and blood flow increases. Too much histamine can cause a problem in itself, especially in the case of allergic responses; hence, the use of antihistamines.

Histamine also attracts other immune-system cells such as neutrophils and macrophages. These white blood cells, in turn, often engage in what is essentially a form of paracrine signaling of their own. Depending on their type—and on their "prey"—neutrophils can consume from several dozen to many hundreds of invading bacteria and thousands of viruses before eventually succumbing to their toxins. But even then, in their death agony, they make a contribution, secreting small proteins known as interferons. Interferons, as their name implies, interfere with something: the spread of viruses.

They do this in two ways. First, they attract additional white blood cells to the site of the invasion or infection. Second, they protect as yet uninfected cells by inducing them to make proteins that interfere with viral replication if the viruses do succeed in entering those cells.

Interferon refers to a class of broad-spectrum antiviral substances, which, when first discovered, raised great hope as a potential cure for a range of viral diseases. Not until the 1980s did genetic engineering permit the production of adequate quantities of interferon to test its efficacy. Although not the wonder drug originally hoped for, interferon has shown promise in treating hepatitis-C viruses, the papilloma virus that causes genital warts, and a rare form of leukemia.

The nonspecific immune system generally does a good job coping with the steady trickle of pathogens that assault every body, every day. If the infection is strong enough, however, signals sent out by the injured cells can generate a total immune response, in which the specific side of the immune system comes into action.

It is specific indeed: it can launch a precise and limited attack against a single kind of disease-causing organism, distinguishing it from other organisms that might be similar. At the same time, the immune system possesses three other notable traits: diversity, the ability to mount a comparable response to millions of different types of disease-causing organisms; memory, or the ability to react to invaders more quickly and completely if it has encountered them before; and self/nonself recognition, so that it does not normally attack the body it is supposed to be protecting.

The specific immune system is made up of white blood cells known as lymphocytes. Unlike the white blood cells we have already encountered—neutrophils and macrophages—these lymphocytes often remain fixed in certain parts of the body. They are divided into B-cells, which differentiate in the bone marrow, and T-cells, which do so in the thymus gland. (B-cells were named not for bone marrow but for the bursa of Fabricius, an organ found in chickens, where they were first discovered.)

In one of the most famous speeches of the twentieth century, Winston Churchill both described and helped arouse British defiance of Nazi Germany during World War II when he intoned, "We shall fight on the seas and oceans, we shall fight . . . in the air, . . . we shall fight in the fields and in the streets, . . . we shall never surrender." Under the command of B- and T-cells, the specific immune system is prepared to fight invaders that have breached the body's other defenses and are freely circulating in the blood, and if that fails, it is prepared to continue fighting even after the body's cells have been entered. The first kind of specific defense is known as the humoral response, because it relies upon the production and release of chemicals: *humors* in an older usage referred to the different kinds of body fluids. The second—fighting in the streets (that is, within invaded cells)—is called the cell-mediated response; we'll get to it shortly.

The Humoral Response

Humoral responses are the work of B-cells, which produce antibodies to destroy an invader. An antibody is a molecule created in response to exposure to an antigen, and an antigen is that part of an unwanted invader recognized by the immune system (*antigen* is simply a shortened form of "antibody-generating"). Like a hand in a glove, a particular antibody responds specifically to the chemical traits of one antigen.

There is a saying that if you give someone a fish, you feed him for a day; teach him how to fish, and he is fed for life (assuming, that is, there are enough fish). Immunity can be like that. It can be "given," in which case the recipient passively acquires it, and benefits for a time. Or it can be "acquired," in which case it lasts a long time, often for life.

Passive immunity is like being given a fish. It happens when someone receives antibodies produced elsewhere: the most common example is the transfer of antibodies from mother to embryo, first through the placenta, and then via nursing (especially during the first few weeks). Passive immunity of this sort provides short-term protection, as with injections of gamma globulin to protect against hepatitis A.

To be immunized for life, however, one must learn to fish, acquiring active immunity, the ability to produce one's own antibodies. This happens when the antigen invades a body, either as part of a malevolent pathogen or as a weakened form intentionally introduced to serve as a vaccine. In the simplest case, an active humoral immune response is generated when the antigen binds directly to a receptor on the surface of a B-cell. When this occurs, the B-cell will start to divide, forming two different kinds of daughters: one gives rise to so-called memory cells, which essentially sit around waiting for the next exposure to the same antigen. More than two thousand years ago, the Greek historian Thucydides, writing about the deadly plague of Athens, commented that those who had survived the disease could safely tend the sick, because "the plague never attacked the same person twice." They had acquired active immunity, through their memory-minded B-cells.

Memory cells can sit around for years, even decades, bearing a helpful grudge, brooding on a precise, detailed recollection of prior affronts, ever ready to pounce at the next repetition. Following initial exposure to an antigen—and without any existing memory cells to react—antibodies usually do not reach protective levels for days or even weeks, which is sometimes too late. But the second time, response is nearly instantaneous. Thanks for the memories!

Next, let's follow the other B-cell descendants, known as plasma cells, which circulate in the bloodstream and secrete antibodies. Antibodies are polypeptides—simpler than proteins, more complex than amino acids—that travel around the body and bind to the antigen (the foreign invader). They are Y-shaped molecules, the tips of each Y formed of a specific molecular pattern that recognizes and binds to a particular antigen. Collared by its counterpart antibodies, the antigen is then deactivated in any of a number of different ways, all of which result in the invader being more easily destroyed by macrophages. Antibodies can neutralize invaders by blocking their ability to do harm, by making them clump together, or by causing the antigens to precipitate from the blood, eventually to be filtered out by the kidneys. One final possibility is that the antibodies can activate a system called "complement," which consists of a battery of about twenty different kinds of proteins. Attracted by the antibodies, these complement proteins can coat an antigen source, making it adhere more readily to a white blood cell, which destroys it. Alternatively, in some cases complement proteins attach to the membrane of the foreign cell and literally bore holes in it.

Most often, the humoral response involves a larger cast of characters. Often a "big-eating" macrophage starts the process by gobbling up some of the invaders. When it does so, the pathogen's antigens move to the outside of the macrophage as part of a special molecular array called the major histocompatibility complex, or MHC. This MHC/antigen mixture is then displayed on the surface of the macrophage that overcame it, not as a signal of triumph, but as a kind of "wanted" poster. Essentially, it alerts lymphocytes to the detailed appearance of the malefactor.

Here's how it works. The "wanted" poster, attached to the macrophage that vanquished it, is recognized by a helper T-cell. This connection between victorious, antigen-displaying macrophage and helper T-cell appears to be made largely in the lymph nodes, which is one reason they become swollen during an infection. The helper T-cell becomes activated when the macrophage secretes wake-up chemicals known as interleukins. (Interleukin is an appropriate name for molecules that provide communication within the immune system; it literally means "between white blood cells.") The activated helper T-cell responds by producing a different interleukin, which stimulates "virgin" B-cells, those antibody makers we just described but which have not yet seen action. By short-range paracrine signaling, the helper Ts stimulate the virgin Bs to form those friendly plasma and memory B-cells, which in turn either wait around as memory cells in case they're called upon later or secrete antibodies that help destroy the invaders. There are billions of B-cells; each one produces antibodies for one type of antigen and only that antigen.

Whether macrophages and helper T-cells are involved with the humoral response depends on the nature of the antigens: some require the involvement of helper T-cells—and are therefore called T-dependent—while others do not. Either way, helper Ts don't produce circulating antibodies, although they help virgin Bs to do so. A common misconception is that the presence of a particular antigen somehow causes the body's B-cells to produce the appropriate antibodies. This is not quite true. Antibodies are not custom-made on-site to "fit" each antigen. Rather, a healthy person normally has about 100 million different types of antibodies circulating at any one time, each type produced by a slightly different B-cell. The result is that most antigens find the perfect Goldilocks antibody—not too big, not too small, not too positive, not too negative, but just right—to bind with chemically. This binding not only decommissions the antigen, it also stimulates the B-cell to reproduce quickly, a process called "clonal selection," since it results in a clone of the appropriate B-cell. The result is a whole lot of antibody: not something new so much as an amplification of what was already there, one of the millions of biomolecular snares that the body had previously set.

This leads, in turn, to another success story of genetic engineering: monoclonal antibodies. These are antibodies secreted by identical copies—clones—of a single, specified type of B-cell. Since antibodies are precisely designed to attack a particular invader, and one B-cell produces only one kind of antibody, a long-standing dream has been to produce large quantities of such antibodies, which could act as a "smart bomb" or "magic bullet," precisely targeting a particular invader, whether bacterial or viral. It is easy enough to generate small amounts of nearly any antibody: just inject an antigen—say, a strain of bacteria—into a laboratory animal. Among the recipient's billions of B-cells there are likely some with precisely the correct antigen receptors to cause them to respond by producing just that antigen.

A special difficulty, however, has been that B-cells do poorly when cultured in a laboratory, so it is difficult to obtain enough antibody to do anyone (other than the animal who produced it) much good. Therefore, part of the trick in making monoclonal antibodies has been to fuse the appropriate B-cells with, ironically, a cancerous lymphocyte, producing a "hybridoma," which reproduces with the enthusiasm characteristic of cancers in general, while also secreting large amounts of one particular substance: the desired antibody! Monoclonal antibodies have made home pregnancy tests possible, as well as screening for prostate cancer; they are also used in some cases to provide passive immunity.

The Cell-Mediated Response

Like the Royal Navy and the Royal Air Force, which were prepared to fight off attackers before they landed, the humoral response destroys invaders, such as bacteria and viruses, when they are outside individual cells. But antibodies are helpless against viruses or bacteria that have gotten inside a cell. This is where the cell-mediated response comes in. It is like the street-fighting infantry; its job is to continue battling against those pathogens that have already made their way into the organism's cells. Most commonly, the invaders attacked by the cell-mediated response are either cancer cells or viruses that are only active inside their host's cells. The cell-mediated response also uses helper T-cells, in much the same way that the humoral response activates B-cells, but this time the helper T brings a different lymphocyte into play: the cytotoxic or "killer" T-cell.

When the body becomes infected with an invader, the free antigens will contact the B-cells directly if they are T-independent, or they will likely be engulfed by a long-armed Mr. Tickle–type macrophage, which will present the invader to the helper T-cells and also stimulate them with interleukins, as already described. In addition to contacting B-cells, however, the activated helper T-cells use yet another brand of interleukin, to call upon virgin killer T-cells. These respond by dividing, whereupon, much like activated B-cells, they form memory and active killer T-cells. (In immunology, a B- or T-cell is a virgin if it has not had "immunologic intercourse" with an antigen.)

When a cell is infected by an invader, it automatically displays some of its invader's antigens on its surface, much as a macrophage does, with its antigenic "wanted" poster. When an infected cell does this, however, it is sending out a distress signal, a kind of flag indicating that it is under attack, while also displaying the chemical characteristics of the attacker.

Activated killer T-cells "read" this message and use it to bind to the infected cell. When they do so, the T-cell secretes a molecule known as perforin, which perforates the membrane of the infected cell, causing it to break and die. This reflects a coldhearted but adaptive calculation that once a cell has been lethally infected, it is as good as dead, and so the best response by the rest of the body's cells is to sacrifice it quickly, if doing so ensures that other, healthy cells will not be similarly infected. It is analogous to quarantining infected individuals, or even whole cities, in the event of a lethal plague, "writing them off" if necessary to save the larger population.

To some extent, killer T-cells will also recognize and destroy cancer cells much as they do an infected cell, because cancer cells also tend to display antigens that are distinct from their normal counterparts. Immunotherapy for cancer involves various ways of boosting the body's immune system, notably the T-cells. One of the newest possibilities is to develop a vaccine, produced by exposing the body to its own cancer cells (which have been inactivated by radiation). Don't be misled by the word *vaccine*; although a true cancer vaccine is a long-range hope, current expectations for cancer "vaccines" are that they might help provide a cure rather than prevent the disease.

Cancer immunotherapy, if successful, could offer a major advantage over other treatments. Unlike radiotherapy, chemotherapy, or even, to some extent, surgery, immunotherapy is likely to take advantage of the immune system's exquisite specificity, thereby targeting and killing cancer cells and only cancer cells. Cancer immunotherapy is undergoing testing and development especially for malignant melanomas (skin cancers that are often lethal, as opposed to the less dangerous squamous-cell carcinomas). Although, as we have seen, cancer cells in some circumstances are normally attacked by natural killers, macrophages, and killer T-cells, this assault is often less vigorous than one might wish; the reason, presumably, is that cancer cells, although "transformed," are still one's own cells, so in many cases they apparently aren't different enough to evoke a full-fledged immune response.

On the other hand, patients receiving organ transplants—and thus, immunosuppressant drugs—are at significantly greater risk of cancer. This points to the role of a healthy immune system in policing the body and removing "normal" cancers as they arise. The idea in cancer immunotherapy would be to fool the immune system into thinking that the cancer is, in a sense, more different from other cells than it really is. This is being attempted by administering a combination of cancer antigens and interleukins. Good luck to it.

We mentioned earlier that HIV especially targets the immune system. Now we can pursue this further. HIV is particularly likely to infect macrophages and helper T-cells. Attacking a victim's macrophages is like a criminal gang assassinating the cop on the beat, while destroying the helper T-cells is like blowing up the central police station, decapitating both the humoral and cell-mediated responses. This, in turn, explains recent interest in using interleukin as a means of treating AIDS. Since the role of helper T-cells is to activate B-cells as well as killer T-cells, it might be possible to compensate for the loss of HIV-devastated helper-Ts by administering the interleukin that helper-Ts would otherwise secrete, if they were able.

It has also been noticed that some people possess a kind of genetically based "natural immunity" to AIDS. Further intriguing: such immunity is especially frequent among certain populations, notably Caucasians, and relatively absent in others, such as black Africans. One interesting possibility, as yet unproven, is that a degree of immunity to the same (or a similar) genetic trait developed when much of Europe was being ravaged by bubonic plague, which also attacks the immune system. In the early stages of plague (also known as "the Black Death"), the disease bacteria *Yersinia pestis* home in on macrophages, as does modern HIV. Conceivably, a disproportionate number of individuals surviving the plague might have been those carrying a mutation that prevented the plague bacillus from gaining entry into immune cells. For its part, HIV gets into macrophages via two proteins that are normally present in the surface of macrophage cells; the "HIV-resistance gene" destroys one of these entryway proteins, leaving the HIV particles locked helplessly outside. It is at least possible that a similar mechanism conferred a degree of immunity upon Europeans who would otherwise have succumbed to plague and whose descendants now enjoy a comparable immunity to AIDS.*

Molecular-clock dating suggests that the HIV-resistance gene appeared around 700 years ago, just when plague was first devastating Europe, and long before AIDS. It remains to be seen, however, whether the descendants of people immune to plague are also immune to AIDS, and if so, why.

Despite the ravages of AIDS and other diseases and malfunctions, the fortunate truth is that the immune system does a remarkably effective job of policing the internal security of *Homo sapiens* and destroying what doesn't belong. But in a sense, cancer cells and those infected with pathogens *do* belong, in that they are (or were) part of us. This helps raise an important question: Why doesn't the immune system attack itself? Or the rest of our body? As we shall see, the question "Who am I?" is not simply an existential quandary for philosophers or theologians. It has life-or-death ramifications.

Self versus Nonself

Toleration is a virtue, especially when it means acceptance of diversity. And yet, despite the immense variety of different cell types within the body, the cellular diversity of every multicellular creature really amounts to variations on a theme. That theme is genetic uniformity,

---

* *There are precedents for this sort of thing. The best-known case is sickle-cell disease, a genetic illness whose persistence had been a mystery given that the sickle-cell gene is debilitating and sometimes lethal in double dose. The key is that in single dose, the sickle-cell gene conveys benefit against malaria. This accounts for the maintenance of the sickle-cell gene, otherwise so deleterious, in populations historically exposed to malaria. Cystic fibrosis has recently been shown to be another example: in double dose, the gene is lethal, typically causing death by age two. But in single dose, it conveys a benefit against typhoid fever.*

based on the fact that all cells making up a body are—or are supposed to be—genetically the same. In short, the immune system is fundamentally intolerant and conservative, deeply opposed to change and diversity. It's okay to be a bone, muscle, gland, or skin cell, so long as each of these cells is at heart—that is, at its DNA—the same. Any departures, whether caused by pathogens or cancer, are threatening.

The cell-mediated immune system works on behalf of this deeply exclusionary principle. It is a kind of hard-nosed Immigration and Naturalization Service, jealously guarding the body's borders and coming down hard on any aliens. It is also a powerful mechanism by which the body destroys cells that are somehow abnormal. In addition to cells infected by viruses or transformed by cancer, however, it will destroy normal cells if they are from a different individual. If you were to graft a bit of skin from one person to another, the foreign skin will be treated as an undesirable alien, a foreign invader to be destroyed, because it will be perceived by the immune system as "nonself." Going back to our INS metaphor, it isn't a citizen, and it doesn't have a visa. Even a mother's immune system will not allow her fetus's tissue to remain anyplace in her body other than inside the placenta, where it is safe from being rejected.

But why doesn't the body's immune system normally attack the "self" antigens present in its own cells? In this case, the crucial steps come about as an embryo is developing. It appears that some B- and T-cells do in fact produce antibodies to other, normal body cells. (Recall that antigens don't literally create new antibodies; rather, they "choose" among the immense variety already present.) But these overly zealous defenders—that is, those which, by chance alone, possess receptors for antigens present on the body's own cells, and which would therefore attack normal body tissue—are destroyed, in a process not yet understood. The outcome, however, is clear: a healthy, developing embryo is left with only those immune cells that do *not* produce antibodies to itself.

This recognition of self/nonself is perfectly understandable and utterly essential when it comes to defending the body against unwanted incursions. But it also means that organ transplants and tissue grafts can be very difficult to accomplish, unless the tissue donor is an identical twin of the recipient (if so, then the antigens present on each donor cell are the same as those of the recipient). Drugs have been developed that suppress the immune system, allowing the tissues to "take." Such treatments are doubly troublesome, however, in that anyone with a suppressed immune system is dangerously vulnerable to genuine pathogens, and also because such suppression often does not persist, in which case the transplanted tissue is eventually rejected.

The best method, therefore, is to introduce tissue that will be accepted by a healthy, functioning immune system; that is, to achieve a match between the molecular markers of the donated tissue and those that are recognized as "self" by the cell-mediated immune system of the recipient. Fortunately, a cell's entire molecular makeup does not comprise its relevant antigens. Only a comparatively small number of marker genes are involved; these comprise part of the major histocompatibility complex, or MHC, discussed earlier in regard to the humoral immune response. There are about twenty genes that make up this MHC, and as many as one hundred possible forms of each gene. So, matching the donor's MHC to the recipient's MHC can still be a daunting prospect. It is, in fact, virtually impossible that two persons, chosen at random, will provide an MHC match. Sometimes close relatives fit the bill, but not always.

Fetal tissue offers a possible solution in some cases. The cells of a six- to eight-week-old human fetus have generally differentiated enough to be useful to a recipient but have not yet developed "self" labels and hence are unlikely to be rejected by the body's immune system. Also, recall from Chapter 1 that one of the intriguing long-term goals for genetic engineering is to introduce markers for some of the common human MHC antigens, so that, for example, pig kidneys might eventually be disguised as human kidneys of a particular MHC pattern, and thus be safely transplanted into someone who needs them and whose immune system will not distinguish between the newcomer and self.

Interestingly, some cancer cells secrete a hormone, human chorionic gonadotropin, that is normally produced by a developing fetus and that prevents the mother's immune system from destroying her developing child. (After all, a fetus is a different individual from the pregnant mother, and its antigens could otherwise be expected to induce a response similar to the body's rejection of transplanted foreign tissue.) It would not be surprising, therefore, if cancer cells have stumbled on a similar way to evade detection and rejection. If so, then by disrupting this process it might be possible to help the body to identify a genuine enemy within.

Sometimes, recognition of self/nonself can even be a problem within an intact animal, without any cancer or foreign material present. This is the realm of autoimmune disorders, in which the cell-mediated immune system begins to attack parts of the individual's own body. Such disorders are numerous, and it appears that many debilitating diseases whose causes had long been unknown are actually examples of pathological "friendly fire." Rheumatoid arthritis, for example, is a relatively common disease in which antibodies and T-cells accumulate in joints such as finger, wrist, knee, or hip, attacking the cartilage and causing pain, deformation, and reduced movement. Some forms of kidney disease are also autoimmune ailments, as are systemic lupus, multiple sclerosis, muscular dystrophy, and myasthenia gravis.

Rheumatic fever provides an interesting medical problem because it resembles an autoimmune disease but isn't quite. It is caused when the body's immune system produces antibodies in response to an infection of streptococcus bacteria, commonly in the throat (strep throat). As it happens, molecules found in the valves of many a normal, uninfected heart offer an antigenic "target" that is similar to the outside of streptococcus bacteria, and so accumulated strep antibodies can also attack and damage the heart. This is why it is important when treating strep throat with antibiotics to continue the medication for the entire prescribed period and not stop when the symptoms have gone away: a small number of surviving bacteria, too few to cause trouble in themselves, can nonetheless trigger a troublesome immune response that does far more damage than the streptococci themselves.

Then there are allergies, the most commonly encountered example of immune-system hyperactivity. Allergies are hypersensitivities to certain substances, resulting in reactions typical of the inflammatory response, including redness and swelling, as well as possibly asthma or hives. An allergen, such as dust, pollen, mold, or certain types of foods, provokes an allergic response by acting like an antigen. After exposure to the allergen, the body's immune system begins a humoral response, with plasma cells producing designer antibodies so that the next time the person is exposed to the allergen, the allergic response kicks in right away. In this manner, a person can become "sensitized" to an allergen, so that the second or third response is more severe than the first.

The responsible antibodies are a specific kind called IgE (Ig = immunoglobulin), found on the surface of mast cells. When the allergen attaches to the IgE antibody, these mast cells release histamine and other substances, leading to an inflammation. Histamine causes blood vessels to dilate and also increases their permeability, which results in swelling. Most often, allergens come into contact with mast cells displaying IgE antibodies in the nose and sinuses, so the released histamine commonly has its most immediate effects in runny noses and contracted airways (asthma).

In severe cases, allergies can lead to so-called anaphylactic shock. This occurs when the mast cells release so much histamine in response to an allergen that the dilation of the blood vessels causes the blood pressure to drop sharply. This can lead to death within a few minutes but is easily reversed by a shot of epinephrine, which increases blood pressure by reconstricting the affected blood vessels. Bee stings, for example, are a common source of anaphylactic shock, killing many more people annually than does snakebite. Penicillin allergy is another insidious cause of life-threatening anaphylaxis. In general, allergic reactions are not this serious, however, and are readily controlled by taking antihistamines, which prevent the release of histamine from mast cells.

In the above examples, it is the hyperactivity and excessive sensitivity of the immune system that causes problems. It should be clear by this point, however, that the immune system is in most cases a very important ally to the bodies of mammals such as ourselves. It is necessary for defense against all sorts of pathogens that are constantly trying to destroy us. As with breathing and the beating of our hearts, we are usually unaware of what our immune system is doing for us. When something goes wrong with such defenses, however, the result can be devastating.

Immunodeficiency is a catchall term to describe any defects in the immune system that reduce one's ability to fight off invaders. We have already discussed AIDS. There are other types of immunodeficiency that deserve attention. For example, an inherited disorder called severe combined immunodeficiency (SCID) results in a similar shutdown of both the cell-mediated and humoral responses, much like AIDS. People with SCID can only survive if they live inside a plastic bubble that keeps out all pathogens—which is, in the long run, impossible. It is hoped that eventually SCID will be cured by gene therapy.

As we age, our immune systems become less efficient. One possible explanation for aging, in fact, *is* the decline of the immune system. There also seems to be a real connection between the nervous system and the immune system. After all, lymphocyte-producing organs are deeply penetrated by nerve fibers, so it is not fanciful to suggest that stress, or even mood, can influence health. One revealing bit of research compared the immune system of college students under two different conditions: after a vacation and during exams. Students just returning from vacation produced more interferon as well as more effective natural killer cells.

It may well be that people who are happy stimulate their immune systems more than those who are sad. Laughter, it has long been said, is the best medicine. That may be an overstatement, but anything that keeps the immune system up to snuff—so long as it doesn't cause it to go overboard—is good medicine indeed.

PART 2

MEDIUM RANGE

Familiarity, we are told, breeds contempt. Although most of us are not literally contemptuous about the details of our aliveness, we are strangely inclined to take ourselves for granted. Yet true familiarity is rare. Most of us know our automobile or television set better than our own brains, our reproductive hormones, or what happens when we crave or digest food.

This would be more understandable, perhaps, if knowing ourselves required some kind of special indoctrination into abstruse mysteries. But like Yogi the rock, which the *Martian Sojourner* bumped into, evaluated, and then made the subject of extensive reports, there is much in the nature of human beings that is beneath the surface but nonetheless there to be examined. In a sense, Yogi was unique among Martian rocks, not because it was of a special class but simply because it happened to be in the right place at the right time. Human beings, similarly, are in the right place at the right time. As living things, as animals, as mammals, as primates, we are also worth getting to know.

In the next three chapters, we will look under the skin in an attempt to get to know the less visible parts of ourselves.

THE BRAIN AND BEHAVIOR:

On the Matter of Mind

Famed DNA researcher and Nobelist Francis Crick has turned his attention from genes to brains, from heredity to mind. One result is a book titled *The Astonishing Hypothesis*. For Dr. Crick, this "astonishing hypothesis" is that "you, your joys and your sorrows, your memories and your ambitions, your sense of personal identity and free will, are in fact no more than the behavior of a vast assembly of nerve cells and their associated molecules." The human mind, in short, is the result of nerve cells doing their thing, oozing minuscule droplets of chemicals and flashing tiny sparklets of electricity, prodding and tickling other nerve cells into similar action and in the process somehow generating thought and consciousness. We agree with this astonishing hypothesis, and we think that by the time you've finished this chapter, you will, too.

It would in fact be far more astonishing if "mind" came from something other than brain, if brain came from something other than the nerve cells that comprise it—if, in short, anything other than genuine, down-to-earth material processes were responsible for wonder, dread, doubt, elation, despair, and all those ideas, sensations, and self-awareness that fill our heads.

Don't get us wrong. There is indeed something amazing about the thought that thought itself, as well as ideas, hopes, dreams, love, hate, perceptions of beauty or despair, that all these mental and emotional pyrotechnics come from material stuff such as molecules and electrical charges. But in fact, the material world is all we know and all we are likely to know. A sunset, too, is amazing, as is the unfolding of a flower or the structure of a snowflake; the fact that they are also material does not make them any less wonderful. But there is nothing hypothetical about them.

No one yet has spotted a nonmaterial efflorescence of consciousness, flashing and fluttering about like Tinkerbell, just as no one has ever identified a soul. (And, unlike the case of Tinkerbell, no amount of believing or hopeful clapping of hands seems likely to change things.)

Whatever else we may be, we are all stuck with our materiality. So let's enjoy it and, insofar as possible, understand it. In short, we are about to ask your nerve cells to contemplate one of the great wonders of the universe: themselves.

Neurons: The Matter of Mind

If you were miniaturized and sent on a journey inside a brain, like in the science fiction movie *The Fantastic Voyage*, and if you were a neurobiologist, you would know immediately that you were surrounded by nerve cells (also called neurons). But you wouldn't be able to tell if they were human nerve cells or if you were traveling among the neurons of a bear, a bat, or a beaver. At the microscopic level, the similarities among neurons greatly outweigh the distinctions. As far as can be told, neurons in different species both look and work similarly. The differences between the mind and behavior of hippo and human reside in the fine tuning and organization of those neurons, not in their basic functioning as nerve cells. For now, we will be talking about the neuron without specifying what mammal—indeed, what vertebrate—it belongs to.

Cells generally are perfect embodiments of Bauhaus architectural theory: form follows function. If you are expected to absorb stuff, as in the lining of the intestine, you're going to have lots of tiny projections so as to enjoy the largest possible surface area. If you have to squeeze into narrow capillaries, as red blood cells do, you're going to be smoothly rounded (if, instead, you have sharp, irregular edges, you are likely to get clogged and clumped, as in sickle-cell anemia). And if your job description calls for sending messages from place to place, as with nerve cells, you can expect to be elongated, not unlike a telephone cable, as well as sensitive and excitable.

Although there are many different kinds of nerve cells, all of them are basically signaling devices: sensory neurons, which convey information from the environment (about such things as light, sound, touch, pain, position, presence of chemicals, and so forth); motor neurons, which send information to the muscles or other organs about what action to take; and, most mysterious, so-called interneurons, which intervene between the incoming and the outgoing, and which organize and integrate all these signals, while also producing thought, sensation, learning, awareness, emotions, and so on.

"Between the idea and the reality/Between the motion and the act/Falls The Shadow," wrote T. S. Eliot in *The Hollow Men*. Actually, this description of The Shadow aptly describes the neuron. And so, we might continue the poem as follows: "Between the conception and the creation/Between the emotion and the response/falls The Neuron."

What is the neuron?

Whatever else they may be, neurons R us. To be sure, we are also composed of bones and blood vessels, lungs and livers, but in a very real sense, it is not so much our bodies that constitute our selves as it is our minds. We *are* in our brains. So neurons are worth a careful look. They come in many shapes and sizes, but the basic pattern involves a lot of spidery arms, known as dendrites, at one end and a single, long cablelike structure, called an axon, at the other. Dendrites are multibranched structures (*dendro* means "tree" in Latin) twisting and stretching in many different directions, bringing information into the cell, not unlike the complex branches of a real tree. They are the nerve cell's in-baskets. A single neuron is likely to have many dendrites, in what looks like a devilish tangle. In some cases, these dendrites closely resemble the crown of a tree; in others, more like a dense mat of twisty, serpentine tendrils. The slender, single-stranded axon (meaning "axle" or "axis" in Greek) carries the outgoing message along the nerve cell itself; a bunch of these axons bundled together is what we mean by a "nerve." The axons going to the human leg may be a meter or so in length, making their cells the longest in the body.

Going back to the tree analogy, if the dendrites are multiple, wavy branches at the top, the axon is the single, slender trunk, or perhaps a taproot. Between dendrites and axon, or—rarely—off to the side, is the nerve-cell body, complete with nucleus, mitochondria, and all the usual apparatus of cells generally. As it happens, neurons aren't really very much like trees at all, especially because a tree's bark is pretty much dead, whereas the "skin" of the nerve cell—more accurately, its cell membrane—is where the action is.

Neurons live to communicate, and they do so by sending signals, known as nerve impulses. These messages travel in via the dendrites and out via the axon, and their language is electrical. Although it is tempting to think of an axon as an electric wire, it isn't, and a nerve impulse is in fact crucially different from an electric current. The difference is this: an electric current consists of electrons traveling through a conductor, roughly like water coursing through a garden hose, whereas a nerve impulse is like water moving in a different way, up and down like a wave instead of horizontally, like a flow of current.

Think of an ocean wave. As it "passes" a given point, water isn't literally flowing in a linear stream (although of course this can happen, too, as with the Gulf Stream or the Japan Current). Rather, a wave consists of water molecules that move up and down, vertically, then stay more or less in place while the energy passes on, horizontally, to the next water molecules, which do the same thing. A water wave doesn't really exist, except as physical excitation that travels in a given direction like the graceful snakelike wiggle you can produce in a long string if you shake it vigorously.

A nerve impulse is just such a wave, consisting of electrical changes across a nerve-cell membrane, instead of the vertical excursions of water molecules or of a shaken string. More technically, it is a "wave of depolarization." This implies that there must first be polarization, and there is.

Start with a resting neuron. The first thing you discover is that it is resting in name only, because it takes quite a lot of energy to maintain this state. If you were to stick an electrode just inside such a nerve cell and another one just outside, you would find that the inside is negatively charged compared with the outside—in other words, it is polarized, like a battery. The difference is surprisingly great, around 70 millivolts, which is about $\frac{1}{20}$ the voltage of a standard AA battery.

This crucial finding was made by British physiologists A. L. Hodgkin and Andrew Huxley,* who took advantage of the fact that among invertebrates, nerve impulses travel more rapidly when the axon is thicker. Consider, for a moment, a large squid: its nerves have to send impulses a very long way, to the tips of its tentacles, and they have to send these messages in a big hurry if the squid is to succeed in keeping out of the jaws of its many predators. The axons of a squid are therefore not only long but thick, thus relatively easy to work with.

Hodgkin and Huxley were able, therefore, to insert electrodes into these large motor axons of squid, which subsequently became the workhorses of much neurophysiological research. As it happens, giant-squid neurons function very much like the nerve cells of mammals—which might give pause to calamari connoisseurs but has also done wonders for the scientific unraveling of nerve impulses. The researchers were able to measure the electric charge across the squid nerve-cell membrane during rest as well as during the passage of a nerve impulse.

How does a neuron become polarized? We speak of a political argument as being polarized if it is sharply divided, usually into two opposing camps. So it is with nerve cells. In effect, negative charges are accumulated inside the nerve cell, while positive charges build up outside. But since cell membranes are permeable (that is, they have holes) and opposite charges attract, you might expect that positively charged ions would tend to move in, while negatives move out, reducing the polarization until there is no voltage difference between inside and outside; that is, until the battery is dead. Several things keep this from happening, however. For one, the negative charges inside the nerve cell are mostly due to protein molecules, which are too large to squeeze through the membrane's relatively small holes. And for another, these holes aren't just passive openings. The cell membrane spends considerable energy keeping out the positive ions (especially sodium) that would otherwise rush in. In addition, potassium ions, which are also positively charged, tend to leak out of the resting cell, propelled by the osmotic crush of other potassium ions.

Andrew Huxley, who shared a Nobel Prize for his findings, continued a remarkable Huxley family tradition of excellence: he was the grandson of renowned evolutionist Thomas Huxley and brother of biologist Julian and writer Aldous.

The cell had previously expended energy pumping this extra potassium in, which further contributes to an unstable situation.

So, the "resting" nerve cell is really more like a loaded gun. It "fires" when sodium ions—the ones that had earlier been kept at bay—rush in. One way of describing this effect is that the ion channels that control the flow of these charged molecules are "voltage-gated," meaning that different ones selectively open and close in response to changes in the voltage difference across the membrane. In any event, within milliseconds, the inside of the neuron depolarizes, switching from negative to neutral, with this change passing like a wave along the membrane. (Typically, the depolarization is so great that it actually swings briefly the other way, becoming positively charged for a short time, compared to outside the cell.)

About as quickly, the neuron returns to normalcy. Its membrane begins actively pumping out sodium while also stuffing smaller amounts of potassium back in. This so-called sodium-potassium pump shoves three (positively charged) sodium ions out of the cell for every two (also positively charged) potassium ions it pushes in. As a result, the cell's inside is once again made negative compared to its outside; in other words, the membrane is promptly repolarized by the hardworking sodium-potassium pump.

The sodium-potassium pump and the various ionic gates are expensive to operate. By some estimates, they cost the body about one third of its total energy expenditures. But by all estimates, it's worth it.

Immediately following all this electrochemical coming and going, pumping and diffusing, the membrane is unable to fire another nerve impulse—also called an action potential—for at least a few milliseconds. In a word, it is tired. More technically, it undergoes a brief "refractory period," when it will not respond to any further stimulation. At this point, the neuron is "absolutely refractory." A bit later, it will respond only to an exceptionally strong stimulus; it is "relatively refractory." Eventually, it becomes chemically refreshed; its voltage-sensitive gates are prepared once again to open and close on electrochemical command, and the neuron is ready to respond as usual.

In human speech or in music, pauses are as important as the sounds. Similarly, with nerve impulses, the period of membrane nonresponsiveness is valuable. There is a silver lining even in this apparent refractory cloud: since the cell membrane cannot respond during the absolutely refractory period, nerve impulses can't backtrack on themselves. They can only be transmitted in one direction along an axon. Unlike a rock dropped into a pond, for example, which generates waves spreading outward in all directions, a nerve impulse is a one-way street. And like one-way streets, this keeps the "traffic" moving smoothly.

We said that a nerve impulse isn't like a regular electrical current, with electricity flowing along the axon; rather, it is produced by sodium flooding into the neuron, then being rapidly pumped back out, while potassium is secondarily pumped back in. This briefly restores the cell's "resting potential," but shortly thereafter, the wave of depolarization progresses to the adjacent, polarized region of the membrane. In this way, the action potential travels along the nerve. It does not weaken as it travels because it is generated as it goes. In a sense, therefore, it is like a burning fuse, which carries a wave of rapid oxidation along its length. Once a fuse is lit, its "impulse" is as strong at the end as at the beginning, because it feeds on the chemical energy stored in the fuse itself. Similarly, the nerve impulse feeds on the energy stored across that peculiar living battery known as the nerve-cell membrane.

Think of a gun. After firing—and unlike a fuse—it can be quickly reloaded, ready to fire again. And like a gun firing, a nerve impulse is all or nothing. Just as you can't shoot a bullet at half speed, a nerve impulse either fires, full speed, or doesn't fire at all. Firing requires, say, a half ounce of pressure on the trigger. One quarter ounce is no more effective than one sixteenth; a full ounce produces the same effect as three quarters. In the case of a nerve impulse, "pressure" on the "trigger" is electrochemical rather than physical and is measured in the amount of depolarization produced in the cell membrane: the greater the depolarization (the greater the number of positive ions entering the cell), the more likely that the neuron will fire.

This introduces another important concept: thresholds. A threshold for a gun is the amount of pressure on the trigger needed to fire it; for a neuron, it's the amount of depolarization needed for it to fire. Different neurons have different thresholds. If the threshold is exceeded, the neuron "goes off" and the action potential or nerve impulse is generated. If not, the understimulated neuron falls back to its resting potential.

Thresholds can add up. Think of many different fingers squeezing on the same trigger. Each one may be below threshold by itself, but the accumulated effect might do the trick. With its many dendrites receiving input, each neuron acts as a kind of adding machine, tallying up the effects of multiple inputs. When the total depolarization, coming in via all the dendrites, exceeds the threshold for that neuron, its action potential is triggered and the resulting wave of depolarization travels along the axon, where it contacts the next neuron "downstream," knocking on the door of *its* dendrites.

In most cases, nerve impulses travel faster than would be possible by waves of depolarization alone. Unlike the axons of squid, for example, which increase the speed of their nerve impulses by growing thicker (similar to reducing the electrical resistance in a copper wire by increasing its diameter), the axons of humans—and of mammals in general—are differently adapted for high-speed transmission, as follows. Special cells wrap around the mammalian axon, encasing it in a fatty insulating material known as myelin, which is impervious to ions or electric charges generally. Each myelin-containing cell grows around the axon, so that when seen in cross section, the myelin wrapping looks remarkably like a jelly roll. (Myelin is white and shiny and is responsible for the characteristic appearance of nerves, or at least, of certain nerve fibers, as opposed to the "gray matter," so conspicuous on the outside of the brain, which is made of nerve-cell bodies.)

Unlike the rubber insulation of an electric wire, this myelin wrapping is not continuous. At regular intervals of about a millimeter, bare unwrapped "nodes" are left. The action potential actually jumps from node to node, in rapid, "saltatory" conduction. Here's how it works: An action potential is produced at a given node, but because of the myelin insulation next door, this local depolarization cannot spread to the adjacent membrane. The built-up electric charge then essentially jumps to the *next* place where depolarization can occur, the next node. As a result, myelinated nerves conduct impulses very quickly, up to 100 meters per second; without myelin, the maximum rate is only about 4 meters per second.

Several centuries ago, when pirates terrorized the Amalfi Coast of Italy—and cell phones weren't available—residents devised a means of rapidly sending an alarm when a marauding ship appeared. Bundles of dry wood were placed at conspicuous points along the coastline. Once a single watch fire was lit, it was seen by the sentries, who in turn lit their fire. Thus, the message jumped from point to point, far more rapidly than the fastest rider.

Without myelinization and the ability of nerve impulses to jump along, effectively bypassing the lengthy membrane and taking an express route instead, nerve conduction would be much slower and less efficient. Everyone has experienced the difference. Stub your toe, for example, or hit your thumb with a hammer, and you get an almost instantaneous message that you've injured yourself. At the same time, there is typically a gap of a few seconds, during which you may say to yourself, ruefully, "Uh-oh, soon this is really going to hurt!" By which time, of course, it does. The initial sensation, telling you that something is wrong, is carried to your brain by fast, myelinated axons. The details—the specific nature of the pain itself—arrive more slowly, via unmyelinated fibers.

Consider what happens when the myelin sheath is disrupted, as happens with multiple sclerosis. In this autoimmune disease, the body destroys its own myelin, which in turn wreaks havoc with nerve signals, resulting in difficulty performing simple, coordinated movements. As is so often the case, however, the more we learn about something, the more complexity is revealed. Thus, multiple sclerosis has long been considered a classic "demyelinating disease." Although the loss of myelin is unquestionably involved in multiple sclerosis, recent findings have also shown that large numbers of neurons themselves are often severed, which makes them unable to transmit impulses and also contributes to their death.

Meanwhile, back at a normal nerve-cell membrane, things are yet more complicated. Not only are there many different fingers squeezing each neuronal trigger—many different inputs coursing through the dendrites of each nerve cell—but some of these fingers are pulling in the other direction! For a neuron, this means *increasing* the electrical polarization of the membrane. Recall that a resting neuron is electrically negative compared with the outside, and that a nerve impulse is a wave of *de*polarization. Moreover, this wave is initiated if the membrane is depolarized beyond its threshold. If, for example, the threshold is 20 millivolts, then a 25-millivolt depolarization will cause the nerve to fire. A jolt of 15 millivolts will not. (On the other hand, two 15s or three 10s will also do the trick.) But at the same time, if the inside of the resting neuron is made yet more negative—*hyper*polarized—then it will be more resistant to reaching the threshold.

Return now to our hypothetical neuron with a 20-millivolt threshold. Many different signals are coming in via the dendrites. Let's say that the total is 25 millivolts toward hyperpolarization. The neuron's interior has therefore been made 25 millivolts more negative than it was just before. As a result, it now requires a total of

45 millivolts of *de*polarization to cause that neuron to fire (the original threshold of 20 plus the additional 25 to which the resting potential has been reset). We can say that the nerve cell has been somewhat inhibited, as our make-believe gun would be if someone were pulling against the trigger.

There are, in fact, inhibitory impulses, those that hyperpolarize, as well as excitatory impulses, which depolarize. This is how each nerve cell acts as an adding machine, in addition to being a gun or a fuse: it constantly sums the excitatory and inhibitory impulses that act upon it, then either generates an action potential if the threshold is exceeded or remains silent if it isn't. Certain nerve cells in the cerebellum receive input from as many as 25,000 other cells upstream, some of them excitatory, others inhibitory, and still others silent at any given time. Although this is extreme, the average, run-of-the-mill neuron has to integrate messages from hundreds, even thousands, of other neurons, some urging it to "go for it," others demanding restraint, in a constantly shifting background pattern of electrochemical murmurings. And this is just a single nerve cell! Now consider that there are about 100 billion nerve cells in the human brain, many of them connected in complex ways to many others, some excitatory and some inhibitory, forming all sorts of cascades, loops, and reverberating circuits, converging on a small number of executive neurons, or diverging to make themselves heard (or felt, or smelled, or remembered) by other neurons working in other systems.

Neurons are great gossips, constantly talking to one another, or trying to. Most of the time these murmurings are sotto voce; their messages aren't passed on. A single whisper—the input from one dendritic spine, however insistent—normally isn't enough to arouse an action potential. At any given moment, for any given neuron, there may be thousands of whisperings from its incoming dendrites, some urging action, but also several thousand that are quiet, and maybe a few thousand more that are inhibitory. Every incoming impetus toward depolarization declines with time and distance; it must be reinforced, either by input from many other neurons (what is called "spatial summation") or by repeating itself over and over, in very rapid sequence ("temporal summation"), to have any chance of evoking the Big Response.

Consider, next, an especially important region of the neuron, the axon hillock, a sometimes discernible bump located where the axon originates from the cell body. These axon hillocks are the vote counters. They add up the information coming in and "decide" whether or not the neuron will respond. Note that neurons have a very high signal-to-noise ratio. When they don't have anything interesting to say, they keep quiet. When they do, it's usually important ("there's a lion chasing me" or "what a cute member of the opposite sex"). So, when the neuronal gossip is at a low level, the axon hillocks keep a lid on things. Only when something is worth saying will neurons expend the energy needed to transmit the message.

Not surprisingly, axon hillocks differ in their sensitivity, which introduces the whole question of individual differences. Give someone's nerve cells a slightly lower threshold, with axon hillocks that are somewhat more likely to fire, and the person is, in turn, more likely to be jumpy, irritable, or sensitive. (Among its various effects, it appears that testosterone reduces the threshold for axon-hillock responsiveness.) Add up all these effects, across literally billions of neurons, complexly interconnected and mutually responsive, and you have gone a long way toward distinguishing what it is that makes each of us an individual.

It's important to keep in mind that "irritability" doesn't mean being cranky or high-strung; it simply refers to the likelihood that a neuron will fire. Since vast numbers of neurons are connected in all sorts of ways, the fact that certain ones are more likely to respond and others are less so means that certain connections are more or less likely to be made. The result? For one person, a particular smell may conjure up a thought; for another, it evokes a different thought; for a third, a special emotion, and so forth.

It is probably no exaggeration to describe the brain as the most intricate device in the known universe. According to Sir Charles S. Sherrington, an early pioneer of neurophysiology, the human brain is a "sparkling field of rhythmic flashing points with trains of traveling sparks hurrying hither and thither." Or, in an unforgettable image, it is "an enchanted loom where millions of flashing shuttles weave a dissolving pattern." Out of this dissolving, shifting harmony of electricity and chemistry, of ions flowing in and out, being pumped hither and yon, there somehow emerges that elusive chimera: thought.

What Turns On the Neuron?

We now know that neurons get stimulated and that out of the sum of all this stimulation (positive and negative, excitatory and inhibitory, depolarizing and hyperpolarizing) they either fire or remain silent. What is actually responsible for those crucial depolarizations?

Several things. For instance, neurons can be highly specialized to fire in response to various forms of energy, especially the rat-tat-tat of the outside world knocking on our door. These are the many different kinds of "sensory receptors" scattered throughout the body but especially concentrated at the head (because that's where a whole bunch of other neurons are also to be found, comprising what we call the brain, whose job is to integrate all this information and then decide what to do about it). Rods and cones, for example, are nerve cells that respond to light energy, sending impulses that make their way to particular regions of the brain where they are perceived as vision. The inner ear contains its own special kind of nerve cells, which respond to vibrations causing the eardrum to jiggle, and whose accumulated messages are eventually perceived as sound. Sensory neurons in the nose are attuned to chemicals, especially the volatile ones that evaporate readily and are carried through the air; they send information that is ultimately identified as smell. There are similar tales to be told about taste, touch, pressure, pain, position, and so forth. Filled with electrochemically encoded information, rather than sound and fury, they signify everything—everything that we know and can ever know about the world, both outside and inside.

Other life-forms, incidentally, receive information unknown to us, except through the use of instruments. There are freshwater fishes—especially those inhabiting turbid rivers where visibility is limited—that are exquisitely sensitive to electrical fields and able to navigate obstacles and even identify prey by the fact that a tiny insect larva, for example, conducts electricity better than a submerged rock. A common aquarium fish achieves this feat, but to do so it must refrain from the full-body undulations that characterize the movements of most other fishes, which would disrupt the precise electric field it is creating; hence, it keeps its body rigid and is known as a knife fish. Bats bounce high-frequency sounds off moths, homing in on their victims via aerial sonar. Bees can see ultraviolet light (which is why so many flowers are white to us but many different shades of white to their discerning pollinators). Most mammals rely heavily on information-rich chemical calling cards deposited on convenient landmarks; this explains why dogs are so intrigued by fire hydrants and why going for a walk is such an exciting experience for the most domesticated canine.

Human beings—need we be reminded?—are mammals. Nothing in our description of nerve cells and their function is *sapiens*-specific. And when it comes to our sensory abilities, we are especially unimpressive. In our sense of smell we are notably obtuse, although recent evidence suggests that human beings may perceive more than they consciously realize, especially in the sexual realm: women can distinguish male from female sweat, for example, an ability that increases during ovulation. In addition, the well-known tendency of women living together to synchronize their menstrual periods seems to be mediated by pheromones (external hormones, operating between individuals, instead of within them) that exert a profound but unconscious influence. Human hearing is acceptable but not extraordinary; if anything, our ability to detect sounds is rather deficient in the higher frequencies, so we employ dog whistles but only to call dogs. On the other hand, we have comparatively good eyesight as mammals go, and we're unusual among mammals in possessing color vision. (As a general rule, animals that can see color are those that are brightly colored themselves: birds, most fish, and insects. A naked person is comparatively uninteresting—at least compared with a rainbow—but we generally make efforts to enhance our coloration through cosmetics, clothing, and other adornment.) As far as can be told, however, all our windows to the world operate by the same neuronal rules as those of other living things, regardless of how many such windows there may be, how widely opened, and what prospect they command.

It is remarkable that for all the subjectively experienced difference between, say, hearing a high C and seeing a high tide, between the smell of a rose and the pain of a stubbed toe, the brain receives, processes, and experiences its input via the same electrochemical events: waves of depolarization traveling along nerve-cell membranes. Because of this common currency, it is possible to fool our various specialized nerve receptors. Push gently on your closed eyelid, and you will stimulate your optic nerve and "see" patterns of light and shape, just as a blow to the head can make you "see stars" or cause your "ears to ring."

So much for neurons responding to the outside world. What about connections among themselves? When you hear or see something, neural excitation has traveled to your brain, where it is not only perceived as sensation but may also trigger alertness or boredom, emotions such as joy or fear, as well as ideas, memories, and so forth. This is not the work of a single neuron. As we've already seen, hundreds and even thousands of impulses impinge on a typical "brain cell," with the additive effect of their excitation and inhibition determining whether this cell will fire and thus in turn become one of the innumerable impulses that impinge on the next cell downstream. Nerve cells, in short, do not only receive outside simulation; they also communicate with one another.

The key concept here is the *synapse*, a word invented by Charles Sherrington (he of the "enchanted loom," who also gave us the word *neuron*). *Synapse* is derived from the Greek, meaning "junction" or "making contact," because it speaks to the contact between nerve cells. Ironically, however, a synapse is not a place of direct physical contact. Instead, each synapse consists of a tiny gap—a synaptic cleft—between the axons of one neuron (the one potentially sending a signal) and the dendrites of another (the one receiving it). Join us now as we enter a synaptic cleft.

It is a world of chemicals. Actually, there are some electrical synapses, across which waves of depolarization pass just as they pass along nerve-cell membranes, with the presynaptic cell sparking the postsynaptic cell into immediate and obedient action. In most cases, however, the minuscule synaptic space between axon and dendrites is awash with various and often competing juices, so that transmission from pre- to postsynaptic cell is far from guaranteed.

This tiny gap, about $\frac{1}{600}$ the width of a human hair, is the realm of neurotransmitters, chemicals that transmit a nerve impulse (or rather, the possibility of one) across the interneuronal no man's land. Passing an action potential from one neuron to another is a bit like passing a baton in a relay race, with the neurotransmitter being the baton. (Don't forget, however, that once the chemical baton is transferred, it then travels in a different form: as an electrical wave of depolarization, as already described.) Recall that the voltage-gated ion channels found throughout a nerve-cell membrane are called that because they respond to voltage changes. The dendritic tips at the receiving end of a synapse are different. Instead of responding to voltage changes, they are structured to generate an action potential in response to certain chemical neurotransmitters: they are *chemically* gated.

Many different kinds of chemicals open these gates, because there are many different kinds of receptors, each responding to its own neurotransmitter. When the molecular "Mr. Right" comes along, however, the response is basically the same: changes in the electrical polarization of the receiving cell.

In the early days of brain research, only a small number of these messenger chemicals were known. Today more than fifty have been identified. Some are excitatory, reducing the polarization of the dendrites, thus making it more likely that the threshold will be reached and the postsynaptic neuron will fire. Others are inhibitory, increasing polarization and thus reducing the likelihood of a nerve impulse being propagated. To make things yet more complicated, the same neurotransmitter might be excitatory in one type of synapse (say, in the brain) but inhibitory in another (where nerve meets muscle). You might wonder why synapses make use of such an elaborate system of both inhibition and excitation, when it would seem that simply varying the amount of excitation alone, for example, would do the job. This would be like driving a car with only an accelerator: feasible, perhaps, but less controllable than if you could deftly employ both the brake and the gas pedal.

The axon tips at the end of each presynaptic nerve cell are different from the rest of the cell membrane along which the action potential travels. As far as their cell is concerned, they are the end of the line. Instead of simply depolarizing, they respond to the onrushing electrochemical wave by releasing specified amounts of their particular neurotransmitter. This release is mediated by a number of things, such as how long ago the cell last fired and how rapidly the action potentials arrive. Calcium ions are especially important in permitting presynaptic membranes to release their load of neurotransmitter: stimulated by the action potential, calcium flows in and induces little sacs containing neurotransmitter to be spilled into the synapse.

The neurotransmitter then diffuses across this gap, uniting with special receptor sites on the postsynaptic cell. At the synapse, postsynaptic cells are rich in receptors, poor in transmitters. And for their part, presynaptic cells have lots of neurotransmitters but very few receptors. The result? A synapse is a one-way street, just like the neuron itself. Otherwise, the seeming chaos that results from all this electrochemical tickling, summation, excitation, and inhibition, divided and multiplied exponentially across literally billions of cells and many billions more possible combinations, would be genuine chaos indeed.

(The more neurobiologists learn about synapses—indeed, the more anyone learns about anything—the more difficult it becomes to generalize; everything, it seems, becomes an oversimplification. For example, it is now known that some neurotransmitters inhibit cells *presynaptically* and autoregulate cell function; that is, they secrete chemicals that inhibit themselves! Thus, to some extent, presynaptic cells do have receptors, and to some extent, synapses carry two-way traffic after all. And as we'll see in a moment, *post*-synaptic cells can secrete neurotransmitters that impact the firing of presynpatics. Nonetheless, the above generalization is useful for a functional overview; as far as transmission downstream is concerned, synapses really *are* one-way streets.)

Depending on the neurotransmitter, its quantity, and its effect at a particular synapse, a brief change in membrane potential is produced in the dendrite of the postsynaptic cell. These various little twitches of depolarization and hyperpolarization are called, respectively, excitatory postsynaptic potentials (EPSPs) or inhibitory postsynaptic potentials (IPSPs). In most neurons, the accumulated EPSPs and IPSPs all come together in the axon hillock. And here, depending on the total effect of all these inputs, the postsynaptic cell either fires or keeps quiet.

We have already traced the little electrochemical drama of nerve-cell firing. Therefore, let's return to the synapse, where, in a sense, the cast of characters is first assembled.

After a molecule of neurotransmitter has diffused across the synapse and helped to generate either a brief EPSP or IPSP, one of three fates awaits it. First, it may simply leak out into the intercellular fluid, becoming part of the flotsam and jetsam of the body. Or it may be cleaved by dedicated search-and-destroy enzymes that specifically target a particular neurotransmitter and thereby allow the postsynaptic neuron to resume its resting potential. Or it may be taken out of action another way, reabsorbed by the presynaptic membrane that originally secreted it, ready then to be released once more when another nerve impulse comes down the axonal line. This mechanism is known as presynaptic reuptake, an awkward but descriptive phrase that, as we shall see, applies to some of the most important new psychoactive drugs.

Neurotransmitters for Beginners

There are essentially three kinds of neurons, defined by what they do rather than how they work. (1) Sensory neurons are excited by light, sound, touch, et cetera, and carry that information toward the central nervous system. (2) Association neurons, or interneurons within the brain, bounce signals off other, similar association neurons and in the process perceive, ruminate, cogitate, and otherwise perform the quiet but crucial jobs constituting the "mind." (3) Motor neurons send outgoing signals from the brain to the muscles, telling us to walk, talk, throw a ball, step away from a hot stove, or scratch where we itch.

Synapses are not found only between nerve cells. They also occur where motor neuron meets muscle fiber, at the so-called neuromuscular junction. Here the signal is conveyed by acetylcholine ("ah-sit-eel-KO-leen"), the first neurotransmitter discovered. Austrian physiologist Otto Loewi had long suspected that nerve stimulation was not merely an electrical process but that chemicals were involved too. In 1921 he placed a frog's heart in a salt solution and stimulated its vagus nerve, which, he knew, would slow its heartbeat. Loewi then removed this heart and placed another, unstimulated one in the same chemical bath; it also began to beat more slowly, showing that some sort of chemical, present in the liquid bath, was involved. (It is easy to be confused here. Although acetylcholine is excitatory at most neuromuscular junctions, at the heart it is inhibitory—one of many cases in which neurochemicals are difficult to pigeonhole. It is also interesting to note that the idea for Loewi's simple, brilliant, pathbreaking experiment first came to him in a dream; he wrote it down, went back to sleep, and then, in the morning, couldn't decipher what he had written. The next day, fortunately, the idea came to him again.)

In any event, neurotransmitters are chemicals, and they are crucial for nerve transmission as well as for the normal process by which neurons stimulate muscles to contract. This is how some chemical weapons and other nerve poisons work: they interfere with the crucial functioning of neurotransmitters. For example, South American rain forest inhabitants discovered long ago that certain brightly colored frogs are ferociously poisonous (not in their bite but in their skin, which also explains why they are so gaudily adorned: would-be predators quickly learn not to eat them). These poison-arrow frogs contain curare, a potent chemical that essentially clogs up acetylcholine receptors, so that the victim's muscle cells are made insensitive to the neurotransmitter. If you rub an arrow point against the skin of such a frog, and then shoot it into a deer, a monkey, or a person, paralysis results. Among other things, the victim's diaphragm stops contracting, causing death by suffocation.

Interestingly, there are two kinds of acetylcholine receptors, called "nicotinic" and "muscarinic" because of the other chemicals that have been found to bind to them. As their name suggests, nicotinic receptors respond not just to acetylcholine but also to nicotine, the active ingredient in tobacco, which may account for some of the physiological effects—but not the addictiveness—of smoking. Muscarinic receptors respond not just to acetylcholine but also to muscarine, a chemical derived from *Amanita muscaria*, a deadly poisonous mushroom, which in small amounts also has hallucinogenic effects. Nearly all mind-altering drugs act in some way or other at the synapse, modifying neurotransmitters or their receptors.

Acetylcholine is normally broken up and rendered ineffective by its own private enzyme: acetylcholinesterase. Otherwise, acetylcholine would accumulate in the neuromuscular synapse, overexciting the muscle cells. This provides an opening for yet another group of poisons, those that inhibit the work of acetylcholinesterase. Many insecticides fall into this class, doing the opposite of curare: they block the work of acetylcholinesterase, so the victim suffers what is essentially an acetylcholine overdose, which produces excessive muscular contractions and death from seizures.

Another one of the early neurotransmitters to be discovered is norepinephrine. It (along with its close chemical relative, epinephrine, better known as adrenaline) is especially implicated in helping the body respond to stress. Norepinephrine is primarily removed via reuptake, that is, reabsorbed into the presynaptic neurons whence it was released. Amphetamines and cocaine block this norepinephrine reuptake, thereby prolonging its effect and producing enhanced energy and short-term stress resistance. But stress can only be defied up to a point, beyond which there is a potentially lethal cost. By exaggerating the effect of norepinephrine, the body can obtain temporary respite and excitation, but people can also die from cocaine or amphetamine overdose.

It is now clear that the brain employs a wide array of neurotransmitters, some inhibitory, some excitatory, some long-acting, some short. The most prominent neurotransmitters in the brain are simple amino acids and their products. Among the latter, for example, dopamine works mostly (but not exclusively) to inhibit nerve firing. The nerve cells in certain brain regions, especially one known as the substantia nigra ("black substance," because that's how it looks under the microscope), specialize in producing dopamine. When these cells degenerate—as sometimes happens in old age or after repeated blows to the head, as in the case of the boxer Muhammad Ali—too little dopamine is available to the brain, and the resulting loss of muscular inhibition is observable as uncontrolled tremors, a condition known as Parkinson's disease.

Some Parkinson's sufferers can be aided by the drug L-dopa, which helps the remaining healthy cells make dopamine. But eventually these cells, too, cease to function. A promising experimental treatment for Parkinson's disease involves a kind of targeted rejuvenation: introducing young, healthy brain cells eager to provide dopamine for bodies bedeviled by its deficiency. The catch is that these transplanted cells are derived from aborted fetuses (adult dopamine-making cells are rejected by the host's immune system), and research of this sort is bedeviled by political opposition from those opposed to any use of fetal tissue.

Like most neurotransmitters, dopamine cannot easily be pigeonholed. It is involved in much more than inhibiting muscular twitches. Take schizophrenia, a serious mental illness that can involve delusions and various impairments in normal thought and that afflicts about one person in a hundred. No one really knows what causes schizophrenia. At one level, it seems clear that genes are involved: the disease runs in families, even if genetic relatives are reared in completely different households. So whatever else may be responsible, heredity is, too.

It is also clear that schizophrenia is a brain disease, and furthermore, that it is somehow related to dopamine. Either schizophrenics suffer from excessive secretion or sluggish removal of dopamine at the synapse, an excess of dopamine receptors, hypersensitivity of these receptors, or some combination of these. The result is essentially a dopamine "overdose." More precisely, the culprit may be an improper balance between dopamine and a neurotransmitter known as GABA, discussed below.

Schizophrenic symptoms can sometimes be treated by drugs that block dopamine release, including so-called dopamine "antagonists" such as Haldol, Thorazine, and Mellaril. But not always. And not surprisingly, these medications can also bring on symptoms resembling Parkinson's (which, you'll recall, is associated with too *little* dopamine).

Another well-studied neurotransmitter, serotonin, also has many different functions. It inhibits pain, allows sleep, and appears to contribute to a sense of well-being. In fact, depression, a mental illness even more common than schizophrenia and no less debilitating, appears to be caused by (or at least, related to) a serotonin shortage in the sufferer's brain. One solution: antidepressants, a class of drugs that are increasingly well-known to the public but whose precise mode of function is rather mysterious. Like norepinephrine, serotonin is removed from synapses by being selectively reabsorbed by the neuron that originally secretes it.

Since depression is associated with too little serotonin, it may be that people suffering from depression have too few postsynaptic serotonin receptors, or that the ones they have just aren't responsive enough, or maybe their serotonin is too promptly reabsorbed. This suggests treatment to enhance the impact of serotonin by slowing down the rate at which it is reabsorbed by the depressed person's presynaptic neurons. Slowing down the "reuptake" of serotonin induces the synapses to marinate in more of this particular neurotransmitter, often alleviating the worst symptoms of depression. Accordingly, an important class of antidepressants includes the selective serotonin-reuptake inhibitors, or SSRIs, the best known of which is Prozac.

This is just scratching the surface of a veritable witches' brew of neurotransmitters. Another interesting one is GABA (gamma-aminobutyric acid, related to the chemical responsible for rancid butter), a relatively simple amino acid that is involved in about 80 percent of the brain's synapses, where—depending on the circumstances—it does a little bit of just about everything. Mostly, GABA is inhibitory, reducing the excitability of neurons so that normally, when well-supplied with GABA, they require relatively more stimulation to cause them to fire. The benzodiazepine tranquilizers such as Librium and Valium are known as GABA "agonists" (the opposite of *antagonists*) because they enhance its action; interestingly, they are used both to prevent seizures and to reduce anxiety. Specifically, the "benzos" bind to GABA receptor sites and cause the GABA channels to open more often. At the neurochemical level, the resulting ion influx hyperpolarizes the postsynaptic cell, making it less likely to fire. At the experiential level, people who take Librium or Valium become calmer and more relaxed, less "nervous."

People born with fewer receptors to their own naturally produced benzodiazapines would presumably be more anxious and so may respond well to extra benzos—tranquilizers such as Valium or Librium. In this and all other cases of neurotransmitter function—not to mention synaptic connections of all sorts—the accumulated effect of an extraordinary number of tiny variations seems to explain the fine tuning that makes each of us behaviorally different and unique.

Then there is a fascinating detective story worked out in the 1970s. Candace Pert and Solomon Snyder, at Johns Hopkins University, were trying to understand how such powerful drugs as heroin and morphine exerted their effects. They found that the brain contains receptors that are specifically sensitive to these complex chemicals. This posed something of a mystery, however, because heroin and morphine, like opium and marijuana, are produced by plants.

Why should the human brain be outfitted with receptors that respond to neurotransmitters that they don't normally encounter? Pert and Snyder reasoned that perhaps there are additional neurochemicals, made by the brain itself and resembling heroin and morphine. Sure enough, they discovered that the brain indeed produces certain natural opiates, which they called endorphins (from "endogenous morphines"—*endogenous* means deriving from within the body, as opposed to *exogenous*, coming from outside). Endorphins appear to be responsible for such self-gratifying sensations as the "runner's high," as well as the reduced sensitivity to pain and the brief but intense enhancement of physical capacity that is experienced by people who perform remarkable feats on the football field or the battlefield, even when seriously injured.

Apparently the brain responds to heroin, morphine, and the like because it is normally wired to respond to its own endorphins, produced by the brain itself and used as neurotransmitters under extraordinary circumstances or when a brief jolt of endogenous anaesthetic is called for. (Endorphins are not only called upon by athletes or war heroes, however. During sex, for example, most people are relatively insensitive to mild pain. Shortly afterward, they may become aware of a persistent ache, perhaps from an elbow or knee that is a bit contorted, but which, during the "ecstasy of love," was masked by endorphins performing a useful erotic service.)

Endorphins may also explain why heroin addicts often suffer such distress when they stop using the drug. Since heroin mimics the body's natural endorphins, the addict's brain mistakes the former for the latter. With endorphin receptors ringing off the hook, the brain essentially says to itself, "I must be making lots of endorphins," so it cuts back on its own production. Going "cold turkey" leaves the addict's brain with an opiate deficit: no more heroin, and hardly any natural endorphins, either. The resulting pain is called withdrawal.

As we have seen, neurons talk to one another. As in a human conversation, if the presynpatic neuron speaks very softly, its postsynaptic interlocutor strains to listen, becoming more sensitive to the signals it is receiving. And when a neuron talks too much or too loudly, the postsynaptic neuron is likely to stop listening; it becomes less responsive to the given neurochemical. This may also have relevance for drug accommodation and addiction: used to getting lots of a particular chemical, the postsynaptic neuron becomes less responsive. Having turned its hearing aid down, it requires a louder yell—more of a "fix"—in order to create the same initial effect.

One of the striking things about many neurotransmitters is what small and simple molecules they often are. At least two simple inorganic gases even get in on the act: carbon monoxide, chemically CO (a deadly poison in even moderate quantities), as well as nitric oxide (NO), whose interesting vasodilating effect we examined earlier. Unlike other neurotransmitters, these gases are not stored in presynaptic neurons, waiting to be released into the synapse when an action potential arrives. Rather, they are cooked up on demand.

In addition, the simple gases NO and CO—although genuine neurotransmitters in the brain—do not directly affect mood or thought. For example, nitric oxide is released as a result of sexual thoughts or stimulation, but it does not cause these feelings in itself. Others, such as dopamine and serotonin, have a more direct impact on subjective experience.

Even in these cases, however, it is important not to oversimplify and expect that there is a simple one-to-one correspondence between brain chemicals, thoughts-and-emotions, and behavior. Despite the dizzying variety of neurotransmitters, there are even more thoughts and moods. On the other hand, recall that just four nitrogenous bases in DNA, by specifying a mere twenty amino acids, can dictate all of the immensely variable proteins that make up a human body. Given the large variety of neurotransmitters (which greatly exceeds A, T, G, and C), the possible combinations increase exponentially. Add to this the immense variability of synaptic inhibition and facilitation, as well as the nearly infinite arrangements of the various synapses themselves, and the possibilities are literally mind-boggling—as well as mind-creating.

Finally, we need to mention that some—perhaps most—neurotransmitters do more than open or close their designated chemical gates on the far side of a synapse, a direct effect that is sometimes called the first-messenger system. Many neurotransmitters also participate in additional modes of operation, collectively termed the second-messenger system or, more accurately, systems, because there are many of them. Research is only beginning to uncover the details, but already the number of relevant biochemical entities is staggering. The distinguishing characteristic of second-messenger systems is that they exert their influence *inside* the cell, specifically at the level of the cell nucleus and enzyme-producing nucleic acids, rather than at the cell membrane alone. Not surprisingly, therefore, second-messenger effects are slower than those produced by first-messenger transmitters, since they are responsible not for whether or not a nerve cell fires but rather for modulating its pattern of sensitivity, growth and development, and overall responsiveness. Second messengers, in a sense, set the stage for the more immediate "go/no-go" decisions that are eventually made at the axon hillock of the post-synaptic cell. It is safe to say that we will be hearing quite a bit from second messengers in the future.

Little Green Men in the Machine

In the hilarious movie *Men in Black*, an extraterrestrial alien is murdered. He looks, initially, like any other middle-aged (albeit dead) man, but in the morgue it is revealed that the victim lacks internal organs and is actually a robot. The real alien is a little green man who resides inside the "person's" head, from which he had been directing the body's movements. It is an image with an old history, dating back at least to René Descartes. This brilliant seventeenth-century mathematician, philosopher, and anatomist was quite insistent that whereas the human body is merely a machine, the mind is something else altogether: immaterial and partaking of the divine.

Descartes did not go so far as to suggest that there was an otherworldly little green man inside everyone's head, but his notion of "dualism"—between mind and matter, body and spirit, brain and soul—lives on.

Descartes is also the author of what is probably the most famous sentence in Western thought—"*Cogito ergo sum*": "I think, therefore I am"—which he proposed as the cornerstone of a philosophy to be founded on incontrovertible truth. (Ambrose Bierce modified this to "*Cogito cogito ergo cogito sum*": "I think I think, therefore I think I am"—adding that this was as close to certainty as philosophy seems likely to get.)

Descartes's view has been criticized by many others, including, not surprisingly, the majority of neuroscientists, who haven't yet seen anything resembling little green men inside anyone's brain. If there really is a "ghost in the machine," it is remarkably elusive; after all, unlike the vast Himalayas, reputed to harbor Abominable Snowmen, or mile-deep Loch Ness, with its supposed underwater monster, the cranial vault is a comparatively small place in which to hide. And it has been pretty carefully explored.

At the same time, it is hard not to be at least somewhat sympathetic to Descartes. To put it differently, *cogito ergo whom*? Who is this "I" who am? And where does he reside? The preceding pages have necessarily left out a lot of detail, but they pretty much exhaust the basic neurophysiology of mental processes. Certainly, there is no special molecule uniquely tagged with consciousness and found only in human beings. (For that matter, there is no molecule unique to our species, except for our nucleic acids, and these are unique to every species.) Second, *Homo sapiens* is not home to any special, magical brain region or any unique physiological, electrical, or chemical process unknown in other mammals. Imagine the hullabaloo if it were!

In the nineteenth century, the German physician Carl Vogt proclaimed, "The brain secretes thought just like the kidneys secrete urine." The self-proclaimed "vitalists" of his day "refuted" Dr. Vogt by asserting that thought is not a fluid. Game, set, match. But what about this proposition, as true as anything we can imagine: the brain generates thought, like the kidneys generate urine? Or, to turn Descartes on his head: *Sum ergo cogito*: I am, therefore I think. Because each of us "is" made of neurons and their crucial chemicals, we are able to think.

It may seem strange that the details of human thought are not more open to human thought itself. Why can't we just close our eyes and concentrate, commanding our brain to turn inward and tell us just what is going on there between our ears?

Here are two possible reasons, not entirely separate. The first might be called the Ministry of Culture paradox. It is well known in Warsaw, Poland, that the best possible view of that city is from the top floor of the Ministry of Culture, because it is the only place in Warsaw from which one cannot see the Ministry of Culture building: a huge, ugly example of Stalinesque architecture at its worst. Similarly, perhaps we cannot "see" the workings of our own brains because we are literally too close. But on the other hand, if there were some biological payoff to perceiving the actual electrochemical processes by which we think, it seems likely that we would have that ability, just as we are "blessed" with the capacity to detect a toothache or a stubbed toe.

This leads to a second reason why we must resort to oscilloscopes, microelectrodes, and other paraphernalia if we want to probe the workings of our own minds. The human brain has not evolved for the purpose of providing accurate information about the world, neither outside our bodies nor inside our skulls. Rather, its evolutionary purpose is to promote the long-term reproductive success of its bearer. Fortunately, perhaps, this purpose is often met by being objectively accurate: telling us, for example, when and where a large animal is approaching, or a potential mate, or whether the river is rising or the nearby volcano is about to blow its top. (Also, with suitable adjustments, all the information on which the insights of "science" are based.) But unless there is a distinct payoff in doing so, don't expect your brain to tell you how *it* works, any more than it will inform you about how your kidney manages to filter your blood. For evolution, it is enough that your brain does work, just as your kidneys do.

This doesn't mean that we are altogether clueless when it comes to connecting brain function on the one hand with, on the other, the subjective experiences of perception, sensation, thought, and emotion. We've already described, for example, some of the effects of chemicals such as dopamine and serotonin. And many people are familiar with other psychoactive drugs, whether "mind-expanding" ones such as hashish, LSD, and mescaline, or "mind-contracting" ones such as alcohol, heroin, or barbiturates. The mental is chemical.

The mental is also electrical, as revealed, for example, by an electroencephalogram (EEG), which is basically a record of the overall electrical activity of the brain. This pattern of electrical activity differs between sleep and wakefulness, among other things, and during the different stages of sleep. The brain-wave information gathered by an EEG represents in a sense the sum total of the brain's electrical activity; as such, it is much less precise than the "behavior" of individual neurons. But it can be useful in diagnosing certain brain tumors, as well as diseases such as epilepsy (which result from excessive nerve firing in certain brain regions). In addition, a blow to the head can produce mental confusion, lasting minutes, hours, even a lifetime, which is not surprising since physical injury can disrupt the complex electrical pathways upon which the brain's normal functioning depends.

Connecting Brain and Mind

Mind and brain: you can't have one without the other. Here are some examples.

First, consider learning and memory. We learn when we modify our behavior as a result of prior experience. And when we remember something, we subjectively experience it. The most elegant experiments in this regard were conducted by Eric Kandel and his colleagues at Columbia University, working with a marine mollusk known as a sea slug. When undisturbed, this rather simple animal extends its siphon, bringing in water that flows over its gills. Touch the siphon, as a predator might do, and it is quickly withdrawn. Do this repeatedly, however, and the animal stops responding; it has learned that having its siphon tickled is not a preface to being attacked.

Sea slugs, not surprisingly, make use of a very simple arrangement of neurons in order to learn this scintillating bit of behavioral brilliance: sensory nerves from the siphon send a signal that the siphon has been touched. Each presynaptic sensory neuron connects via a single synapse with a postsynaptic motor neuron that, when adequately depolarized, induces the sea slug's muscles to withdraw the vulnerable siphon. When the sensory nerves are repeatedly stimulated, they simply produce less and less neurotransmitter, so that eventually, the motor nerves, on the other side of the synapse, stop firing. Voila! The sea slug has learned something. And so have we.

To make matters even niftier, it is possible to counteract this simple form of learning, known as habituation, by its opposite: sensitization. By applying a mild electrical stimulus to a much beset-upon sea slug's tail at the same time that they touched its siphon, the researchers were able to *enhance* the siphon's withdrawal. This is due to a third neuron that adds its two cents' worth of neurotransmitter to the presynaptic (sensory) neuron, thereby producing an enhanced effect when it is appropriate: being touched at both siphon and tail suggests that the sea slug is under increased threat.

There have been many efforts to analyze human behavior, in all its complexity, as being merely the accumulation of simple reflexes. The name Pavlov should ring a bell; also psychologists John Watson and B. F. Skinner, among others. Reflexes and simple "conditioned" behavior undeniably exist and have been studied extensively, at the level of neurons and neurotransmitters as well as through the actual behaviors in question. But reflexes and conditioning are generally seen as inadequate to explain the immense sophistication of what people do, or at least the more interesting things that they do. Nonetheless, the likelihood is that human memory as well as thought itself consist of short-term electrochemical firing, probably forming some sort of reverberating circuits among the neurons involved.

We all experience short-term memory, such as quickly "learning" the name of someone we just met—and then forgetting it almost as quickly. On the other hand, different kinds of memories (our name, address, telephone number, etc.) are firmly entrenched and can only be disrupted by serious trauma. It seems reasonable that a different mechanism is involved in such long-term memory, possibly involving the microanatomy of the brain, such as the growth or acti-vation of additional pre- or postsynaptic connections. It is easy to imagine fine neuronal tendrils, straining to reach out to one another and eventually achieving a "memory" when they succeed. But so far, no one has been able to demonstrate that the growth of specific nerve connections causes—or results from—specific memories. It does seem increasingly clear, however, that when we learn, certain synapses, already in place, become more likely to fire.

If you stimulate a synapse once or twice, you won't have any effect on it. But after regular, frequent stimulation with microelectrodes in a laboratory, it becomes easier to cause the postsynaptic neuron to fire, even weeks later. Thus, neurons develop what is called long-term potentiation (LTP), and it looks wonderfully similar to "laying down a memory," as existing synapses become quicker on the draw. The "better" the memory, in a sense, the easier access we have to it, which, at the level of neurons, appears to mean that the downstream neuron(s) are more readily stimulated. It appears, therefore, that at the neuronal level, learning involves "potentiating" those connec-tions already present.

It's a strange notion, suggesting, first, that there is a synaptic cir-cuit for every memory. But this is no stranger than neurons medi-ating thought, emotion, and memory in the first place! It is also strange in its implication that every memory may in some degree have preexisted in each person's catalog of synapses, just waiting to be potentiated. Even this is conceivable, however, given that we are talking about not only billions of neurons, but that each neuron is typically connected to thousands of others, many of them via thousands of different synapses; in short, there may be as much micro-anatomical subtlety in our brains as there is cognitive subtlety in our minds. It's no coincidence.

Of the many neurotransmitters, glutamate seems especially involved. It causes "up regulation"—that is, the presence of glutamate causes the production of more glutamate receptors, which in turn makes the neuron with the receptors more likely to fire. Glutamate itself is excitatory, which explains in part the effect monosodium glutamate, or MSG, has on some people (and why it shouldn't be given to kids: it's too neuro-excitatory). Aspartamate, the main ingredient in NutraSweet, is similar, which raises questions about excessive use of that product.

In the course of normal learning and long-term potentiation, some neurotransmitters violate the rules: known as *retrograde* neurotransmitters—because they go in the "wrong" direction—they travel from the postsynaptic cell (receiver) to the presynaptic (sender) and cause the latter to send more glutamate. The result is yet more LTP. Alcohol, incidentally, can inhibit LTP, which is why alcohol overdose causes loss of memory, or alcoholic blackout. So can high levels of stress—which has complex hormonal effects—as well as a severe concussion.

Most of the things we "learn" don't stay with us very long, but some do. This is where long-term potentiation comes in. Something must happen when a memory is transferred from short-term to its long-term form. In fact, there is strong evidence that this involves a specific part of the brain, known as the hippocampus.

A famous case is that of a man known as H.M. (to retain his privacy). In an effort to control his severe epilepsy, H.M.'s left and right hippocampus were removed. Following surgery, H.M. had normal memory for everything that had happened to him before his hippocampus-ectomy, but he was unable to transfer any new experiences into long-term memory. For example, H.M. behaved normally with someone he had just met; but if this person went away for more than a few minutes, he was completely forgotten, as though they had never met. H.M. could read a page, but by the bottom, he would forget what he had read at the top. His epilepsy was gone, but H.M. was condemned to live in a perpetual state of now.

Human memories, of course, can be marvelously complex. They often consist of more than simple, unitary recollection, far more involved than the sea slug's memory of its siphon being prodded. Here is a famous passage from Marcel Proust's fictional masterpiece, *Remembrance of Things Past*, in which the narrator's rich and complex memories—specifically, of his childhood in the village of Combray—are evoked by a sip of tea and a small taste of madeleine cake:

> Many years had elapsed during which nothing of Combray . . .
> had any existence for me, when one day in winter, my mother . . .
> offered me some tea, a thing I did not ordinarily take. . . . I
> raised to my lips a spoonful of the tea in which I had soaked a
> morsel of the cake. No sooner had the warm liquid, and the
> crumbs with it, touched my palate than a shudder ran through
> my whole body. . . . And suddenly the memory returns. The
> taste was that of the little crumb of madeleine which on Sunday
> mornings at Combray . . . my aunt Léonie used to give me,
> dipping it first in her own cup of real or of lime-flower tea. . . .
> Once I had recognized the taste . . . , all the flowers in our
> garden and in M. Swann's park, and the water-lilies on the
> Vivonne and the good folk of the village and their little
> dwellings and the parish church and the whole of Combray . . .
> sprang into being. . . .

Like the rest of us, Proust had rich long-term memories stored in his brain, capable of being evoked by the right stimulus. Does it diminish his accomplishment, or our own, to acknowledge that these long-term memories can be brought to life by neurotransmitters, initiated by taste receptors? Regardless of one's answer, the *fact* of such memories—their capacity to be evoked and their existence as physical entities—is not seriously in question.

Evidence comes from the work of pioneering Canadian neurosurgeon Wilder Penfield. He performed brain surgery on patients suffering from severe forms of epilepsy. Most epileptic seizures are preceded by characteristic sensations; Penfield hoped to control these seizures by destroying precisely that part of the afflicted person's brain that generated them. To find the right spot, he used a mild electric current to stimulate various regions of the patient's brain, trying to evoke the all-too-familiar preseizure sensation. (For all its endowment with nerve cells, the surface of the brain has no pain receptors, so Penfield was able to explore the cerebral cortex of conscious patients, who then related what they were experiencing.)

And experience things they did! As many of his patients described it, having various brain regions excited by Dr. Penfield's electrodes was not simply to remember an earlier event, but literally to feel that it was happening again, right there in the operating room. Long-ago conversations were relived, in precise detail. Old friends were revisited, musical events reexperienced. Had Marcel Proust been on the operating table, it presumably would have been possible to achieve with a tickle of electricity the same effect as his renowned tea-and-madeleine. Also to the point: Proust's memories must have been encoded in his cerebral cortex long before he transferred them to paper. (Only about 5 percent of Penfield's patients actually experienced the "memories" described above, and their accuracy has been questioned. Nonetheless, stimulating so delicate a structure as the human brain with something as crude as a randomly placed electrode is a bit like trying to perform precise needlework with a pair of ax handles; the fact that any mental experiences were generated at all is powerful testimony to the reality that the brain stores information.)

Penfield's remarkable results lead to the question of localization of function within the brain. The nineteenth-century French surgeon Paul Broca, witnessing the effects of battlefield brain injuries, reported that a patient who had suffered damage to a small region on the left side of his brain was unable to speak. This region has since been identified as "Broca's area" and is associated with the motor pattern of speech. People suffering injury to this area are unable to verbalize normally, and yet they have no difficulty reading or understanding the speech of others. By contrast, injury to another region, Wernicke's area, can destroy the ability to comprehend written or spoken language, although the sufferer remains able to produce the basic sounds of speech and to imitate the rhythm of normal talking. In the first case, people can make sense of language, without being able to produce it; in the latter, they can produce something like language but cannot do so sensibly.

In addition to a basic distinction between "motor" and "sensory" areas, different parts of the brain are involved specifically with smell, speech, hearing, skilled movements, taste, reading, vision, and so on. Persons with brain damage due to stroke, accident, or other causes often reveal that in some ways, at least, the brain is partitioned in remarkably precise ways. On the underside of part of the cortex, for example, resides the ability to recognize faces. One patient, with an injury to his left frontal cerebral lobe, had only one symptom: he could no longer name fruits or vegetables. (This led a science writer, describing the case, to ask whether the brain has a produce section.)

Next among the ways in which mind and brain go together is personality. In a sense, one's personality or character is every bit as elaborate, if not more so, as one's memories. Personality, or "temperament," is influenced in complex ways by an interaction between genetic background and personal experiences. But however it is laid down, it seems clear that certain fundamental aspects of human personality reside in the brain. The living—or formerly living—proof for this assertion is a fellow named Phineas Gage.

The remarkable story of Phineas Gage is renowned in the annals of neurobiology. In 1848, he was working as a foreman on a railroad construction project. One of his jobs was to prepare sections of rock for demolition, using a metal rod to tamp explosive into holes drilled in the rock. He must have tamped too hard, because the blasting powder exploded and the rod—a yard long and 1½ inches in diameter—was propelled out of the hole like an elongated bullet. It entered Gage's left cheek from below and passed out through his skull (and brain), landing some distance away. Miraculously, he survived; in fact, he recovered quickly.

But the word *recovered* is problematic here. For after the accident, Phineas Gage wasn't the same person. Whereas earlier he had been a sober, responsible family man, successful in his work and highly regarded in his community, the new Phineas Gage was irresponsible, prone to foul language and a quick temper, unable to hold down a job or to delay gratification. One side of his brain's frontal lobe had been obliterated; subsequent studies have confirmed that this region is responsible for, among other things, generating behavior that is careful, socially appropriate, and suitably inhibited.

People who lack social propriety and have difficulty with impulse control are unlikely to have lost massive regions of their brains in spectacular, traumatic accidents à la Phineas Gage. However, it is not unlikely that something is amiss in their brains. Perhaps they have experienced microstrokes in brain regions otherwise concerned with inhibition, or, as part of normal mammalian variability, maybe they are a wee bit underdeveloped in their frontal lobes. This is not to deny that experience also can play a crucial role. Our point is that when neuroscientists ask *how* experience influences behavior, they are likely to come up, eventually, with answers in terms of brain chemicals, synapses, and neurons. (Even when "experience" is the culprit, it is bound to operate via the brain.)

The issue of localization of function leads to one of the most perplexing and challenging findings in neuroscience: the matter of cerebral lateralization and the "split-brain" phenomenon. It has long been known that there is a crossover in the way the brain interacts with the body. Thus, we speak of the cerebral hemispheres because the most highly developed part of the brain, the cerebrum, is divided into two halves, looking a bit like your two fists, clenched and held together. The left side controls the right side of the body, and vice versa. Thus, if a stroke damages the left side of your brain, the result will likely be right-sided paralysis. For right-handed people, motor regions in the left hemisphere are dominant. Or, as some left-handers like to point out, they are the only ones truly in their right minds!

Human beings are unusual among animals in experiencing handedness, and even more so in being primarily right-handed (rats and parrots, for example, are also "handed," but righties and lefties are equally common). Aside from the question of handedness, however, the two cerebral hemispheres are not quite mirror images of each other. The "left brain" is especially verbal, mathematical, and almost computerlike in its use of logic, whereas the "right brain" deals more with spatial relations, and also with emotion, music, and art.

Normally, the right and left sides communicate with each other via a thick band of connecting fibers known as the corpus callosum. But beginning in the 1960s, a radical surgical treatment for severe epilepsy provided an opportunity to probe the distinctness of the two hemispheres, leading to the astounding finding that to some degree, people are composed of two brains inhabiting the same body, a kind of Siamese twins of the mind. (From the hippocampus of H.M. to Wilder Penfield's electrical evocation of memory to the famous split-brain research we are about to describe, neuroscience owes a remarkably large debt to epilepsy.) Some patients had their corpus callosum severed, so that their right and left cerebrums could no longer "talk" to each other. Roger Sperry and his colleagues at Cal Tech noticed that at first glance, these people seemed normal in their mental functioning. Sophisticated research showed, however, that in such cases, a left-sided person and a right-sided person were secretly cohabiting.

For example, when such individuals held a key in their left hand and also looked at it, they could identify it and use it to open a lock.

If they were blindfolded, they recognized the key by feel and could use it normally; however, when holding the key in their left hand, these split-brain subjects were unable to call it a key. Remember, the right hemisphere in most people is nonverbal, and tactile information identifying the object as a key was transmitted only to this, essentially mute brain half. With the corpus callosum no longer intact, there was no way for the verbal left brain to tell the nonverbal right brain that the correct word was *key*. (A story is told about the time Sperry described his experiments to a group of Cal Tech scientists during a dinner party at his home. As his distinguished guests were leaving, physicist Richard Feynman shook Sperry's hand and commented how much he had enjoyed the evening; Feynman then did a little shuffle, extended his left hand, and added in a peculiar, high-pitched voice, "And I had a good time, too!")

It is generally true that the right brain is more intuitive and the left more rational. However, the distinction does not appear to be as absolute as earlier reports had suggested. In addition, people can sometimes compensate for damage to either side, so that with training, the right brain can develop some language skills and the left, linear reasoning. And following injuries due to strokes or auto accidents, women are generally able to recover language ability more rapidly than are men. In some ways at least, it appears that the female brain is genuinely different from the male brain. It is less compartmentalized, for example, so loss in one region can more readily be offset by activity in another.

For many years it was thought that after basic growth and development is complete, the human brain does not add neurons; it only loses them. Other species, however, have long been known to perform the trick. Birds, for example, grow new neurons as they learn new songs, and rats do virtually the same thing when exposed to an interesting and complex environment. So-called neurogenesis (growing new neurons) has been reported among monkeys, and moreover, it turns out to be dramatically slowed as a result of stress. Finally, a study of middle-aged human cancer patients used a chemical label to distinguish new brain cells (those generated after the label was introduced) from preexisting cells. The results showed conclusively that even in human beings, adults do in fact produce new brain cells . . . at least in some cases.

Neurogenesis could eventually offer long-term hope for someday developing recovery regimes for victims of Alzheimer's disease, Parkinson's, strokes, and other injuries that result in loss of brain cells. It would be no small trick to induce brain cells to replace others that have died, but such interventions have, at least, moved from being mere science fiction to the realm of the theoretically possible—or, shall we say, not impossible.

Brain Imaging

Which brings us, finally, to "imaging," new ways of seeing what's going on inside a healthy brain. For a long time, the intact, functioning brain was something of a black box, a mysterious, unknown, and largely unknowable "something" that has inputs (experience, nutrition, injury, etc.) and outputs (thoughts, feelings, behavior, etc.) but that could only be accessed directly in cases of accident or surgery. This is why the new world of brain imaging is so exciting, because it provides a noninvasive way of opening the brain's black box without literally opening the skull.

First, a new range of techniques known generally as tomography (*tomos* is Greek for "slice," and *graphe* means "drawing"). Fortunately, no literal slicing is involved. So-called CT scans—for "computerized tomography"—involve taking a series of thin, cross-sectional X-ray slices. The patient is moved slowly through the CT machine while an X-ray source circles about, getting many different views from slightly different angles. The resulting three-dimensional image is more meaningful than a simple, two-dimensional X ray but is still somewhat bedeviled by the "shadow" effect of the bony human skull. X rays are especially good at seeing bone, because they are blocked by bone tissue and pass through the soft stuff, which is why they are used to diagnose fractures. But in the case of brain studies, it is the soft tissue that is of special interest.

Next, MRI, or magnetic resonance imaging. This technique relies on the fact that living tissue contains lots of water, and the hydrogen atoms within water molecules will align themselves in the same direction when exposed to a powerful magnetic field. If these atoms are then jolted with radio waves, which kick them out of alignment, they spring back again almost immediately, in the process giving off their own very faint radio signals. An MRI scanner detects these signals and converts this rapid, resonating dance of hydrogen atoms into a detailed picture.

Imaging technology can generally be used throughout the body (MRI, for example, is good at visualizing possible knee injuries) and is especially rewarding when focused on the brain, which, because it is so delicate, had previously been impervious to biology's prying ways. Moreover, surrounded by the thick human skull, the brain hasn't shown up especially well on X ray. Because MRI is attuned to water—well-represented in soft tissue such as the brain—and passes readily through hard, relatively dry stuff such as bone, it provides unusually good brain scans.

Although MRI gives a good picture of soft tissue, the result is a static image that conveys immense detail about structure but nothing about function. Enter the current star of brain imaging: the PET scan, or positron emission tomography. Before being PET scanned, the subject is given an injection of biologically active molecules, usually glucose, that have been labeled with a harmless radioactive isotope. The isotope, in turn, emits positrons (positively charged), which collide with electrons (negatively charged) within each cell, producing gamma rays. The PET-scan machine essentially records the different levels of radiation that have accumulated in different parts of the brain. The more biologically active a cell, the more glucose it metabolizes, and the more radiation. For easier interpretation, different levels of radiation are converted into different colors on a computer screen. The resulting image of the brain's activity resembles a weather map, except instead of precipitation or wind velocity, it depicts amount of metabolic activity.

PET scanning allows neurosurgeons to detect the presence and exact location of brain tumors, which dramatically appear as "hot spots," since tumor cells metabolize more vigorously than normal tissue. PET scans have also revealed substantial differences, for example, in the brain activity of normal people and those suffering from Alzheimer's disease or schizophrenia. A surprising result of PET scans has been the finding that the cerebellum, previously thought to be involved in noncognitive events such as balance and coordination, also "lights up" during other, more "cerebral" activities.

Most interesting, for our purposes, is that PET scans reveal some of the actual brain patterns—of energy use, not electrical impulses—associated with thought itself. (Thinking really *is* work! At least, it really does use up energy.) In one oft-cited study, for example, subjects were asked to perform four different mental activities associated with words: listen to words pronounced by the experimenter; look at a list of words but keep quiet; read words out loud; and come up with simple synonyms. These different verbal tasks produced distinctly different PET scans, showing that the brain functions differently depending on what it is doing. In a sense, this seems a tautology: How could the brain do different things *unless* it functioned differently each time? What is remarkable, however, is to see these differences and to know that they are correlated with distinct metabolic activity in different regions of the brain.

We mentioned evidence for male-female differences in localization of brain function. By use of PET scans, researchers found significant differences in the neural activity of men and women performing the same mental task. For example, when men were asked to look for words that rhyme, their brains became active in Wernicke's area, in their left cortex; in women, the area of activation was broader and more diffuse and involved both left *and* right hemispheres.

Brain imaging has been growing rapidly, not unlike a cerebral tumor, only more benevolent. For example, the original MRIs, described earlier, are essentially static, whereas the latest in off-the-shelf MRI technology essentially combines the noninvasive aspects of static MRI with the action-oriented PET scan. So-called functional MRI provides a continuous picture of brain metabolism (and thus, activity), by recording differences in oxygen level within the brain. Working brain regions, just like working muscles, use more oxygen. With this procedure, researchers compared brain activity in dyslexic and normal people when both groups were asked to perform a rhyming task. They found that there is a distinct neural system called upon when people read: individuals suffering from dyslexia have a disconnect in this wiring between their auditory and visual brain areas, making it difficult for them to associate written words with their auditory counterparts.

Seeing is believing. Maybe the next step will be to *see* believing.

Just as nineteenth-century philosophers found it easy to dismiss Carl Vogt's analogy of the brain and kidneys, it is also too easy to bypass the physical reality of neurons and the brain by taking refuge in "emergent properties," arguing that we are interested in outcomes, not inputs, in the whole rather than its parts. We do this, for example, when we acknowledge that atoms exist but comfortably leave their details to the physicists. After all, someone interested in, say, the construction of chairs is more likely to be concerned with the "emergent properties" of wood, paint, or glue than in their atomic structure. Fair enough. But no one, not even the most ardent furniture-maker, takes the existence of such emergent properties— or the fact that in the case of chairs, the whole is greater than the sum of its parts—as reason to go around speaking in hushed tones about the inexplicable spiritual and mystical import of wood, paint or glue, even for the most ornate Louis XIV chair. Admittedly, the furniture of our minds is more complex and downright awe-inspiring than any chair, which is why we have written this chapter about brains and synapses, rather than barstools and sofas. But the fact that something is "emergent" does not necessarily mean that it is removed from the realm of scientific explanation.

In the closet scene of *Hamlet*, Queen Gertrude accuses that melancholy Dane of "seeing things," namely, his father's ghost, chastising that it "is the very coinage of your brain. This bodiless creation. . . ." In most productions of *Hamlet*, the audience too sees the ghost, although Queen Gertrude evidently does not. This makes for an interesting conundrum: Is the ghost real or merely a "bodiless creation"? Our point is that either way, whether the prince was really seeing his dead father or something conjured out of his fevered imaginings, Hamlet's actions, perceptions, and thoughts are nonetheless the "very coinage" of his brain. And ours.

RELEVANT REPRODUCTION:

Sex and the Human Animal

Sex: just about everybody does it, just about nobody understands it. "Fools give you reasons," we learn in the musical *South Pacific;* "wise men never try." Well, we are going to try.

First we'll look at why sex happens at all; the answer may surprise you. Then we'll consider a few other questions, such as what is the meaning of male and female, and why are there two sexes (rather than three, or three hundred)? We'll get into the sweaty details of reproduction itself: how it works, how it can be prevented. Finally, we'll look at the birth process and follow a developing person on the journey through life, through puberty and into parenthood. In a very real sense, the story of sex *is* the story of life, for almost all of us.

Even though, as we'll see, some living things are asexual, the shared biological reality of human and nonhuman animals is probably nowhere as clear as in sex and reproduction. Consider the hard-core existence of males and females, the even harder-core facts of sexual intercourse, fertilization, implantation, pregnancy, the gill slits and fishlike tails of all vertebrate embryos, the uterine contractions of birth, as well as lactation afterward; if you're at all inclined to doubt the animal, biological nature of *Homo sapiens*, you need only think of these things to know your place: smack in the middle of life itself.

There is a certain poetic justice here; reproduction, after all, is about continuity, not only of generations but of generation itself, across all living things.

Why Sex?

What a stupid question! Doesn't everyone know that sex—at least, at its biological core—is for reproduction? Indeed, resolutely nonsexual interactions between a young man and woman can come across as absurd. In 1952, a broadcast on Moscow radio contained this red-hot scene between two young workers at a collective farm: The young woman, a tractor driver, sighed, "How wonderful it is to work on such a beautiful moonlit night, and do one's best to save fuel!" To this, her male counterpart ardently exclaimed, "The night inspires me to overfill my quota by a higher and still higher percentage!"

Equally ludicrous is the pronouncement of the influential Hungarian psychoanalyst Sandor Ferenczi that "the purpose . . . of the sex act can be none other than an attempt on the part of the ego . . . to return to the mother's womb." Clearly, sex serves reproduction, and yet biologists know that there are other and easier ways to make babies. Lots of animals reproduce quite well, thank you, without sex. There is even some reason to suspect that they may do better than the rest of us, who are so hung up on sex that we can't breed any other way.

It may seem counterintuitive, but the fact is that when it comes to reproducing, sex is just not a very good method. Consider just a few of its drawbacks. You have to find a suitable mate, and this is not always easy. You have to coordinate your activity with your mate's; this, too, is not always easy. Sometimes, moreover, courtship and mating result in injury, especially to the smaller, weaker, or less fervent party (generally, but not always, the female). Not only that, but courtship usually involves making oneself conspicuous, which increases the risk that predators will take advantage of a temporarily lovestruck pair; there are, for example, bats that specialize in picking up the signals of courting frogs. The mate of one's dreams may also be carrying hidden (recessive) or not-so-hidden genes that are hurtful when combined with your own. He or she may also have hidden agendas, conducive to personal selfish interests but not yours. And then, of course, there is the problem of sexually transmitted diseases, prominent in human beings but in no way limited to them.

This list seems daunting enough, but even these liabilities seem minor compared to the huge *genetic* cost sexual reproduction entails.

You might think that despite its problems, sex is obviously suitable for reproduction, for the same reason that elderly people are generally grateful for a birthday: it beats the alternative. But the alternative to sexual reproduction is not *non*reproduction, but *asexual* reproduction, a process that has much to recommend it. Remember mitosis, that delicate dance of DNA whereby one cell becomes two? It is quick, relatively simple, and—for the genes involved—highly efficient. If a cell can reproduce that way, why can't whole organisms? In fact, they can and do. Strawberry plants, for example, reproduce by forming "runners" that lead, eventually, to new individuals. Hydras and many different kinds of animals reproduce by "budding," with new individuals simply developing from outgrowths of the old. Many invertebrates, including common water fleas and garden aphids, reproduce via parthenogenesis, that is, by the development of unfertilized eggs (the Greek *partheno* means "virgin," so parthenogenesis is pretty close to "virgin birth"). Even some vertebrates, including a few fishes and reptiles, have gotten into the act; no birds or mammals, however.

When a plant or animal reproduces asexually, it is guaranteed that all its genes will be represented in every one of its offspring. With sexual reproduction, each parent must literally share the genetic payoff with its partner, since mother and father only get to bequeath an egg or a sperm, each of which carries half of the parent's initial genetic endowment. The result is that sexual reproduction carries with it, right off the bat, a 50 percent evolutionary tariff compared with the payoff awaiting the amoeba, who simply cranks out exact duplicates of itself.

Sex emerges, therefore, as one of the great mysteries of life. As D. H. Lawrence asked, "Why were we crucified into sex?/Why were we not left rounded off, and finished in ourselves . . . ?"

Don't get us wrong. For all mammals, including human beings, sex is *the* way to reproduce. (It will be a long time, if ever, before cloning becomes the norm.) And for most of us, sex is more ecstasy than crucifixion. At the biological level, pleasure—whether the satisfaction of a full belly or a good night's sleep, even a good bowel movement—is how our body rewards us for doing something that is ultimately in its interest. Conceivably, we might find budding or parthenogenesis to be excruciatingly joyous . . . if reproducing that way were in our long-term evolutionary interest. But neither we nor any other mammal has been "left rounded off, and finished in ourselves." Why not?

Biologists haven't yet decided. Some even remain glumly convinced that sex as a means of reproduction is a losing proposition, that people and many other living things have been led down a regrettable evolutionary garden path, so that we are stuck with making the best of a bad situation. Others—and they are in the majority—recognize the immense genetic cost of sex but argue that it is compensated for by one particular, equally immense benefit: producing variable offspring. In this view, the primary asset of asexual reproduction is also its major liability. By producing identical copies of oneself, the argument goes, it is possible for an asexual breeder to reap an immense evolutionary benefit, but at great potential cost if and when the environment turns sour. Sexual reproduction, by contrast, entails the mixing and matching of genes from two different parents, so that the resulting offspring are always somewhat different from either progenitor.

Just as a field of genetically identical wheat is vulnerable to a newly arrived parasite, genetically identical offspring are vulnerable—every one, equally so—to being wiped out by climate, competitors, or any other vagary that comes their way. In agriculture, it is crucial to have a number of different crop varieties on hand, so genetic diversity can insure that among the available seed, at least some resistant strains will be represented. Similarly, when it comes to reproducing, asexual parents are (almost literally) placing all their eggs in one basket, a strategy that may yield short-term dividends but courts long-term disaster. Current thinking among evolutionary biologists is that the payoff of sexual reproduction may be especially intense because of parasites and pathogens. These are ever-present in the living world, afflicting not only human beings today but also our primordial ancestors hundreds of millions of years ago, when the "decision" to reproduce sexually was made.

The idea is simple enough. As disease-causing organisms begin to specialize in a particular type of prey, there is benefit in taking evasive action. Keep making offspring that are identical to yourself, and the viruses, bacteria, et cetera, that prey upon you will fatten themselves on your children as well. But if you bob and weave, offering a moving target, it might be possible to keep a step ahead of your enemies. All this wouldn't be necessary, of course, if those enemies were standing still, but since they, too, have discovered sex and the genetic variability it confers, everyone has to keep at it or be overwhelmed. Evolutionary biologists call it the "Red Queen effect," after a character encountered by Alice in Lewis Carroll's *Through the Looking Glass*: because her world was always moving, it was necessary to run just to stay in place. And, as the Red Queen pointed out, "To get anywhere, you have to run twice as fast!"

In a sense, it would be all too easy to give up on sex, to gather the evolutionary advantages of substituting a 100 percent genetic return for a mere 50 percent—until the final reckoning.

So, ironically, it may be that when people "indulge" in sex—so often criticized as being shortsighted and dangerously enmeshed in immediate gratification—they (or rather, their genes) are actually planning ahead, and laudably so, sacrificing short-term gains for long-term security.

The Meaning of Male and Female

So much for sex itself. But why males and females? After all, if the biological significance of sexual reproduction is that it creates diversity by combining the genes of two different individuals, why do those individuals have to be from what we call the opposite sex? Why not do it like paramecia, which have dozens of "mating types," with an individual from one type free to exchange genes with any individual from another? Better yet, why not have any two individuals exchange genes and reproduce that way? (Actually, paramecia reproduce asexually, by dividing in two; they and a number of other one-celled creatures practice the ultimate in nonreproductive sex: exchanging genes but without reproduction.)

More than two thousand years ago, in Plato's *Symposium*, Aristophanes offers up an explanation. It seems that once upon a time there were remarkable creatures known as Androgynes, which sported four arms and four legs, as well as the sexual apparatus of both male and female. Great was their power and glory, but they grew insolent. Zeus responded—as was his habit—with thunderbolts, splitting each offending creature into two, every one equipped with two arms, two legs, and one paltry set of genitals, either male or female. (Fortunately, the Androgynes abated their insolence at this point, for if they had not, Zeus was prepared to split them yet again, which would have resulted in beings who hopped about on one leg, with sex organs that can scarcely be imagined.)

According to this version, our search for emotional and carnal union is simply our continuing attempt to reestablish our prior, androgynous wholeness, each of us seeking to reunite with his or her missing half. That's why there are two sexes.*

Biologists take a different view.

---

* *Plato's tale also comes with the helpful suggestion that there were originally three kinds of Androgynes: male-female, male-male, and female-female, thereby explaining the origin of male and female homosexuality as well.*

To understand it, we must first ask another question that may seem silly but isn't: What do we mean by male and female? For biologists, maleness and femaleness are not defined by external physical traits. It is not penis or vagina, breast or beard, color or costume that determines the difference between male and female. Rather, it is the type of gamete—egg or sperm—that conveys the crucial difference between bulls and cows, stallions and mares, men and women. Eggs, which are large and relatively few in number, are produced by females, whereas sperm, which are small and abundant, are the defining characteristic of males.

Another way of looking at it: males and females are each specialists, in making either vast numbers of tiny gametes—sperm—or a comparatively small number of large, well-nourished eggs. We'll see later that this difference is truly momentous, with implications for anatomy, physiology, and behavior. But in a sense it begs our earlier question. It still seems possible to have a species composed of individuals who reproduce sexually (that is, they exchange genes) but are not recognizably male or female, who simply produce nondescript, middle-sized gametes, which fuse to produce offspring. Such an arrangement might actually have made life easier all around, since if everyone were sexual-reproductive fair game, the number of potential mates would be essentially doubled!

But things didn't work out that way, for any species. As to *why*, we are back in the realm of speculation, but here is a reasonable possibility, based on a computer simulation. Researchers put together a model consisting of many individuals, each producing slightly different gametes: some large, some small, most of them in-between. Because they took less energy, the smaller gametes were produced in larger numbers, and vice versa. Small gametes had the advantage of being more abundant and able to move more rapidly; large gametes had the advantage of extra nutrients, which gave them a head start once fertilized. Very soon, small gametes were competing with other small ones for access to large gametes, just as throughout the natural world males typically compete with other males for access to females. Not only that, but as the computers cranked out generation after generation of theoretical creatures and assessed their reproductive success, it quickly became apparent that there were two optimum strategies that were rewarded (equally) in this brave new world: make lots of quick, little, competitive gametes or a comparatively small number of large, well-endowed ones. In short, be male or female, not in-between.

The Three Great Mysteries of Female Sexuality

Before plunging into how reproduction works, we'd like to continue exploring long-term evolutionary questions ("why sex?" and "why two sexes?") by addressing the three great mysteries of sex. Interestingly, these mysteries are characteristic of females (the sexuality of men is comparatively straightforward) and involve traits that are unique, or almost so, to human beings. They are also deeply biological as opposed to spiritual or intellectual. Thus, someone looking for evidence of *Homo sapiens* as uniquely created in God's image and the apple of His eye would probably not leap to draw attention to concealed ovulation, the female orgasm, or menopause.

First, concealed ovulation. Most female mammals are not exactly secretive about their readiness to release a fertile egg. Female dogs or horses in heat are unmistakable. Visitors to the monkey house at any zoo need only glance at the red, swollen behind of a female baboon or chimp to know these primates are ovulating. Not so for human beings. With all our intense self-knowledge and personal evaluation—from the philosophical to the medical—we as a species are paradoxically hard-pressed to determine, short of high-tech intervention, exactly when we ovulate. Yet few biological processes are more fundamental. Given that human ovulation is so important, why is it so secret?

Here are some possibilities, all variants on the theme first enunciated following the senseless slaughter of World War I, when British prime minister Lloyd George commented ruefully that war is too important to be left to the generals: in some way, it appears that ovulation is too important to be left to the conscious awareness of the people doing the ovulating!

For one, concealed ovulation may be an evolutionary strategy that functioned to keep prehistoric men in close touch—literally—with their mates. Since human infants are unusually helpless at birth, it would have paid mothers to retain the assistance of a devoted father. According to this view, if ovulation were clearly signaled, as in so many other species, then men might be free to abandon a consort as soon as she stops ovulating. Concealed ovulation therefore may have essentially forced men to maintain a relationship throughout the female's cycle. This, in turn, could have led to greater confidence of paternity and thus greater willingness to help care for the dependent offspring.

Another possibility argues in precisely the opposite direction. It is based on the observation that infanticide—killing of infants—is quite common among free-living animals, and it typically involves males killing offspring who are not genetically their own. This hypothesis is based on the further observation that concealed ovulation is intimately connected with a lack of estrous, or "heat." If ovulation is to be concealed, it cannot be given away by a dramatic increase in sexual receptivity; as a result, women are free to be sexually receptive at any time. Thus, they may copulate with more than one male, thereby giving each of them the impression that he may be the father of the offspring, which in turn makes each potential father less likely to kill the baby.

Another, related idea, emphasizes the "heatless" nature of women's sexuality. It suggests that natural selection may have promoted an absence of estrous in *Homo sapiens* as a way of giving women more control of their reproduction. Mammals in heat are not noteworthy for their choosiness in selecting a mate. Women are.

It has also been suggested that concealed ovulation evolved as follows. From as early as prehistory, women were aware that childbirth was painful and often dangerous. Accordingly, those women who were able to detect their own ovulation may have endeavored, often successfully, to avoid sexual intercourse at that time. The result would have been natural selection favoring those women who could *not* tell when they were fertile: that is, whose ovulation was concealed. Ignorance led not to bliss but to breeding.

Next mystery: female orgasm. Male orgasm is easy to explain. For men, reproduction requires the discharge of sperm (that is, ejaculation), and it is not surprising that this biological act should be accompanied with pleasurable sensations (that is, orgasm), just as human biology has wired our species so that it feels good to eat when hungry, sleep when tired, and so forth. But orgasm is not needed for women to reproduce. Unlike cats or rabbits, for instance, which are "induced" ovulators, releasing their eggs in response to stimulation they receive during mating itself, human beings ovulate spontaneously. Many women reproduce without ever experiencing orgasm. Maybe the female orgasm is a nonfunctional adjunct to male orgasm, just as male nipples are a nonfunctional vestige of their female counterparts. Or maybe it is biologically functional after all, as a way of providing the woman with feedback as to the quality of her lover and their relationship. (In many mammals, the copulation of dominant males tends to be relatively prolonged, whereas that of subordinates is more likely to be hurried—and thus, perhaps, less satisfying to their mates.) Or perhaps contractions associated with orgasm make fertilization more likely.

Whatever the answer in this case, the phenomenon itself seems less noteworthy in recent years; this is because it used to be thought that women were unique among female mammals in experiencing orgasm, but new findings have made it clear that orgasm, or something very much like it, takes place in several primate species. Nonetheless, its evolutionary rationale (technically, its "adaptive significance") remains unknown.

Finally, the third great mystery of female sexuality: menopause. Again, the male situation is relatively uncomplicated. Men continue making sperm throughout life, just as other mammals do. With increasing age, the abundance and possibly the viability of sperm decline, but despite talk about "male menopause," there simply is no such thing, at least, nothing as clear-cut and biologically mandated as occurs among women. To our knowledge, there is only one other mammal species—the pygmy pilot whale—in which female ovulation ceases, more or less abruptly, when the individual in question still has a significant amount of living yet to do. As with concealed ovulation and the female orgasm, the reason for menopause remains obscure, but there is no shortage of hypotheses. (For all their supposed intellectual caution, scientists love to speculate.) One idea that seems especially cogent is known as the "grandmother hypothesis."

Bear in mind that from an evolutionary perspective, the purpose of reproduction is to project genes into the future. But this is not achieved only by producing children. Or another way of looking at it: children are merely the first step toward evolutionary success. (Nor are they the only step; we'll revisit this issue in Chapter 9). To be "selected for," genes must be catapulted as far as possible into succeeding generations. Thus, children represent a triumph for natural selection, but if they die prematurely or do not reproduce, the triumph is short-lived, as they are an evolutionary dead end. Better yet, therefore, to be assured grandchildren, and even more so, great-grandchildren, ad infinitum. But let's stick with grandchildren for now.

Add that pregnancy falls upon women, not men, and that it falls with some weight. Pregnancy and childbirth are risky, with the risk increasing as the woman ages. There is even a downside for the fetus as well, since older women are more likely to produce genetically defective offspring. Now, add the coup de grace, that among large-brained mammals such as *Homo sapiens* (and pygmy pilot whales?) aged individuals probably have something of value to contribute to the eventual success of their grandchildren, whether assistance in childbirth, child care, food gathering, advice and information, and so on. The result? As they age, women may well reach the point at which they do more, on average, to further the success of their genes if they stop reproducing and start being available to help with their grandchildren. In short, menopause.

This hypothesis has been strengthened in recent years by the finding that among several groups of nontechnological people, notably the Aché, hunter-gatherers of Paraguay, the presence of grandmothers increases the calories consumed, on average, by their grandchildren, which not only improves their health and survival but also frees up Grandma's adult children to make more children.

But, you might say, what about modern women who don't have grandchildren or even children? And what of women whose children and grandchildren live far away, and who are therefore unable to act as a grandmotherly caregiver, except perhaps on rare occasions? Why should they undergo menopause? Keep in mind that most aspects of human biology were established hundreds of thousands of years ago. Modern times, even recent centuries, are just a drop in the bucket compared with the hundreds of thousands of years during which natural selection operated on our ancestors. Thus, when looking for evolutionary explanations, the crucial causes are generally factors that operated in the past, regardless of whether they seem important now. (And, of course, no one who has ever been born is descended from anybody who did *not* reproduce!)

The Development of Sex Cells

Now we'll get into the nitty-gritty of reproduction, beginning with the creation of eggs and sperm that will lead to a developing person, or any other mammal. These sex cells differ from normal cells in a number of ways, the most important being that they have only half the genetic material of all other cells in the body.

Recall that a normal cell is diploid, that is, it contains two sets of each chromosome and therefore two sets of every gene. One of these sets comes from the mother and one from the father, prepackaged, delivered, and combined at fertilization. If the egg and sperm that merged, eventually becoming a human being, were themselves diploid, then the fertilized egg and any cells descended from it would be tetraploid (diploid times two). And in turn, if such a tetraploid individual produced gametes that were genetically the same as his body cells, the next generation would be octoploid, and so on, with the number of chromosomes doubling every generation, clearly an untenable situation.

To prevent this, part of gametogenesis—the making of gametes—involves a so-called reduction division, after which the resulting eggs and sperm contain only one set of chromosomes. All human gametes, therefore, are created haploid: that is, with half the "normal" number.

For both eggs and sperm, the process whereby cells go from diploid to haploid is called meiosis (derived from a Greek root meaning "to lessen or diminish"). Recall that mitosis takes one diploid cell and splits it into two diploid cells, identical to each other and to their immediate ancestor. Meiosis is similar but splits one diploid cell into four. These four "offspring" are all genetically identical to one another, but each has only half the genetic material of the starting cell.

Meiosis takes place very much like mitosis, with similar phases of prophase, metaphase, anaphase, telophase, and cytokinesis. The major difference is that during the first meiotic division, pairs of chromosomes line up at metaphase, one member of each pair going, intact, into each of the two daughter cells, without the sister chromatids splitting apart as they do in mitosis. As a result, each of these cells contains just one representative of each set of chromosomes, rather than the full set produced by mitosis. And not only is the number of chromosomes reduced by half, but the genes they carry are randomly rearranged, since one member of each chromosome pair, chosen at random, appears in each egg or sperm.

The crucial work of meiosis has been accomplished after the first meiotic division. There are now two cells where there had been one, and each cell is haploid. The second round of divisions is essentially mitosis, resulting in four cells, each with one copy of each type of chromosome. For example, if the parent's DNA could be represented by the genes AaBbCcDd (one dominant and one recessive gene for each pair), then after meiosis is complete the resulting gametes will be an essentially random assortment of the following sixteen different types ABCD, ABCd, ABcD, AbCD, aBCD . . . all the way to abcD and abcd. Even in this highly simplified example, a tremendous amount of genetic diversity is in the making, since any of sixteen different male gametes can combine with any of sixteen different female gametes, creating $16 \times 16 = 256$ different genetic permutations.

Meiosis occurs in the ovaries for eggs and in the testes for sperm, but beyond this, eggs and sperm follow very different developmental paths. By the time she is born, each female human being already has, within her own tiny ovaries, all the eggs that she will eventually release. After puberty, one of these cells develops each month into a full-blown egg, potentially ready to be fertilized. This is part of the menstrual cycle, which we'll discuss later. Males, by contrast, are constantly making sperm. Compared to eggs, which mature for years, even decades, sperm have a very brief shelf life. Each ejaculation releases between 150 million and 350 million of these tiny pollywogs, which are replenished daily.

Spermatogenesis, the development of sperm, takes place within the testes, inside the seminiferous tubules. There a reproductive reservoir of "primordial germ cells" are undergoing almost constant mitosis, so there is a steady supply of new spermatozoa. Some of these freshly minted cells begin to go through meiosis and eventually form sperm, but others continue dividing so the supply is never used up. After meiosis, each primordial germ cell has given rise to haploid cells called spermatids. In order to become functional sperm, these need to undergo a number of different changes, among them the development of characteristic structure, including a long, thin, lashing tail and a tiny head. Inside the head is a nucleus and a small cap, called an acrosome, which is necessary to achieve fertilization. Directly behind the head is a mitochondrion that provides the energy needed to move forward, followed by the tail. All the other organelles found in most cells have disappeared; the resulting tiny sperm is basically a delivery vessel for the nucleus and its DNA, combined with a propellant (the tail and mitochondrion) to get the nucleus to the egg, as well as a key (the acrosome) to let it in. Sperm are much larger than viruses, and unlike viruses, they are unquestionably alive. Like viruses, however, they are highly stripped-down structures, consisting of just nucleic acid and a way of moving from one place to another.

The egg, on the other hand, is enriched with many other nutrients and organelles and is much larger than a sperm. Although only about the size of a pinhead, it is one of the largest cells in any mammalian body. As a midterm fetus, a girl may have as many as 2 million primordial germ cells inside her ovary; this is whittled down to about 700,000 at birth and around 300,000 by puberty. Only a small percentage of these get to develop more, until each cell is suspended in prophase I of meiosis. The eggs remain in this state until primed by hormones at puberty. At this point, one cell each month undergoes its first meiotic division, producing two cells. But cytokinesis in developing egg cells is uneven; the result is one large cell and one small one. The latter, called a polar body, is eventually reabsorbed, but the larger cell can be released from the ovary and, if fertilized, will produce a new human being. During her lifetime, a woman will ovulate about 400 eggs from the millions she starts out with.

Fertilization and Development

Over time, a single cell must form itself into a complicated being made up of about a trillion cells and more than a hundred different cell types. Development is not simply a matter of one globular cell making copies of itself, because then we'd be left with a large blob of undifferentiated cells, not a human being at all. This is the subject of embryology and differentiation, and it is a marvelous and exciting tale. But first there is the greatest of all boy-meets-girl dramas: sperm meets egg.

Fertilization is a big event in every life, perhaps the biggest. After an egg is released from the ovary, it begins its journey down the oviduct, also called the Fallopian tube, which leads from the ovary to the uterus. If, during this time, the woman has sexual intercourse, sperm begin to migrate through the vagina and uterus, eventually, perhaps, reaching the oviduct. There is only a very slim possibility that any one sperm will come into contact with the egg of its dreams. Of the hundreds of millions of sperm that are released upon ejaculation, only several hundred reach the area of the oviduct where the egg and sperm must meet.

For men, a sperm count of fewer than 50 million per cubic centimeters of semen generally means infertility. This is probably because the success rate is appallingly low, at least from the sperm's perspective. Attrition is immense, in part because the pH of the vagina is somewhat acidic, good for the egg but bad for sperm. In addition, some sperm are poorly formed, with two tails or a crooked neck: even in a healthy and reproductively potent man, up to 40 percent of sperm may be structurally abnormal. Moreover, many don't swim straight, or fast enough (slower than 20 micrometers per second is a problem), or with enough persistence: if a sperm were the size of a man, it would have to swim about seven to ten miles! To make matters worse, much of this is upstream, since the cilia lining the oviduct are beating the other way, helping to push the egg along to a rendezvous that may never happen. Not only that, but having traversed the uterus, half the sperm go up the wrong oviduct!

Aldous Huxley seemed to appreciate the fact that every human being represents a unique biological contingency:

> A million million spermatozoa,
>
> All of them alive:
>
> Out of their cataclysm but one poor Noah
>
> Dare hope to survive.
>
> And of that billion minus one
>
> Might have chanced to be
>
> Shakespeare, another Newton, a new Donne—
>
> But the One was Me.

When that sperm that was to become Aldous Huxley (or one half of him) encountered his mother's egg, it released a host of different enzymes from its acrosome, that mysterious little pouch at the head of the sperm. These enzymes help the sperm tunnel through the egg's outer coating, made up of two layers and boasting some of the loveliest names in biology: first is the corona radiata, a "radiating crown" surrounding the egg and derived from the nourishing cells that surrounded it while in the ovary, and then the zona pellucida, a "clear zone" of jellylike material immediately outside the membrane of the egg itself.

The labors of Snow White's prince, slashing through seemingly impenetrable thornbushes (not to mention dragons) are trivial compared with the travails of sperm seeking to penetrate the egg's defenses. Perhaps the passage is made difficult so that only the best "man" will win. In any event, many are called but few are chosen. Most sperm will typically be trying to be among the latter, although unbeknownst to them, they are not wholly competitors: the combined action of acrosomal enzymes from numerous sperm appears needed to weaken the surrounding corona and zona. When one finally breaks through, however, rapid chemical changes occur within the egg that prevent any other sperm from getting in, while stimulating the egg to finish meiosis II.

At this point, the sperm's nucleus migrates to the nucleus of the egg, and they fuse. This fusion creates a cell called a zygote (from a Greek word meaning "yoked together," as opposed to "yolked," which, of course, is true of every egg). Each zygote, composed of haploid egg and haploid sperm, is therefore diploid; it has two sets of every chromosome, one from the mother (the egg) and one from the father (the sperm). After fertilization, what's left of the sperm—and it isn't much—is simply absorbed. As a result, a sperm contributes only its share of DNA to the zygote. It does not provide nutrients or any organelles to the developing baby. Each zygote has the basic intracellular anatomy of other cells—endoplasmic reticulum, lysosomes, and especially mitochondria—all of which are contributed by the mother, not the father. In the case of mitochondria, this is particularly significant, since they possess their own small amounts of DNA, passed directly from mother to child without the intervening genetic sieve of meiosis. Since the DNA in a mitochondrion comes only from the mother, geneticists can trace an individual's maternal lineage with great confidence by focusing on this mitochondrial DNA (whose function is not understood).

Within twenty-four hours after it is formed, the zygote begins to divide. These divisions are simple mitotic ones, as described in Chapter 3. Throughout the next few days, it keeps dividing, but the entire mass remains the same size. (Each cell, therefore, gets smaller even as the number of cells increases geometrically.) During this time, the glob of cells migrates into the uterus, and by day six or seven it has implanted into the surrounding tissue. Occasionally, the migrating mass of cells will implant before it gets to the uterus, perhaps inside the oviduct. The result is an ectopic ("out of place") pregnancy, which can be lethal to mother as well as offspring if not aborted surgically.

Once implanted, the zygote is an embryo, and it passes through a number of different stages early in its development. By day four, when it is about sixteen to thirty-two cells, it is called a morula, from the Latin word for "mulberry," which it resembles. In the ensuing days, the morula changes from solid ball to hollow, becoming a blastocyst, which consists of both outer cells and an inner mass. These cells appear undifferentiated; eventually, they become specialized, giving rise to the more than two hundred different cell types—muscle, bone, blood, et cetera—that constitute a full-fledged human being. In 1998 two different research teams reported that they had succeeded in plucking such undifferentiated, multipurpose "embryonic stem cells" from the inner mass of blastocycsts as well as "germ cells" from embryos (cells that also escape early differentiation) and culturing them in laboratories, where they seem able to continue dividing indefinitely. This raised the exciting—and, for some, frightening—possibility that such cells could eventually serve as a kind of multipurpose spare-parts system, for regenerating lost or defective cells in heart, bone marrow, or even brain tissue.* (The idea is to alter these stem cells so that they match the recipient's immune system, then introduce them into a patient, after which they would follow the local signals of their new host, providing custom-designed replacements as needed.)

During normal development the human blastocyst implants in the wall of the uterus, and shortly afterward four different essential membranes develop around the embryo. These are produced from cells that were originally part of the cell mass, and although some of them become part of the rapidly developing embryo itself, others simply feed and support it. One of these membranes, the yolk sac, will eventually form parts of the digestive system and the embryo's future germ cells (either eggs or sperm). Of the remaining membranes, the innermost is the amnion, a fluid-filled sac protecting the embryo from sudden movements and from drying out. Next is the allantois, which eventually will support structures such as the umbilical cord. The final membrane is the chorion, which for the first three months continues secreting the hormones necessary for the pregnancy to continue. The placenta also develops. Through it, nutrients and oxygen are exchanged between mother and embryo.

The developing mass of cells is called an embryo until about eight weeks into the pregnancy, or during most of the first trimester (three months). During this time most of the structures that will eventually make up a baby are becoming pronounced. Soon after implantation, the embryo begins a process called gastrulation, in which the inner mass of cells rearranges itself in a complicated pattern to form the basic tissue types.

---

* This research involved cells from fertilized eggs that would otherwise have been discarded, obtained during treatment of infertile couples, as well as tissue from already aborted fetuses. Because federal funds have been banned for most fetal-tissue research, it was supported entirely by private, corporate money, raising the possibility that a major biomedical breakthrough will be limited to for-profit exploitation.

The word *gastrulation* comes from the Greek for "stomach" or "pouch," because among most vertebrates it involves a dramatic invagination of the previous embryonic stage, the blastula. If a blastula is like a soft, hollow rubber ball, gastrulation involves a finger pressing into the ball, producing a multilayered structure—a gastrula—with a pouch leading to the outside where the finger is pressing. Cells migrate past a linear structure known as the "primitive streak" and submerge themselves to produce additional cell layers.

One of the first things to develop in a human or any other sort of vertebrate embryo is the nervous system. During a process called neurulation, which immediately follows gastrulation, the basic structures of the nervous system are put into place. A complicated series of cell migrations and foldings extends the primitive streak into a tube that will eventually become part of the spinal column, and tissue begins to develop that acts as a primitive brain. Surrounding the neural tube there arise a series of paired segments called somites; these will eventually form the main bones and skeletal muscles of the future baby. Even at this early stage in the development of the embryo, serious birth defects can become apparent, such as spina bifida, when the neural tube fails to develop correctly.

During gastrulation and neurulation, the embryo grows at an incredible pace, so that by the end of the fourth week it is more than five hundred times its original size. At this point, the basic setup is clear: small heart, buds of tissue where the limbs (arms and legs) will eventually develop, a primitive brain, and a neural tube. Also some features that will eventually be lost, such as a tail. These are common to all vertebrates during early development and indicate our shared animal nature as clearly as can be imagined. Indeed, at this stage of development it is very difficult to tell different types of mammals apart. Four weeks after conception, a human being looks very much like a rat, an elephant, or even a turtle or a fish at the same embryonic stage.

During approximately the next four weeks, the embryo begins to lose some of its general features and become more distinctly human. It also starts revealing itself as male or female. If the parental sperm contained a Y chromosome, then a gene on this chromosome becomes activated, causing testes to develop, and the embryo eventually becomes a boy. On the other hand, if the sperm contained an X chromosome—and thus lacked the Y chromosome's male-determining gene—ovaries develop. Female structures are thus the "default" option, the "natural" developmental path, while male structures only develop if the embryo is prodded in a specifically male direction.

Other features begin to arise at this time, including the heart, limbs, fingers, toes, umbilical chord, eyes, and ears. At the end of the first trimester, the embryo has acquired most of its organ systems, is usually identifiably male or female, and is just over an inch long. Now it acquires a new name: a fetus.

The second trimester, from about three to six months, is also full of changes. Mostly, however, the fetus grows, so that by the end of the sixth month it is about 9 to 10 inches long. Features begin to appear more completely, and muscles arise to move these features. The face becomes more distinct, identifiably different from other fetal faces, and usually the sucking reflex kicks in, along with different types of facial movements. The fetus begins to kick and squirm. Eyelids and eyelashes begin to form.

During the third trimester, the fetus's eyes open and the final touches are made to prepare for life outside the uterus. The lungs become coated with a chemical called surfactant, which will enable the baby to breathe, and systems develop that allow the baby to maintain a normal body temperature. When a baby is born prematurely, incomplete development in these areas is the biggest obstacle to its success outside the uterus, and doctors have to be very careful with medical support to enable it to survive.

Even in an increasingly overpopulated world, it is hard not to be positively disposed toward a fertilized egg that successfully navigates the myriad shoals of possible developmental disaster to arrive, finally, at a finished—or rather, a beginning—human being. It is even difficult to gainsay the optimistic confidence expressed by Walt Whitman in *Song of Myself*, along with the self-congratulatory certainty that the whole world was organized around producing the particular individual that is oneself:

> Before I was born out of my mother generations guided me,
> My embryo has never been torpid, nothing could overlay
>   it. . . .
> Vast vegetables gave it sustenance,
> Monstrous sauroids transported it in their mouths and
>   deposited it with care.
>
> All forces have been steadily employ'd to complete and
>   delight me,
> Now on this spot I stand with my robust soul.

Birth and Newborns

After approximately nine months in the uterine oven, the baby is fully "cooked" and ready for the outside world. Infants born as young as twenty-three weeks have survived with intensive medical care, although this is rare. By the ninth month of gestation, the immediate survival rate for newborns jumps to 95 percent. The birth process is induced by a complex interplay of many different hormones, both from the mother (her ovaries and pituitary gland) and from the fetus. These hormones begin to stimulate contractions of the uterus, and also induce the placenta to make more hormones, causing even more contractions.

There are three phases of labor. First comes dilation of the cervix (the border between vagina and the uterus), which stretches and becomes very thin, providing the fetus with a pathway out, into the vagina and eventually the open air. The second is the actual delivery of the baby, when extremely powerful uterine contractions move the baby along. The third stage is the delivery of the placenta, making the birth complete. The umbilical cord is cut, and the baby takes its first breath of air.

After birth, human babies are ready for something unique to mammals, indeed, the defining characteristic of every mammal: they nurse. During pregnancy, different hormones have been priming the mother's breasts, causing them to grow. Soon after birth, they produce an important fluid called colostrum, which contains essential nutrients and antibodies. With the sensation of suckling, a signal is sent to the mother's pituitary to secrete the hormone oxytocin, which earlier was involved in the onset of labor and now stimulates milk production. (Sometimes nursing women experience mild labor pains as their "let-down" reflex and its hormonal underpinnings begin to take effect.)

Interestingly, the size of a woman's nonlactating breasts have no relation to the amount of milk she can produce. Most female mammals, in fact, have very small breasts, roughly equivalent to the male's. They grow when she is nursing, then shrink afterward. *Homo sapiens* is unusual in that nonlactating females have prominent breasts. No one knows why (this might be considered yet another mystery of female sexuality, except it isn't clearly sexual). Zoologist Desmond Morris suggested that prominent breasts were a kind of buttocks mimicry, useful for stimulating the male during intercourse. After all, human beings commonly make love face-to-face, whereas most other mammals copulate "doggy style," with the female's buttocks more prominently displayed to the male. We are skeptical, in part because mammals such as gorillas that—like humans—engage in face-to-face mating do not have especially prominent breasts, and mostly because it seems unlikely that evolution would have to resort to such efforts to assure male sexual enthusiasm.

On the other hand, it is entirely possible that relatively prominent nonlactating female breasts evolved as a kind of deception, whereby females were able to "promise" future nourishment to the offspring, a promise that prehistoric males may well have found especially alluring, especially if they lived in an environment in which malnutrition or even starvation was a risk for infants. In any event, the size of nonlactating breasts is determined by the amount of fat deposited there; it has no relationship to the glandular tissue employed in making milk (hence the deception). A small-breasted woman will have her bra size grow considerably during pregnancy and especially nursing but then shrink again afterward. Breast size does not influence breast function, but it may be that men have been fooled into responding as though this were the case!

Twins and Multiple Births

Let us return to pregnancy briefly to explore some of the unusual things that can happen. One is the occurrence of twins, triplets, quadruplets, quintuplets, sextuplets, septuplets, or even, as illustrated recently in the United States, octuplets! Some mammals, including most rodents, dogs, and cats, typically produce litters. For others—horses, cows, whales, bats, and most primates, including human beings—singleton births are the norm.

Looking first at natural reproduction, there are two main ways in which a multiple birth can take place, producing siblings that are either "fraternal" or identical. If the former, the babies are no more alike than two siblings born many years apart, except that they share a uterus, a placenta, and a birthday. (Obviously, the term *fraternal* is misleading, since fraternal twins can be two girls, or one girl and one boy—a case of rampant sexism.) Identical twins have exactly the same genetic makeup, since they arise from union of the same egg and sperm. When more than two babies are born at once, they may be a combination of fraternal and identical siblings.

Identical siblings occur when the fertilized egg, very early in its development, separates: from one, two. If these two cells become physically separated, they are able to form completely independent embryos. These cells divide again and again, beginning development just as if they had been two separate zygotes all along. If separation occurs after the zygote has divided repeatedly, it is possible, once more, for these embryos to develop separately. As a result, a zygote can divide quite a few times, then fall apart and begin to create new embryos. It is theoretically possible for identical triplets, quadruplets, or even more "lets" to develop, although it happens rarely. We know of one mammal species, the thirteen-banded armadillo, that regularly produces identical quadruplets; with this exception, identical births are very rare among mammals. (In the highly unusual cases of identical human triplets, the likelihood is either that one of four identical quadruplets aborted spontaneously and was too small to be noticed or, after producing identical, single-celled twins, one of these may have separated yet again.)

From all this we can infer that each of the first few cells created after fertilization is capable of acting like its own zygote; it is, in scientific terms, "totipotent," with the potential of giving rise to cells that are able to differentiate, totally, into any tissue or organ. It was thought for some time that mammalian cells are totipotent only in the earliest stages of development. Recent experiments with cloning, however, have shown that even cells from a fully formed adult can, if appropriately stimulated, become totipotent and give rise to an entirely new individual. It is one of the more remarkable findings of the late twentieth century, and it requires a rewrite of the developmental-biology textbooks (see Chapter 1).

Fraternal siblings are more frequent and more pedestrian. Each month, one human egg ripens, not unlike a fruit; indeed, fruits are neither more nor less than the ripened ovaries of plants. It isn't clear exactly how one egg decides—or is chosen—to mature and be ovulated and how others are inhibited from doing the same thing at the same time. This pattern of choice and inhibition operates across the ovaries as well: if an egg is maturing in the left ovary, this will temporarily inhibit any in the right. Occasionally, however, two or more may mature simultaneously, although whether from the same ovary is also a mystery.

If these plump and juicy fruits of the womb are then released, and if they each encounter a sperm, more than one embryo may develop.

There has been a dramatic increase in the number of multiple births recently because of fertility drugs, which promote the maturation and release of multiple eggs, so-called superovulation. In these cases, the siblings are fraternal.

When the uterus supports more than one embryo, there is extra stress on both the mother and the infants. Often these babies are born prematurely, or one or all may be miscarried. Even in the best of times, pregnancy is physiologically (and sometimes psychologically) stressful, and often maternal resources simply aren't adequate to meet the needs of multiple babies. It appears that in somewhat more than half of all twin conceptions, one is naturally aborted. Thus, many singletons are unknowingly the only surviving representative of an early set of twins.

Abortion and the Meaning of Birthdays

The debate over abortion has generated huge amounts of radiation, most of it, unfortunately, in the infrared rather than the visible spectrum. Often it seems to boil down to when, precisely, life begins: if at conception, then abortion is murder. Others argue that life does not begin until birth, or—as the Supreme Court holds—when the fetus is "viable," that is, able to survive outside the uterus.

One might think that the question of life versus nonlife would be easily resolved, that, like a Supreme Court justice's famous description of pornography, we may not be able to define it, but we know it when we see it. And yet, as our previous discussion of prions and viruses should emphasize, the issues are far from clear—even without the complication of heated emotions and theologically driven "certainties." Like it or not, the moment life begins is debatable. Conception starts a process that will eventually *result* in a life. But in our judgment, life itself has not begun at this point: equally, the gametes that join are *already* very much alive, produced (as we have described) by primordial germ cells that underwent meiosis. Traced this way, the beginning of life—every life—must be counted as having occurred about 4.5 billion years ago!

But isn't that begging the question? No one seriously considers that the egg that is shed monthly or the hundreds of millions of sperm that are released in a single ejaculation should each be buried ceremoniously. A baby tooth, kept under the pillow and magically transformed by the tooth fairy into money, receives—and probably deserves—more respect. What, then, about a single fertilized egg? Does the conjunction of DNA from sperm and egg warrant that much more fuss? Does the two-cell stage, or the four-cell? Is something with the *potential* to become complicated and clearly alive really alive when it is so small, so simple, so unformed and undifferentiated? Gastrulation, implantation, and all sorts of developmental milestones remain to be achieved, and yet there is a seamless biological continuity between something exceedingly simple, somehow living, and yet not really a "life," and—at the other end of nine months' development—unquestioned humanity.

Perhaps a better point for the beginning of life would be when the fetus is capable of sustaining itself outside the womb. But does that mean that a fetus with recognizable features and a beating heart is not alive, even if it cannot maintain independent life? And what about those rapidly advancing medical procedures that allow younger and younger fetuses to survive?

Some people—especially abortion opponents—have begun arguing that a person's age should be measured from conception, not from birth. Perhaps the current definition of a person's birthday as the anniversary of his day of *birth* is arbitrary. Perhaps someone is not really "alive" until after puberty; this is, after all, the approximate age at which most religions consider people to be adults and fully responsible for their actions (bar mitzvahs and bat mitzvahs in Judaism, confirmation in Catholicism, etc.). Or, as some teen-agers might argue, maybe someone is not truly alive until he or she can drive, or is legally responsible for his or her actions at 18. Some people even claim to have been "born again." Were they less than alive before? The morality and legality of abortion is for society, not science, to determine.

Abortion is not new. Throughout history, desperate women in virtually every society have attempted to interrupt pregnancy in one way or another, nearly always at great risk to health and life. In modern times, even as abortions have become highly controversial, they have also become safer than ever before—so long as they are performed aboveboard, in legitimate medical facilities rather than at home or in back alleys. First-trimester abortions can be done surgically or chemically. Surgical abortions physically disrupt the implantation of the embryo in the uterus. Chemical abortions, such as those using the pill RU-486, also impede implantation in its early stages. RU-486 is a molecule closely related to the hormone progesterone, which normally functions to maintain the uterine lining. RU-486 blocks progesterone, causing the lining to slough off—as it does during menstruation—taking the zygote or embryo with it.

Puberty

A growing child undergoes almost as many changes after birth as before. The first few months are filled with extraordinary feats of information gathering as the brain develops and the baby learns. Many interactions during the first two years of life begin to shape the child's personality and way of dealing with the world. Language develops, and the baby acquires social and physical skills. And, of course, there is growth and maturation. Through all these changes, the young child's reproductive system stays relatively quiet. Young children can't reproduce. A baby boy, for example, can get an erection but has no developed sperm. Although a baby girl's ovaries have all the cells necessary to form eggs, none are mature. Around age 10 to 15 for girls, and 12 to 16 for boys, this changes dramatically, and children make the leap to sexual maturity, children (at least reproductively) no longer. Puberty is one of the most awkward life stages, and surprisingly little is known about what actually goes on during this time.

In a typical female, puberty usually involves a large growth spurt, breast development, the appearance of hair in her armpits and genital area, and changes in body shape, as the hips become more pronounced. About this time, she will usually begin menstruating. This indicates that she is making eggs and that her ovaries are releasing them, with the accompanied preparation of the uterus for implantation. In the absence of pregnancy, the uterine lining is periodically shed, causing the bleeding characteristic of menses. In the typical male, puberty usually begins with a huge growth spurt; his voice deepens; hair develops under his armpits, in the genital area, and (more gradually) on his face. He begins to acquire male patterns of muscular and physical development, including broader shoulders and increased general muscle mass. In addition, he may start having "wet dreams" or ejaculations while asleep, and these will begin to contain viable sperm. Once a girl begins to get her period and a boy begins to make sperm, they are biologically capable of becoming parents, on their way to becoming sexually mature.

Much mystery still surrounds the body processes that initiate puberty. Although its onset is variable, given enough time, nearly everyone gets there. What is the role of signals, from outside as well as inside? It appears that for females at least, the amount of body fat is crucial in triggering puberty. It has been known for many years that a girl who is extremely skinny is likely to have irregular periods or none at all. Young dancers and athletes with very little body fat often do not undergo puberty until they stop strenuous exercising, and some women athletes stop menstruating until they relax their training.

In recent years researchers have identified a likely mechanism for this trigger. It involves a hormone called leptin, important for the regulation of body fat. Leptin is secreted directly by fat cells, after which it travels in the bloodstream to the brain, where it announces how much fat is present in the body: the more leptin, the more fat. Inclination for exercise, metabolic rate, and food intake appear to be adjusted accordingly. A number of studies have proposed that in mice and rats leptin also influences the reproductive system. Prepubertal mice dosed with leptin are more likely to go into puberty earlier, as though their brains have been fooled into thinking that their body has accumulated enough fat to proceed with reproduction. Not surprisingly, there has been considerable excitement over leptin's potential manipulation in human beings. If leptin works in *Homo sapiens* as it does in rats and mice—as a signal whereby the body tells itself that its fat level is high—dosing with leptin might suppress appetite and even help induce exercise. Interfering with leptin receptors could also provide a means of contraception.

Other than this, however, no one knows what, physiologically, causes people and other animals to enter puberty and become sexually mature. Presumably, there is some sort of internal clock, but thus far no one has heard it ticking.

The Sex Hormones

Sexual maturity requires a complex interplay of hormones and body parts, all working together to provide a man with viable sperm, a woman with functional eggs, and both of them with the inclination and ability to get the two together. As we have seen, a hormone is any messenger molecule secreted from one tissue that travels in the blood to another tissue, where it causes some kind of response. Hormones can be steroids, which enter the receiving cell and actually travel into the nucleus, where they directly affect the transcription of genes. Or they can be peptides or proteins, which hook on to specialized receptors on the target cell's outside, then cause changes in a series of molecules that influence genes that are transcribed inside the cell (the second-messenger system). In either case, very small amounts of hormones can cause very large effects; hormones are also the main way sexual characteristics are conferred upon developing and developed animals of all sorts, including men and women.

Sex hormones regulate themselves by one of two feedback mechanisms: negative or positive. Negative feedback, the most common type, occurs when too much of something causes a chain of events that eventually limits its production. A house thermostat uses negative feedback: when there is too much heat in the house, the thermostat turns off the furnace. When there is too little, it turns it on. As a result, the temperature is kept approximately the same at all times because the output feeds back, negatively, into the control mechanism.

Similarly, in the body negative feedback can keep the concentration of chemical substances approximately constant. If there is too much, it produces less; if there is too little, it produces more. For every chemical regulated in this manner, there must be a set point— a range of acceptable concentration, analogous to the range of acceptable temperatures inside a house—associated with a control mechanism that can stimulate the target organ to produce more or less of the chemical in question.

By contrast, positive feedback occurs when the presence of something causes a chain of events that eventually *increases* that substance. It is the original vicious circle, feeding upon itself. Like an avalanche, in which a little bit of falling snow induces more and more snow to cascade down a mountain, positive feedback keeps increasing the amount of the substance within a body.

Sex hormones are secreted by only a few different tissues. Within the brain, the hypothalamus secretes gonadotropin-releasing hormone, or GnRH, which in some ways is the ultimate initiator of the other sex hormones. (If in fact leptin functions to jump-start the reproductive system, as described above, it presumably acts in the brain by influencing the neurons that secrete GnRH.) GnRH travels to the pituitary gland, situated in the brain just below the hypothalamus. The pituitary secretes many different hormones; for the reproductive system, the most important are luteinizing hormone (LH) and follicle-stimulating hormone (FSH). These chemicals operate in both males and females, but via different patterns.

Male Hormones

For males, sex hormones are controlled mostly by negative feedback. Here is how it works in a normal male, after puberty. Specialized neurons in the hypothalamus secrete GnRH in response to low testosterone levels in the blood. (It is useful to think of the hypothalamus—in both males and females—as always inclined to secrete GnRH; depending on the control mechanisms that are operating, this tendency is either accentuated or inhibited.) GnRH, in turn, stimulates the pituitary to secrete LH—which in men is sometimes called interstitial cell–stimulating hormone, or ICSH—and FSH, which could as well be called sperm-stimulating hormone, but isn't. In the case of men, this dynamic duo of hormones, chemically the same in both sexes, travels via the blood to the testes, where they encounter several different cell types (each named for its discoverer). Leydig cells, which are found between the tubes in which the sperm are made, respond to the LH signal by making and releasing testosterone. Sertoli cells, which are found inside the tubes, respond to FSH and the testosterone secreted by the Leydig cells to help the developing sperm. (FSH is necessary to initiate sperm development during puberty, but no one knows if it is necessary in an adult male.) In this way, the secretion of GnRH eventually causes the secretion of testosterone, which then stimulates sperm formation as well as the development and maintenance of secondary sexual characteristics that appear during puberty. But how does a male body know when enough testosterone has been produced?

This is where negative feedback comes in. High levels of testosterone inhibit the release of GnRH, just as low levels stimulate it. There is a second method, also an example of negative feedback but a bit more complicated because it involves another hormone, inhibin. Inhibin is secreted by Sertoli cells when the sperm count is very high, that is, when testosterone levels are going through the roof and must be decreased. As its name suggests, inhibin is inhibitory. Specifically, it inhibits the secretion of GnRH, just like testosterone itself, but it also inhibits the secretion of FSH from the pituitary. In this way, the body has developed three different mechanisms of negative feedback to control the amount of testosterone and the number of sperm being produced: testosterone and inhibin's actions on the hypothalamus to decrease GnRH secretion, and inhibin's actions on the pituitary gland to decrease FSH secretion.

Thus, the male hormone pattern is fairly simple. Negative feedback enables a man to have a constant sperm supply, so he can ejaculate at any time with millions of viable sperm. Thus, the male sexual "cycle" isn't really a cycle at all; rather, it's a relatively constant plateau, maintained in its steady state by negative feedback. In females, however, things are more complicated, more interesting, and more fluctuating.

Female Hormones

Ovulating is very different from squirting sperm. Since each egg produced is a VIP, making a special appearance only once a month or so, all of a woman's hormones have to be balanced at this time to make sure her egg gets an appropriate reception. Specifically, if her egg meets a sperm and is fertilized, her body must be prepared to handle and support the pregnancy. For men—and male mammals generally—sexual relations may be the beginning and end of parental involvement. But for women—and female mammals generally—sexual intercourse is just the start.

Humans are almost unique among mammals in that they have menstrual cycles instead of estrous cycles. With estrus,* there is only a limited, specific period in which the female is sexually receptive. The rest of the time, she is sexually inactive and her hormonal levels are pretty constant. With a menstrual cycle, by contrast, women are sexually receptive almost all the time, although they are actually only able to conceive during a small window of time surrounding the release of an egg. On the other hand, you could say that women, unlike other mammals, are *never* sexually receptive, in that they do not undergo anything comparable to "heat," in which they will encourage sexual advances from nearly all comers. Actually, there is some evidence that women are most sexually stimulated around the

---

*Estrous *is the adjective;* estrus *is the noun.*

time of ovulation, but this is still unlike the "heat" that is so dramatic and characteristic of mares, for example, or bitches.

Another difference between menstruation and the estrous cycle is that the uterine lining, which becomes vascularized and thickened during both types of cycles, is reabsorbed into the female's body in the case of the estrous cycle but shed during a woman's monthly period.

The menstrual cycle results from an extremely complicated interplay of many different hormones. Generations of women have called it "the curse," but in fact it is, if not a blessing, a sensitive masterpiece of biological control. The initial pattern is reminiscent of the male hormonal system, in which the hypothalamus releases GnRH, which causes the pituitary gland to release LH and FSH. Within the ovaries, LH and FSH stimulate the growth of eggs and the production of estrogen, much as they stimulate testosterone and sperm production in the testes. But from here the female pattern differs markedly from its male counterpart.

To follow the process, it is necessary to understand how the ovary prepares to ovulate—that is, to release a fertile egg. Eggs in the ovary are surrounded by many different cells; the combined egg/cell structure is called a follicle (from a Latin word meaning "small bag"). The development of these follicles eventually leads to ovulation, while their decay after ovulation helps generate the female hormonal cycle. Early follicles are "primordial," which is the state of the eggs that a girl carries until she reaches puberty. After puberty, one primordial follicle matures each month, as other cells form around it and the zona pellucida begins to develop. At ovulation the follicle, now mature, ruptures and the egg is released into the oviduct. The empty follicle forms a new structure called the corpus luteum ("yellow body," because that's how it looks), which eventually degenerates if there is no pregnancy. In the event of pregnancy, the corpus luteum remains, secreting another important hormone, progesterone.

Those hormones released by the pituitary—LH and primarily FSH—cause the follicles to develop. Cells surrounding the follicles respond by secreting estrogen. The initially low concentration of estrogen in the blood increases, repressing the secretion of LH and FSH via our old friend, negative feedback. As estrogen levels inch up, they stimulate the uterine lining; its blood flow increases, and the lining thickens in response. As the follicle develops, the rate at which it produces estrogen increases; just before the follicle is ready to rupture, estrogen production rises dramatically and then suddenly peaks, causing a dramatic surge in LH secretion from the pituitary, along with a smaller FSH surge.

This differs significantly from the typical system of negative feedback, according to which high levels of estrogen should decrease the pituitary's secretion of LH and FSH. Instead, a blast of estrogen from the ovaries has the opposite effect on the hypothalamus: it changes the secretion pattern of GnRH, which seems to cause the surge in LH. Again: low levels of estrogen *inhibit* LH secretion by acting on the pituitary, while high levels of estrogen *stimulate* LH secretion by acting on the hypothalamus. The resulting LH surge is a dramatic example of positive feedback, or a biologically benevolent vicious circle, because increasing levels of estrogen eventually cause estrogen to increase even further.

The LH surge is for a reason. It causes the ovarian follicle to rupture and the egg to be released into the oviduct. After this, the follicle, left behind in the ovary, becomes the corpus luteum in response to LH stimulation. And this corpus luteum keeps up estrogen levels as well as secreting that hormonal newcomer: progesterone. At this stage of the cycle (roughly the midpoint), negative feedback takes over and rising levels of estrogen and progesterone finally slow down the release of LH and FSH. But LH was responsible for creating the corpus luteum and is needed to sustain it. So, in a sense, it carries the seeds of its own destruction: progesterone secreted by the corpus luteum shuts off LH production, without which the corpus luteum self-destructs. It is negative feedback with a vengeance, leading to precipitous declines in all sex-hormone levels—and, for many women, the premenstrual syndrome of cramps, fluid retention, as well as (sometimes) heightened sensitivity and irritability.

Between ovulation and the degeneration of the corpus luteum, the uterus has been preparing for implantation of an embryo. The uterine lining has thickened, blood vessels have grown rapidly, and the cervix has secreted a thin layer of mucus, perfect for conducting sperm up the vagina, through the cervix and the uterus, and into the oviduct, where it just might meet and fertilize an egg. In a nonpregnant cycle, when the corpus luteum degenerates and the levels of sex hormones (especially progesterone) drop, blood vessels in the uterine lining constrict and the lining degenerates, leading to menstrual flow, as the "disappointed" uterine lining falls away and passes out through the vagina amid clots of blood.

The whole business is now ready to start again: the extremely low levels of estrogen and progesterone begin the hormonal cycle once more, as the hypothalamus and the pituitary are released from the inhibitory effects of the two hormones. GnRH causes LH and FSH to be secreted from the pituitary, and a new crop of follicles begins to mature.

But what if fertilization has taken place?

Then the first requirement is to keep the uterine lining intact, or the embryo will be dumped out with the menstrual bath water. So, the early embryo secretes its own hormone, human chorionic gonadotropin (HCG), which acts like LH to maintain the corpus luteum and keep up the production of progesterone, which in turn keeps the uterus as a welcoming environment in which the embryo can implant. And this is why the most notable sign of pregnancy is failure to menstruate.

HCG is so important in maintaining an early pregnancy that it is produced in large amounts, enough so that some of the excess spills into a woman's urine, providing the basis for most home pregnancy tests. Just as where there's smoke there's fire, where there is HCG there must be an embryo producing it. But of course, HCG didn't evolve to provide a convenient indicator for pregnancy testing; it has a far more important role: alerting the rest of the body that an embryo exists. Eventually, the placenta gets the message and takes over the job of producing progesterone, thereby maintaining the uterus.

Birth Control

All these high-flying, fascinatingly fluctuating hormones do more than simply cause the growth and maturation of eggs and sperm. They also provide the drive to mate in the first place. Mammals that we are, sex is a big part of our lives, but people are not always prepared for the responsibilities and burdens that sometimes come from sexual intercourse. Obviously, the most effective method of birth control is abstention. But "just say no" is easy to exhort, hard to do. Other animals use sex as a means of reproduction and only that. Human beings use sex for many things, reproduction not the least. And so, we come to birth control.

Fortunately, abstinence is not the only way of preventing reproduction. We have already discussed abortion, albeit briefly. Most people—whether pro-life or pro-choice—agree that recourse to abortion is, in a sense, an indication of failure. Far better to prevent unwanted pregnancies in the first place.

Birth control, in one form or another, has been around for many thousands of years. But early methods, still sometimes practiced, are not very effective, and some are dangerous. Probably the oldest technique, withdrawal or coitus interruptus, involves removing the penis from the vagina before ejaculation. It is not very effective for two reasons: some fluid containing sperm can be released from the penis before ejaculation (remember, it only takes one sperm to fertilize an egg), and the withdrawal method requires very strong willpower.

A second "natural" method of birth control is rhythm, really a modified form of abstinence. The idea is to avoid intercourse when a woman is fertile, that is, from a few days before ovulation until a few days afterward. A woman can estimate when she is ovulating by carefully tracking her menstrual cycle. In addition, female body temperature increases slightly—about 0.5 to 1 degree Fahrenheit—just after ovulation, so this also can be used to indicate fertility. (By determining when a woman is likely to be ovulating, a couple can also time sexual intercourse to *increase* the likelihood of pregnancy.)

As a means of contraception, however, "rhythm" presents many problems. For example, a woman's cycle is not always reliably rhythmic, and it is often difficult to determine exactly when ovulation takes place. Moreover, viable sperm can linger inside the uterus or oviducts for seventy-two hours after intercourse, so if the timing is off by only a few days, pregnancy is possible. Women are sometimes capable of conceiving shortly after their period; the timing of the menstrual flow is determined by when the last ovulation took place and does not directly determine when the next one will happen. It is even theoretically possible, although unlikely, to get pregnant while having a period, since fertilization typically takes place in the oviduct, not in the uterus. In addition, sexual urges are not determined by the calendar, and therefore this method, like withdrawal, requires more "won't power" than many people can muster.

Another, almost useless procedure for the prevention of pregnancy is douching, or washing the woman's vagina with a chemical rinse after intercourse. Unfortunately, it only takes a few seconds after ejaculation for sperm to migrate up through the cervix and thus out of reach of the most determined douche.

Physical barriers, by contrast, provide some highly effective methods of birth control. The most common of these is the condom, a thin sheath that covers the penis and collects the semen. If used correctly and if they do not break or leak, condoms work. A latex rubber condom is much less likely to break than one made from a natural membrane; moreover, only a latex condom will provide protection against sexually transmitted diseases. Female condoms have also been developed; they fit inside the vagina and cover most of the labia as well. They appear to be effective, although probably less so than male condoms. Both of these methods require an interruption in sexual activity as well as a degree of planning, and are often less than enticing. (Although unwanted pregnancy and sexually transmitted diseases are even less erotically stimulating.)

Other barriers include cervical caps and diaphragms, dome-shaped devices that are inserted into the vagina before intercourse and provide a tight seal around the cervix, preventing sperm from penetrating. This means that a woman must plan ahead for an amorous encounter, although a cervical cap can be left in place for up to three days. Unlike either male or female condoms, both cervical caps and diaphragms are not obvious on the outside of the body and are therefore more discreet. Neither is entirely reliable, however, although when used with a spermicidal foam or jelly they can be quite effective. They don't protect against sexually transmitted diseases, however.

The same must be said of IUDs, or intrauterine devices. These are small plastic or metal structures that come in a variety of shapes, including coil, T, and zigzag. An IUD is inserted by a health care provider into the uterus, where it can stay for several years, depending on the woman's response to it. No one knows exactly how an IUD works, but it seems to disrupt the lining of the uterus, preventing an embryo from implanting. IUDs can work very well for some women, although others develop adverse reactions and there is some risk of infections or perforations of the uterus. IUDs can also cause increased menstrual cramping and longer periods, or be expelled from the uterus, not always with the woman's knowledge. New devices are being developed in the hope of solving these problems, but it remains to be seen whether they will prove effective.

The most widespread type of birth control is the Pill, or oral contraceptive. Its development marks a milestone in science helping women gain greater control over their bodies and their lives. Oral contraceptives use a combination of compounds closely related to estrogen and progesterone to inhibit the release of LH and FSH from the pituitary. With birth control pills, a woman's body is essentially fooled into thinking that it is in the second half of a normal ovarian cycle and that an egg has just been released. At this time, estrogen and progesterone prevent the secretion of FSH and LH, thus making sure that a second egg isn't produced—although because of the pills, a previous egg wasn't produced, either.

Recall that LH and especially FSH are necessary for the maturation of follicles, and therefore for the development and eventual release of eggs from the ovary. Estrogen in low levels inhibits the release of FSH, so when a woman is taking birth control pills, FSH is not released; therefore, her ovaries do not produce ripened follicles and eggs. The progesterone mimic, in turn, inhibits the release of LH. With no LH surge, the egg, even if it were present, would not be released into the oviduct. Estrogen and progesterone work together to inhibit the release of GnRH, resulting in a 99 percent probability that ovulation will not occur.

Now, back to the uterine lining. Under normal conditions, it is maintained under stimulation from estrogen and progesterone. If fertilization does not occur, the corpus luteum—source of most of the estrogen and progesterone—degenerates, after which estrogen and progesterone levels decline abruptly, causing the uterine lining to break down and menstruation to begin. A woman taking birth control pills has no corpus luteum because she has not ovulated; her estrogen and progesterone came from the pills. This is why the standard birth control pill regime calls for taking the hormones for twenty-one days, then stopping them for seven days. During "on" time, synthetic estrogen and progesterone cause the uterine lining to develop much like under normal stimulation. When the hormones are stopped, at day twenty-one, the woman usually menstruates, like non-Pill users whose corpora lutea have stopped producing their own "endogenous"—internally produced—hormones.

Some women use birth control pills to make their periods more regular or lighter, added payoffs that can be gained by dictating the hormonal state of one's body. For example, it appears that the once-a-month menstrual "off-time" experienced by women on the pill is not entirely necessary. Birth control pills can be taken continuously, thereby preventing one's period for up to several months in a row, without harmful effects. (Women should keep in mind, however, that after skipping a month or two, their menstrual cycle may be heavier, with more symptoms of cramping and bloating.) Skipping a cycle may have additional and as yet unknown consequences, so it should be done with care.

Birth control pills also reduce the risk of ovarian and endometrial cancer, and have not—contrary to some popular misconceptions—been implicated in higher risk for breast cancer in the general population. Slightly different combinations of synthetic estrogens and progesterones work better for different people, and although side effects are rare, they are nonetheless potentially dangerous. Prominent among these is blood clotting, especially in the case of smokers, so birth control pills must be taken under the supervision of a physician. In addition, because these pills typically involve rather small amounts of hormones, so as not to disrupt other body functions, missing just a few doses can allow a developing egg to be released, leading to a possible pregnancy.

At least two other chemical/hormonal methods of birth control are now available to women. Depo-Provera is an injection that must be given every three months. Norplant is a surgical implant, usually placed under the skin in the upper arm; it works for five years. Both methods rely on hormones called progestins, similar to progesterone, which provide a high level of protection against pregnancy, primarily by thinning the uterine lining (thus preventing implantation) and thickening the vaginal mucus, making it more difficult for sperm to travel upstream.

The primary advantage of Depo-Provera and Norplant over birth control pills is that they replace the one-pill-a-day regimen, with its risk of human error. In addition, neither uses estrogen, which has been associated with a slightly increased risk of heart disease in some women, and which may also be a bad idea for anyone known to be genetically susceptible to estrogen-dependent breast cancer. But there are also disadvantages: most women using Depo-Provera and Norplant experience irregular menstruation, and after about a year their periods may stop altogether.

With both systems, women also sometimes experience weight gain and possibly fatigue. Also, Depo-Provera and Norplant have a longer-term effect than oral contraceptives. Although this frees a woman from daily concern with contraception, it also necessitates a delay if she wants to become pregnant. Following the last Depo-Provera injection, it usually takes about a year for fertility to resume.

In the case of Norplant, minor surgery is required to remove the implants, after which fertility returns in about a month.

Artificial hormone replacement in women can have other purposes besides birth control. Compared to men, women generally have a lower risk of heart disease before age 55, but after this, they catch up. In addition, as women age, they are more likely to develop osteoporosis, or severe weakening of the bones, leading to increased risk of fractures. These changes apparently result from the decline in estrogen that comes along with menopause, at about age 51. At this time, a woman's ovaries become unresponsive to hormone stimulation from the hypothalamus, so despite a rise in LH and especially FSH as the body tries to kick-start the ovaries, no more estrogen is produced.

This is all perfectly normal and natural. The problem, however, is that in addition to its effects on the reproductive system, estrogen plays an essential role in the absorption of calcium and in maintaining healthy bones as well as favorable cholesterol ratios (low LDL, high HDL). Replacing estrogen in post-menopausal women has therefore become an effective way of reducing risks for osteoporosis and heart disease.

Hormone replacement regimens typically consist of a combination of estrogen and progestins, much like birth control pills, the estrogen aiming to keep up circulating estrogen levels while the progestin prevents overstimulation and thus growth of the uterine lining. Unless a menopausal woman is at high risk for breast cancer, it is generally agreed that hormone replacement therapy is beneficial; indeed, it is often recommended for women who have estrogen deficiencies for some other reason, such as exercise-induced amenorrhea (failure to menstruate), or who have had their ovaries removed, perhaps along with a hysterectomy.

Finally—to return to birth control—there are two methods of sterilization, both surgical and difficult to reverse. They do, however, provide the most reliable method of birth control this side of abstinence. In women, tubal ligation involves severing the oviduct and then tying the cut ends. Women who "have their tubes tied" need to be confident that they don't want children in the future. For men, vasectomy is analogous: the vas deferens—the tube leading from the testes to the penis—is cut, so that no sperm are present when he ejaculates. Vasectomies occasionally fail (the cut tube grows back together), so it is important to monitor sperm counts for some time after the operation. Once successful, vasectomies are essentially hassle-free and permanent; the resulting decrease of about 5 percent in seminal volume is scarcely missed.

There are as yet no male contraceptives analogous to birth control pills for women. Perhaps this is because of the notorious reluctance of men to monkey with their precious genitals, and a corresponding preference to "let the women do it." In our opinion, sexism is indeed at the root of male-female contraceptive asymmetry, but not so much the sexism of male-biased research or even performance, as the sexism of the natural world of which human beings are a part: male mammals (of every species) produce sperm in the hundreds of millions, and although low sperm count generally reduces the chances of fathering a child, the fact remains that just a single sperm could do the trick. The task of guaranteeing zero sperm by anything short of castration or vasectomy is gargantuan, if not insurmountable. By contrast, since women produce just one egg per cycle, and since, moreover, the success of this single egg requires a precisely orchestrated series of events, a zero-tolerance policy is much easier to implement for females. In addition, women offer more potential hormone targets for disruption.

Among the possibilities being actively pursued for new contraceptive devices is a vaccine that generates antibodies against human chorionic gonadotropin, without which embryos cannot implant in the uterus. Also being pursued are possible interference with the acrosome's release of egg-penetrating enzymes, as well as a variety of drugs that might block the binding of sperm and egg membrane.

Finally, there is the question of unintended contraception, via environmental pollution. There is some evidence, for example, that human sperm counts are currently much lower than in the past. According to one study, average sperm levels in 1940 were 113 million per cubic centimeter; by 1990, that number had plummeted to 66 million, which inspired a leading reproductive physiologist, testifying before a congressional subcommittee, to comment that "every man in this room is half the man his grandfather was."

The culprit—assuming this decline is real—appears to be a variety of "estrogenic" pesticides and industrial by-products (including DDT), many of which directly mimic the effects of estrogen. Yet others bind to testosterone receptors, blocking the normal action of male hormones and feminizing males in this way. It is clearly established that at least one estrogen-mimicking pesticide has been responsible for producing feminized male alligators in a Florida lake, resulting in abnormally small penises and reduced breeding success. The impact of such "environmental estrogens" on human beings remains highly controversial, although given that sex hormones choreograph so much human biology, it seems unduly optimistic—if not willfully blind—to assume that something troublesome is *not* taking place.

Baby Technology

Pregnancy Monitoring

As technology develops, the human uterus is becoming increasingly transparent. Before ultrasound, it was impossible to tell the sex of a baby until it was born. Now it is often possible to determine sex within the first trimester. Ultrasound, which uses high-frequency sound waves to provide an image of the fetus that can be seen on a screen, has been around for many years and is often the only monitoring done on a developing fetus. With the onset of new and more powerful techniques, however, it is becoming common for parents to conduct tests on the developing embryo or fetus. These techniques are especially useful for diagnosing genetic disorders.

Two main types of genetic testing can be accomplished in utero: amniocentesis and chorionic villus sampling, or CVS. Amniocentesis is done by inserting a long needle into the uterus and withdrawing some of the amniotic fluid that surrounds the fetus (*centesis* is from the Greek *kente*, meaning "to puncture as with a needle," and is the same root as in the word *center*, referring to the pivoting point used in drawing a circle). The amniotic fluid often contains sloughed-off cells from the fetus, and these cells can be grown in a laboratory. Amniocentesis can be performed around the fourteenth to sixteenth week of pregnancy, and a few weeks later the cell cultures can be analyzed.

In CVS a small part of the chorion, which is the fetal-produced part of the placenta, is removed and cultured. CVS can be done by the eighth week of pregnancy and the results obtained within a few days. Amniocentesis and CVS provide similar information. In either case, DNA is examined to see whether the chromosomes are aligned properly and are all present, a procedure known as karyotyping. There is a downside to both amniocentesis and CVS, however:

because they invade the uterus, they are potentially harmful to the placenta and/or the fetus, and as a result both increase—although only slightly—the possibility of miscarriage.

A few other procedures are sometimes used for testing a fetus. One of these is FISH (fluorescent in situ hybridization), which takes advantage of the small number of fetal cells that cross the placenta and are found circulating in the mother's bloodstream. The DNA from these fetal cells is isolated, then exposed to small, fluorescent DNA segments corresponding to regions of DNA that are known to cause birth defects. If the fetal cells contain these genes, they will glow, or fluoresce. FISH generates results within a few days and is much safer than other procedures because it does not invade the uterus. Another technique, known as fetoscopy, permits observation of the fetus via a thin fiber-optic tube inserted into the uterus, through which a physician can examine the fetus for any developmental deformities. Like amniocentesis and CVS, however, fetoscopy involves physically invading the uterus and thus can cause complications.

Gene therapy may eventually provide a means of correcting genetic abnormalities that are detected via fetal monitoring. Until then, monitoring techniques raise thorny ethical questions: if a seriously debilitating condition is uncovered in the developing fetus, there is, at present, almost nothing that can be done about it. The parents have to decide whether to abort or to prepare for a child with a potentially devastating disorder.

In Vitro Fertilization

Once upon a time, test-tube babies were science fiction fantasy. Now they are a reality. In the United States alone, in addition to about 60,000 annual births from donor insemination, there are roughly 15,000 babies born from in vitro fertilization, or IVF, and at least 1,000 from surrogacy (in which a woman is inseminated with another woman's fertilized egg).* "In vitro" fertilization literally means fertilization "in glass," and although it is hardly romantic, for many couples it not only beats childlessness or adoption but is a whole lot sexier than years of desperately "making love" on demand, only to be bitterly disappointed month after month. (Don't confuse

---

* By contrast, there are only about 30,000 healthy infants given up for adoption annually.

"artificial insemination," through sperm banks and so forth, with the new forms of assisted reproductive technology. Artificial insemination is artificial only in that the sperm are introduced by syringe or pipette instead of by a penis; with in vitro fertilization, eggs are removed from an ovary, and fertilization is accomplished outside the female body.)

The first human being produced by IVF was Louise Brown, born in England in 1978. IVF now offers hope for people suffering from low sperm count, blocked or scarred oviducts, and a wide array of other sources of infertility that, altogether, afflict one couple in six. The basic technique is simple enough, although the procedure is delicate, hence expensive: about $5,000 per try, with a success rate of only about 14 percent at each go-round. On average, therefore, an in vitro baby costs around $40,000.

The woman is given drugs—GnRH followed, if need be, by FSH and LH—to stimulate ovulation. Sometimes, when the problem is simply a recalcitrant ovary, these hormones alone are sufficient to induce ovulation, after which fertilization proceeds in vivo, occasionally leading to multiple births, as with the McCaughey septuplets born to an Iowa couple in 1997. In other cases, seemingly ripe eggs are identified via fiber-optic laparoscopy, and a dozen or more are "harvested" by being sucked into a micropipette. If, after examination under a microscope, they appear healthy and mature, they are stirred in a glass dish along with freshly obtained sperm (if the would-be father's sperm count is low, his ejaculate may be centrifuged first, to increase the concentration). Forty-eight hours later, any developing zygotes will have reached the eight-cell stage, at which point they are harvested once again and gently expelled into a waiting uterus, either that of the egg producer or, if the woman has adequate ovaries but is for some reason unable to undergo pregnancy, that of a surrogate mother, who carries the baby to term. At this stage, zygotes can also be frozen for later implantation. In a sense, Aldous Huxley's *Brave New World* is already upon us.

There are two major variants: GIFT and ZIFT. In the first, gamete intra-Fallopian transfer, eggs and sperm are collected, then placed directly inside an oviduct or Fallopian tube, where, if all goes well, "normal" fertilization takes place with sperm delivered the old-fashioned way. In the second, zygote intra-Fallopian transfer, zygotes are introduced into a Fallopian tube after first undergoing in vitro fertilization. In another technique, developed in Belgium and not yet widely used in the United States, a single sperm is literally injected into a single egg. A fine glass pipette holds the egg in place by gentle suction, while an even finer pipette punctures the egg to introduce a sperm that may be genetically normal but lazy.

Fertility clinics and would-be parents face some difficult dilemmas. Since only a small number of eggs actually "take" when initially exposed to sperm in vitro, and this number is further diminished when any zygotes are reintroduced into a uterus, it is generally advisable to have as many candidates as possible. For example, out of sixteen harvested eggs, only seven might produce zygotes, and of these, perhaps just one pregnancy will result. But what to do if those seven zygotes give rise to, perhaps, four potentially viable embryos? Many couples do not want multiple births; moreover, this type of pregnancy can be dangerous to a woman and it consistently produces a higher percentage of damaged babies, because they are always born prematurely. The human animal is biologically designed for singleton births, occasionally twins, but not litters. Should otherwise viable embryos therefore be sacrificed? If so, which ones? Given the costliness of in vitro procedures, couples not surprisingly are influenced by the success rates of competing fertility clinics. Should these clinics be permitted to run the risk of producing large numbers of medically undesirable multiple pregnancies simply in order to keep up their statistics? (More than 33 percent of "assisted reproductive technology" pregnancies result in multiple births, as compared with about 2 percent in the "unassisted" population.)

Additional ethical questions abound, especially when would-be parents and/or surrogates, after contracting to produce a child, change their minds, perhaps because of divorce or misgivings. For example, which is more important when it comes to motherhood, genes or gestation? Is fatherhood a function of insemination or intention (for example, being a sperm donor versus contracting for someone else's sperm to do the job)?

The ancestors of human beings had sex and reproduced with remarkable—perhaps alarming—success for millions of years. Although doing what comes naturally seems simple enough, we hope to have shown that it really isn't simple at all—but it is altogether comprehensible. At the same time, it is currently becoming safer, more controllable, and yet more complicated than ever.

THE ENERGETIC LIFE:
Food, Fuel, and Fat

What is energy? It resides in a gallon of gasoline, a Big Mac, and the sun. There may be too much of it in a four-year-old, lots of it in a crowd at a rock concert, not enough of it in oneself at the end of a hard day. Technically, energy is often defined as the capacity to do work. In other words, it is what must be present if something is to happen, and after it happens, there is less of it around. Without energy, everything would just sit still, unable to move or grow. In fact, one of the fundamental rules of physics is that nothing takes place unless energy of some kind is expended.

For a car to move, it needs to "burn" gasoline; for a ball to fall, it needs first to be lifted; for a volcano to erupt, spewing lava and debris in its path, the chemical energy inside the earth must have built up the pressure (and be somewhat diminished afterward). If water rushes down a waterfall, it does so only because it originally absorbed the energy of the sun, allowing it to evaporate and become a cloud so that later it could come down to earth in the form of rain or snow and join a rushing river. A tree grows because it absorbs energy from the sun, a termite grows because it eats the tree and uses the energy captured inside it to build more termite, and an aardvark grows by eating the termite, converting the stored energy inside the termite into more aardvark.

We do the same when we eat. Unlike plants or water, animals—whether termites, aardvarks, or people—cannot use the energy of the sun directly to do something active, such as grow, move, or evaporate. That energy must first be captured and transformed into something handier than a bunch of photons, such as the insides of a plant or another animal.

Living things have lots of work to do. Even a spider sitting all week inside its web waiting for prey to wander close enough to pounce gets hungry, although it doesn't appear to be expending energy. The same is true of a person confined to bed or simply watching TV or reading. Hunger lets the body know that energy reserves are getting low and need to be replenished. This is because bodies, even at rest, are working and using energy.

When you're sitting watching a movie, your heart pumps, causing blood to flow. You breathe, that is, your diaphragm rises and falls in a rhythm, causing your lungs to expand and contract, allowing oxygen to enter your bloodstream and carbon dioxide to leave. Your brain is actively shooting out messages all over your body; perhaps you are figuring something out or deciphering words and actions. Your body is probably digesting your last meal; hormones are being secreted and detected by various organs, causing different responses; and your muscles are actively firing, perhaps keeping your head up or causing you to nervously tap your right toe. In addition, different cells are constantly regenerating—dividing, moving around, committing suicide—and your entire body is keeping itself in a state that is dramatically different from the nonliving world around it.

Most of these different acts of work serve one goal: to fight against entropy, the natural tendency of everything on earth to become more disordered. Energy—thus, food—is our ally in this good (albeit losing) fight.

Every animal or plant, once it dies, will become increasingly disordered as it falls apart. When a life stops using energy, it stops being a life. To be a gourmet, therefore, is to make a virtue, or an art form, out of necessity—which may be what George Bernard Shaw had in mind when he wrote that "there is no love sincerer than the love of food."

As a general rule, the natural world finds simplicity easier to maintain than complexity. This is why one of the hallmarks of living things is, in fact, their complexity, which distinguishes a rooster from a rock. It also explains why as soon as something dies, there is a rapid breakdown of complexity into simplicity, and why it requires a constant input of energy to keep anything alive. It takes energy to go from simple to complex (as in growth) and also to keep anything that is complex from breaking down and becoming simpler (that is, maintenance). On the other hand, when things go from complex to simple, energy is generally released.

All animals are good at manipulating energy. Indeed, their survival requires it. Unlike plants, which use the sun's energy to build fancy molecules of sugars and starches from simple carbon dioxide and water, animals keep their own complexity going by ingesting fancy molecules—whether from plants or other animals—and carefully degrading them back into simple stuff, all the while harnessing the energy that is released.

How we animals use and rearrange energy will be the subject of this chapter. Why do we eat what we eat? How is the energy from our foods transformed into energy that can be used by our cells? What about other types of nutritional needs, such as essential vitamins and minerals? How does our body know how much fuel it needs, and why does this system seem to go wrong in some cases (as in obesity)?

Mitochondria

"How do I extract strength from the beef I eat?" asks Walt Whitman in *Leaves of Grass*. "What is a man anyway?" To answer the first question, let's turn to the smallest and, in many ways, the most basic player in the energy game: the subcellular organelles called mitochondria. They are often described as cellular power plants, but this is not strictly correct. They are more like money-changers. (Actually, the same can be said of traditional power plants: they also don't really *make* energy; rather, they change what is already present in coal, oil, or moving water into a form usable by television sets and light bulbs.) Think of the cell's various energy-demanding activities as vending machines, which require quarters, and the various energy-rich food molecules as hundred-dollar bills. Food, even digested food, is simply too clumsy to do the job without first being changed into smaller, more useful coins. These handy pieces of pocket change are molecules known as adenosine triphosphate, or ATP, and they are the nearly universal energy currency. What the U.S. dollar is to international finance—or the quarter is to vending machines in the United States—ATP is to the personal economy of nearly every living thing.

A particularly lovely and well-understood example is the "fire" in a firefly. This glow, technically known as bioluminescence, occurs when the suitably named pigment luciferin, present in the abdomens of fireflies, is acted upon by the enzyme luciferase. The needed energy is provided by our (and every firefly's) friend ATP. It is even possible to generate bioluminescence in a test tube, if the right chemicals, including ATP, are provided. In fact, this reaction is so reliable that it is now used to test for the presence of ATP, as a telltale sign that something alive is nearby. Manufacturers of Coca-Cola, for example, routinely add firefly luciferin and luciferase to batches of soft-drink syrup: if it glows, it is contaminated by bacteria. The same concoction traveled to Mars aboard the *Viking* spacecraft, in a (thus far unsuccessful) effort to detect signs of Martian ATP.

Here on earth, individual cells use energy in their daily activities: communicating with other cells, maintaining concentrations of molecules against the natural flow of osmosis, participating in the cell cycle, synthesizing different substances, moving the fibers that constitute muscle tissue, or (in the case of immune cells) protecting the rest of the body from harmful invaders. Most of the things that a cell does are controlled by enzymes, and those enzymes, in turn, pay for their ability to put things together, take them apart, and generally maintain that tenuous, complicated, nonrandom state called life by having ready access to lots of ATP. Muscle cells use ATP to contract. Nerve cells expend ATP sending messages—and even keeping themselves ready to send messages. Food for thought? Absolutely, but also for construction. Without ATP, new cells could not be built or kept in working order, because the workers would not get paid and necessary materials could not be obtained.

How does it all happen? ATP. It is a worthwhile molecule. As its name suggests, ATP consists of a base known as adenosine, the same building-block chemical found in one of the four bases that make up DNA. In ATP, adenosine is connected to three identical chemical groups called phosphates. These phosphates all have very strong negative charges, and negative charges don't like to be close to one another. As a result, it requires energy to take a molecule of ATP's precursor, ADP—adenosine diphosphate (for "two phosphates")—and squeeze in yet another phosphate, so that three of them are

bunched together in one small molecule of ATP. This energy is provided by the chemical bonds of carbohydrates, proteins, or fats,
without which ADP could not be induced to take up another phosphate and become ATP. The resulting ATP is unstable, however,
because of its mutually repelling phosphates. Think of it as literally
bursting with energy.

This energy is released when one of the phosphate groups is eventually removed, leaving two attached to adenosine. The resulting
molecule is ADP once again. When the first phosphate is removed
from ATP, energy is released in a useful form, energy that can be coupled to most of the different activities that occur in the cell. "Coupled
reactions" of this sort are crucial to the body's energy economy.
They are like a series of meshed gears, with molecules going from
high to low energy providing the power to turn the next gear, which
goes from low to high as it absorbs energy and prepares to do work.

But for this to occur, ATP needs to be produced in the first place.
As already mentioned, no process that uses or gives energy can take
place unless energy is first added to it. In animal cells, the production of ATP occurs mostly inside mitochondria, and it gets the
wherewithal to construct this energy-rich molecule by breaking
down relatively complicated, high-yield foodstuffs into their most
basic, and therefore low-energy, components: carbon dioxide and
water. (In plant cells, chloroplasts have much the same function as
mitochondria, except that instead of breaking down food molecules
for energy, the enzymes absorb energy from the sun and use it to
create ATP.)

Why can't the breaking down of food be connected directly to
the other processes in a cell, leaving ATP out of the picture? Why
don't bodies simply engage in a kind of barter system, directly
exchanging pepperoni pizza, say, for a few new liver enzymes, or a
chocolate-chip cookie for a few hundred million connective-tissue
cells, without bothering to go through the extra step of using pizza
and cookie to "buy" ATP, and then paying for our cellular transactions in that particular coin-of-the-metabolic-realm? The answer is
that food molecules are big, awkward, and difficult to break down
consistently. Needed, therefore, are specialized processes involving a
whole host of different enzymes and molecules, which harness the
energy and keep it in convenient units.

Between daily-maintenance needs and special-activities requirements, the average human goes through a staggering 2.3 kilograms

of ATP every day. It is a bit less surprising if we realize that the production of ATP is what "cellular respiration" is all about, and that cellular respiration and plain old breathing-type respiration are intimately linked. It is in ATP manufacture that oxygen is used, the same oxygen that all mammals obtain by breathing in. And it is in ATP manufacture that carbon dioxide is produced, the same carbon dioxide that all mammals get rid of when they breathe out. Keeping this in mind, along with the large amount of breathing we all do, several kilograms of ATP should not seem so outlandish after all. At any given time, however, the total amount of ATP inside a human body probably amounts to only a few grams. Mitochondria are both fast and efficient at producing enough ATP to keep up with the energy needs of our cells and, thus, ourselves.

What, then, are these mitochondria? They are small structures bounded by a membrane that allows only certain substances to pass in or out. Actually, they have two membranes, an outer, relatively smooth one and an inner membrane that is intricately folded. Between the two is the appropriately named intermembrane space. Inside the inner membrane is the mitochondrial matrix. Specialized reactions occur in specific parts of each mitochondrion.

In addition to the different membranes and enzymes necessary for ATP production, mitochondria also contain some DNA and ribosomes. This DNA occurs in a closed ring, similar to that found in primitive bacteria. No one knows the purpose of this mitochondrial DNA, but its presence leads to an interesting theory about how mitochondria originated. In the early history of life, mitochondria (and chloroplasts that later became part of plant cells) might have been separate, independent organisms. Perhaps as a result of being "eaten" by a larger cell or because they became parasites within such cells, mitochondria found themselves living inside a bigger organism. If these primitive mitochondria were particularly good at creating ATP, a sort of cooperative relationship could have developed, with the larger cell getting its energy needs met and the smaller "cell" perhaps gaining protection in return. As time passed, the two life-forms would have become increasingly dependent on each other, until they eventually evolved into one entity. This so-called endosymbiotic theory can never be proved, but it suggests that the genetic material within mitochondria may be left over from the time when mitochondria made their way independently in the world, perhaps buzzing around while the rest of us were quite a bit more sluggish.

The introduction of mitochondria to animal cells would have been quite important to their ability to extract energy from food molecules. Without mitochondria, the body can still produce ATP but not nearly as quickly or efficiently as inside a mitochondrion. This process, which takes place within the cell's cytoplasm and outside any particular organelle, remains the first step in the breakdown of food. It is called glycolysis. By the time a food molecule is ready to begin glycolysis, it is in the form of a molecule of glucose (*glycolysis* means "the cutting of glucose"). Glucose, of course, is the simplest form of sugar, and although it is not the only food substance that can be "burned" to make ATP, it uses the same pathway as other compounds.

Glucose (and other types of sugars) contains six carbon atoms, in addition to oxygen and hydrogen. The connections among the different carbon atoms are of a higher energy than the connections between carbon and oxygen, and as a result glucose (with lots of carbon–carbon bonds) has considerable stored energy. When these carbon bonds are broken, energy is released. The lowest-energy molecule containing carbon is carbon dioxide, whose carbon is attached to two oxygen molecules. By the end of glycolysis, which is a complicated process involving at least ten different enzymes, glucose has been split into two smaller molecules, called pyruvate, but is still a long way from its lowest energy state. There is still a lot more to be milked from the molecules of pyruvate, and it is the mitochondria's job to do so.

After glycolysis, pyruvate molecules are transported into the mitochondrial matrix, where they encounter the many different enzymes of the Krebs cycle (named for its discoverer). When the Krebs cycle is done, all the carbon atoms of the original glucose molecule are isolated as singletons, converted into carbon dioxide. It would seem that the cycle would at last be finished: the high-energy glucose molecule has been reduced to six molecules of low-energy carbon dioxide. But so far, little actual ATP has been produced. During glycolysis, two molecules of ATP per molecule of glucose were directly generated, and during the Krebs cycle only two more were made. Not only that, but the energy of two molecules of ATP is *used up* when transferring the products of glycolysis into the mitochondrial matrix, so the net gain after the Krebs cycle is a measly two molecules of ATP! This pitiful quantity does not reflect either the full amount of energy within a molecule of glucose or the energy needs of a cell. So where did the rest of the energy go? It was transformed into a simpler carrying system, which requires many more steps in order to produce meaningful quantities of ATP.

This carrying system involves creating two high-energy molecules known as NADH and $FADH_2$, which then travel to the matrix membrane inside the mitochondria, where they dump their extra, high-energy electrons into what is logically called the electron transport chain. This consists of a number of different enzymes that use the energy of the electrons to pump hydrogen into the mitochondrial intermembrane space. The resulting hydrogen-concentration difference between the intermembrane space and the matrix creates what is called a concentration gradient, which is power in the making.

Substances naturally seek to "even out" their concentrations across a barrier, so the hydrogens attempt to move back into the matrix from which they were pumped, much like water in two different but connected tanks that will move from the fuller to the less full tank as the two levels equalize. In the final step, a special enzyme embedded in the inner membrane, called ATP synthase, uses the energy of this hydrogen ion concentration difference between the intermembrane space and the matrix to combine ADP with a phosphate and make ATP. Lots of it.

ATP synthase is thus like a water wheel between two tanks of differing heights that uses the equalizing flow to make electricity, or in our case, ATP. For each original molecule of glucose, ATP synthase creates about thirty-four molecules of ATP. Combined with the results of the Krebs cycle and glycolysis, this brings the grand total of ATP molecules produced per molecule of glucose to about thirty-six: a respectable number at last. But why is such a complicated process necessary to achieve it?

The answer lies in the way chemical energy (stored in the carbon–carbon bonds of glucose and later in the phosphate bonds in ATP) has to be controlled in order to be used. Take, for example, the electron transport chain. Its net result, excluding ATP synthase at the end, is that two hydrogen atoms combine with oxygen to create water. This reaction produces a heap of energy. To appreciate this, consider what happens if you ignite a balloon filled with hydrogen gas. A very large, very loud explosion takes place when hydrogen combines with oxygen in the air to produce water that is vaporized instantly by the heat the reaction produces. (The same thing happened when the German zeppelin the *Hindenburg* exploded after reaching the United States.)

This is fine for a laboratory demonstration, but an explosion is the last thing that a body or even a single cell needs inside its mitochondria. Living things are not designed to withstand such a rapid energy release, although perhaps they could have been, if evolution had had a good reason for it. But any heat energy released in an explosion would be wasted. Unless it is somehow captured, heat simply drifts away. The electron transport chain, by contrast, is a carefully controlled way of ensuring that there are no explosions, that little heat energy is wasted, and that the maximum possible amount of energy is contained and transferred into the chemical bonds of ATP.

Aerobic versus Anaerobic Respiration

So, the body stores the energy it needs by using a series of reactions to break down complicated molecules into smaller ones, harnessing the energy that is released into ATP. There are really two parts to this process: what occurs outside versus what happens inside the mitochondria. This distinction has to do with oxygen and has important practical consequences. Glycolysis, occurring outside the mitochondria, is anaerobic, that is, it runs without oxygen. Cellular respiration, which takes place inside a mitochondrion, is aerobic, which means that oxygen is needed. So when the body is starved for oxygen, it is forced to get by on the meager amounts of ATP produced by glycolysis.

In order for glycolysis to occur there must be a constant supply of the electron carrier NADH, or more precisely, NAD+, which is converted to NADH during glycolysis.* During cellular respiration, oxygen's role is to absorb electrons from the food molecule (to oxidize is to draw electrons away from something). NAD+ does the same job during glycolysis, absorbing electrons from glucose and becoming NADH. When cellular respiration is occurring, NADH is converted back to NAD+ by disposing of those absorbed electrons through the electron transport chain. When this chain is not functioning because there isn't enough oxygen, the cell must somehow reconvert NADH into NAD+ so that glycolysis can continue. This is done through fermentation.

---

*NAD+ *plus an electron produces NADH.*

Fermentation has many different forms, the ultimate goal being to recycle NADH back into NAD+ so it can be used again. Fermentation is very common in yeasts as well as many bacteria. Under anaerobic conditions, such as those used for beer brewing or wine-making, it produces an interesting metabolic by-product: alcohol. Human beings don't make alcohol, whether aerobically or anaerobically. Instead, we and other animals (as well as bacteria and certain kinds of fungi) make lactic acid. When produced by bacteria and fungi, lactic acid can help make cheese and yogurt; in human muscles, however, it is much less useful, causing cramps and fatigue.

Lactic-acid buildup and the resulting cramps usually strike near the beginning of strenuous exercise or whenever the body's activity level is more intense than it is prepared to handle. It occurs when a muscle's need for large amounts of ATP is greater than the body's ability to provide oxygen to it. Under these conditions, the muscle is forced to rely primarily upon glycolysis and thus the production of lactic acid to renew its supply of NAD+. Sprinters, who actually may not breathe for the few seconds it takes to run their race, or other athletes who need a sudden burst of power, find themselves panting afterward as their body catches up aerobically to the sudden decrease in NAD+ supply caused by their anaerobic workout.

Food Processing

Now that we've looked at how food is used, let's step back and talk about the obvious: how food gets where it needs to be. We are all food processors, operating under a procedure that involves four steps, each of which consists of several vital, complicated processes: ingestion, digestion, absorption, and elimination. First comes ingestion: eating, swallowing, and the initial breakdown of food. Digestion is the breaking down of that food into even smaller and more manageable pieces so that during absorption, these particles can find their way into cells where they can participate in the cells' machinery. The final step, elimination, takes place when the waste products are removed from the body.

Ingestion involves getting food from the mouth to the stomach. Even this seemingly simple process has some important implications. For one, the shape of an animal's mouth and the distribution of its teeth tell much about what that creature eats. A carnivore such as a dog or a lion has teeth that are pointed and sharp, good at grasping and tearing flesh. A herbivore such as a cow or koala has teeth that are large and flat, good at grinding tough plant material into smaller pieces. An omnivore such as a human, which eats both animals and plants, sports a mixture of different types of teeth: the incisors and canines, in the front of the mouth, capable of cutting into and tearing meat, and in the rear, premolars and molars, which grind down fibrous plant material. Paleontologists need only discover a fossil jaw to deduce much about the owner's lifestyle.

Once food has entered the mouth and is being dismembered by the appropriate teeth, digestion begins. Even before it leaves the mouth, food has already taken the first few steps toward becoming part of the body, and saliva plays a remarkably big role. Most people think of saliva as mainly a moisturizing balm for the mouth, lubricating the throat so that food can slither down to the stomach. But it also contains some important enzymes that begin to digest food chemically. The most prominent of these is amylase, which converts the starches that you eat, such as those found in pasta, bread, and potatoes, into sugars. As shown by the mighty mitochondria, it is actually food in the form of sugar that gives the body its energy supply (by producing ATP), and saliva plays an important role in making sugar available to the body. Spit deserves respect.

After passing through your mouth, food travels down the throat, helped by regular contractions of the esophagus, into your stomach, where digestion begins in earnest. The stomach can store up to four liters of food, serving as a holding tank for your meal and a gated pathway to the small intestine; it also helps break down food even further. In some animals, the ability of the stomach to store large amounts of food for long periods comes in quite handy: a snake, for example, will typically eat a large meal and then digest it over a course of days, weeks, or even months, depending on the size of the snake and what it eats (the larger the animal, the larger the meal, and the longer the period between meals).

Next stop: digestion. People probably think more about indigestion than about digestion, since it is usually things that go wrong that get our attention. Digestion really does warrant a look, though, not only because of its manifest importance, but also as a useful corrective to the proclamations of various dietary ayatollahs who are so convinced (and often so wrong) about what is right. "You are what you eat," they emphasize, and to some extent, that is correct: from birth on, all of human substance (and of every other animal, too) derives ultimately from what is consumed.

But what we eat does not go directly to our muscles, bones, or nerves. Nearly everything gets digested first; that is, it passes through a reverse assembly process, a "disassembly line," in which it is broken down into its components before being incorporated into human stuff. That is why people are still recognizably human whether they are rigid vegetarians or bloodthirsty carnivores, whether their diet is big on rice, potatoes, cassava, cow's blood, or Big Macs. No matter how much pizza you eat, you will not become a pizza. Rather, pizza becomes you. Of course it matters what you eat, especially if your diet contains toxins, too much of a good thing, or not enough of something essential. Thanks to digestion, however, a fancy French meal and a TV dinner have pretty much the same bottom line.

The digestion that occurs in the human stomach involves mechanical churning and crunching, which chops up pieces of food, as well as two kinds of chemical digestion. The cells lining the stomach secrete large amounts of hydrochloric acid, making the stomach quite acidic and causing food particles to begin breaking down. In addition, the stomach produces an enzyme called pepsin, specialized to cut up proteins into smaller segments called peptides.

In view of these highly acidic conditions and the abundance of enzymes designed for the express purpose of destroying proteins, why doesn't the body digest itself? Because the stomach wall secretes large amounts of mucus, which keeps the acidic mixture at bay. In addition, the cells lining the stomach and intestines divide very rapidly, giving us two new stomach linings every week! (Such rapid cell division, incidentally, makes the gastrointestinal tract very sensitive to radiation and chemotherapy, which target fast-dividing cells; hence, diarrhea and nausea typically accompany radiation exposure as well as many anticancer treatments.)

Under normal conditions, the stomach is further protected by secreting pepsin in an inactive form that only becomes functional when it encounters acid, which is found inside the stomach cavity, not along its wall. These defenses are not foolproof, however, and when the mucus lining of the stomach becomes too thin, ulcers can develop as the acid and pepsin begin to attack the stomach itself.

It has recently been found that a particular bacterium, *Helicobacter pylori*, contributes significantly to human stomach ulcers. Australian researcher J. Robin Warren noticed that these bacteria were associated with people suffering from severe gastritis and ulcers, but it wasn't clear whether *H. pylori* caused the illness or were secondarily attracted to inflamed and damaged tissue. In 1984 Warren's associate, Barry Marshall, bravely drank a concoction containing billions of *H. pylori* and promptly got sick. Other volunteers developed genuine ulcers. Finally, in 1994, the National Institutes of Health recommended antibiotic treatment for ulcers. It appears that "ulcer bacteria" interfere with mucus production, after which the inner surface of the stomach breaks down, rendering it vulnerable to acid and pepsins, which diffuse into the lining and damage the tissue, causing an open sore that bleeds and can lead to secondary infections. Other predisposing ulcer factors include chronic stress, smoking, and excessive use of aspirin. (Many people harbor *H. pylori* without developing stomach ulcers.)

When stomach digestion is finished, the food (now a globular mess called chyme, from the Latin word for "juice") passes through a circular ring of muscle known as the pyloric sphincter into the small intestine. At this point, the food is still inside the digestive system and thus, topologically speaking, still *outside* the body. After all, with a long enough, flexible enough tube, you could trace the twistings and turnings of the human digestive tract, from mouth to anus, without ever going "inside." To gain entry, food molecules must somehow travel from the surface lining of the stomach and the small and large intestines into the body's cells. Ingestion and digestion, therefore, must give way to absorption.

The only substances that are absorbed into the bloodstream through the stomach are water, alcohol, and certain water-soluble drugs, including aspirin. Alcohol is absorbed much faster when the stomach is empty, which is why drinking on an empty stomach enhances its effect.

Along with further digestion, most of the absorption of food particles occurs in the small intestine, basically a tube an inch or so wide and about 10 feet long. (For a long time, when human internal organs were known largely through autopsy and dissection, the small intestine was thought to be longer, because it lengthens considerably after death, when its musculature relaxes). Digestion in the small intestine is accomplished with the aid of secretions from the liver and gallbladder, called bile, and from the pancreas, called pancreatic juice. Bile is a mixture of different salts that act like detergents to isolate small globs of fat and make it easier for the pancreatic juice to digest them. Pancreatic juice contains an array of different enzymes, including amylase (the same as in saliva) to complete the breakdown of carbohydrates; lipase to break down fats (also called lipids); and different kinds of proteases, which cut up proteins and peptides into ever smaller chains, eventually isolating them into individual amino acids. This mixture also contains fairly large amounts of sodium bicarbonate (the same as household baking soda), which neutralizes the hydrochloric acid found in the chyme coming from the stomach and converts the food mixture from acidic to slightly basic.

Once small bits of your meal are sufficiently digested, they are absorbed through the small intestine and eventually make their way to the bloodstream, where they are distributed to the body. The lining of the small intestine has millions of small, fingerlike projections called villi (Latin for "shaggy hair"), containing cells that absorb nutrients. When a sugar, amino acid, small globule of fat, vitamin, or mineral encounters these cells, it is transported inside, where it often undergoes further processing before being secreted into either very small capillaries (blood vessels) or, in the case of fats, into the lymph system. After food has passed through the small intestine, absorption has essentially been finished and it is time to prepare for the final step of food processing: elimination.

The large intestine is actually much shorter than the small intestine, although it is about twice as wide. By this point, any remaining food is mostly waste: leftover fats and proteins that were not completely digested, the cell walls of plants (vegetables and fruits), and water. The large intestine is filled with a huge supply of different bacteria that feed on this waste. They also contribute notably in their own right, producing a variety of vitamins not normally found in our diets, including vitamin K, thiamin, riboflavin, and vitamin B_{12}. These vitamins are absorbed into the bloodstream through the walls of the large intestine. As the leftover food continues its journey down the digestive system, along with a growing mass of dead bacteria, it eventually encounters the rectum, the last 6 inches or so of the large intestine, which expands, signaling the brain that—to modify the common bumper sticker—shit has happened.

Nutrition

"How can you have any pudding if you don't eat your meat!" yells a domineering adult in the Pink Floyd movie *The Wall*. Disappointing it may be, but there is some merit to this parental admonishment urging dinner before dessert. Although it is true that most desserts alone provide enough fats and carbohydrates to keep us going, at least in terms of pure energy needs, there is a host of different nutrients that people need in order to stay healthy, and meat contains most of them—which should not be altogether surprising, since we are meat ourselves.

This is not to say that meat is necessary for us humans, but pudding is definitely not sufficient. Nutrition, the study of what we must eat to fulfill our body's needs, is a complicated and important field, warranting scrutiny. It runs counter to at least one aspect of human psychology, namely, our preference for quick fixes and rapid feedback. The positive effects of good nutrition, as well as the negative effects of bad nutrition, take time to make themselves felt. At the same time, this relatively slow response—at least, compared with the way the nervous system, for example, interacts with its environment—makes nutrition a fertile breeding ground for all sorts of quack cures and idiosyncratic beliefs that remain stubbornly resistant to either verification or disproof. Thus, millions of people remain convinced that they cannot eat certain things or that they must eat other things, that they feel immensely better with certain diets or worse with others—regardless of the biological realities, since the delay between ingestion and effect is generally quite long, leaving abundant opportunity for the imagination to intervene.

In a world so rife with misconceptions, it is interesting to note that most of what holds true nutritionally is the same for all mammals, simply because mammalian bodies, whether person, pony, or porpoise, make similar use of food and nutrients.

Nutritional needs are crucially determined by metabolic rate, the amount of energy that an animal uses in a given unit of time. It is measured in calories or kilocalories. A calorie is the heat required to raise the temperature of 1 gram of water by 1 degree Celsius. A kilocalorie is 1,000 calories, and when dieticians or the backs of cereal boxes refer to Calories, with a capital C, they mean kilocalories. In any case, the caloric needs of any animal depend on its activity. An animal needs fewest calories when it's not obviously doing anything, and the most when it is maximally stressed physically, such as when running hard.

To compare the metabolic rates of different animals or different people, scientists use the minimal or "basal" metabolic rate. This is the amount of energy used by an animal at rest, unstressed, and on an empty stomach. For a human male it is about 1,600 to 1,800 kilocalories per day, while for a female it is about 1,300 to 1,500 kilocalories per day. Keep in mind that this is the energy used by the most devoted couch potato, someone so inactive that he or she isn't thinking about anything, writing in a journal, or even digesting a small meal. Any of these things, not to mention walking across the room, uses additional energy and raises the daily metabolic rate above basal.

In addition, there is great variation in this rate, since more "efficient" bodies use less energy, perhaps allowing them to store more excess energy-providing material (fat) in selected parts of their body. Those with high metabolic rates, on the other hand, may be able to eat voraciously without accumulating excess fat, simply because all (or at least a high percentage) of the energy consumed is used rather than stored.

Fats

Rarely admired, almost never loved, fat is well known and widely shunned, whether detected on one's own body or that of others. It is a white, gelatinous mass that jiggles when you walk and does not seem to fulfill any obvious purpose. Unlike muscle, it cannot be "tightened"—made hard—and used directly for work. It accumulates in different places on different people: on the hips, thighs, breasts, and buttocks of women, and around the stomach in men. Everybody, even the leanest body builders or long-distance runners, has fat. Jack Spratt presumably had less of it than Mrs. Spratt (who allegedly "ate no lean"), but even he was not merely skin and bones.

Fat accumulates in a variety of fat cells, and it is to a large extent the size of these fat cells, not their number, that determines how much fat people have and, thus, "how fat" they are. The bulk of each fat cell is given over to storing lipids, in the form of a so-called fat droplet, and as this fat droplet grows, so does the cell.

It appears that the number of fat cells is determined partly by heredity but also by how much was stored as an infant. In any case, the number of fat cells stays constant for each person's adult life, and although someone prone to obesity *may* have more fat cells than someone who isn't, what really counts is how much fat is in each cell.

Fats are made of three different fatty acids connected to a molecule of glycerol, hence their technical designation as triglycerides. Each fatty acid is a long chain of carbons and hydrogens, which can be either fully saturated, meaning that each middle carbon has two hydrogens attached to it, or unsaturated, meaning that some of the carbon atoms are missing their full share of hydrogen atoms and instead have multiple bonds with adjacent carbons. Saturated fats are relatively linear and can stack together quite closely. Unsaturated fatty acids, on the other hand, have kinks in their structure, which makes them less compressible. The result is that unsaturated fats, such as those found in vegetable and fish oils, are liquids at room temperature, while saturated fats, such as those found in butter and lard, are solid. Saturated fats are considerably more harmful than the unsaturated variety when consumed in excess because they increase blood-cholesterol levels, and cholesterol tends to be deposited in the walls of blood vessels, leading to atherosclerosis and eventually heart disease.

As for cholesterol itself, it turns out that the amount you eat is not terribly important when it comes to determining blood levels, because most of the cholesterol found in your body is synthesized in the liver, where it is used to help make bile. On the other hand, the amount of saturated fat that you consume *is* important. This is because saturated fats increase the amount of cholesterol produced by the liver, at the same time slowing the rate at which it is excreted from the body. Moreover, the cholesterol that saturated fats stimulate is in the form of low-density lipoproteins, or LDLs (the so-called "bad cholesterol"), as opposed to high-density lipoproteins, or HDLs (the so-called "good cholesterol").

Let's talk a bit about lipoproteins. They are complex clusters made of cholesterol, triglycerides (fats), phospholipids (fats containing phosphorus), and protein. The chemical difference between LDLs and HDLs lies in the amount of protein they contain: HDLs are about 45 percent protein, whereas LDLs are only 25 percent. HDL is higher in density because protein is denser than fat. Another way of looking at it: low-density lipoproteins have a higher proportion of cholesterol relative to protein. So there is no such thing as good or bad cholesterol: cholesterol is cholesterol is cholesterol. The difference is how it is packaged in lipoproteins.*

There is also a practical difference. The "badness" of LDLs arises because, in addition to stimulating cholesterol from the liver, they deposit it in the blood vessels. HDLs, by contrast, are good because they bring cholesterol *to* the liver, where it is made into bile and then

---

* There are also "very high," "very low," and even "intermediate" density lipoproteins, but for our purposes, enough is enough.

excreted. Thus, HDLs actually help lower blood-cholesterol levels, while LDLs, whose production is stimulated by saturated fats, raise those levels. The absolute quantities of LDL and HDL are less important than the amount of each relative to the other.

As to any take-home messages regarding cholesterol in the human diet, the scientific jury is still out. Some cholesterol is necessary as part of a cell's structure, for the production of bile, and as the backbone for all the steroid hormones including testosterone and estrogen, but this requires only about 1 percent of the body's cholesterol economy. Too much can be dangerous, especially to the heart. It makes sense to avoid eating large quantities of fats in general and saturated fats in particular. On the other hand, direct ingestion of cholesterol as such is less important than how the body metabolizes it, much of which is under genetic influence. (The fact that different people are born with different propensities when it comes to managing their cholesterol economy is, if anything, a reason for being *more* attuned, not less, to nutrition; if you have a genetic tendency toward high cholesterol levels, it is probably especially important to watch your intake of saturated fats.)

Fats are effective energy storers. Indeed, the structure of fatty acids is surprisingly similar to that of gasoline. Fat holds more than twice as many calories as either carbohydrates or proteins: nine calories per gram for the former and only four for the latter. Because they are "hydrophobic"—that is, they don't like water—they pack lots of calories in a much smaller space than either proteins or carbohydrates, which absorb water. (Water, of course, has no calories.) A cheesecake, therefore, can have about as many calories as a whole truckload of celery sticks.

Fats, like glucose, are broken down and used to make ATP, by employing specialized enzymes to dissolve the bonds holding them together. When this happens, the glycerol part of fat begins the ATP-making process by undergoing glycolysis, while the rest of the fatty-acid molecule enters the mitochondria and is broken down in the Krebs cycle. These processes can also occur in reverse, whereby glucose, which is often obtained first in the form of carbohydrates, is transformed into glycerol and fatty acids, which can then combine to form triglycerides. This set of reactions occurs in the liver and is why eating sugar and carbohydrates can make you fat. Because of their high energy output, fats are often found close to muscle and heart tissue, ready at any time to provide an instant source of energy—but also to clog things up if present in excess.

While being "fat" is not now considered healthy, Mother Earth Goddess statues and even Renaissance paintings suggest that in the past, pleasingly plump was highly desirable, probably because it indicated a likelihood of surviving if there were a sudden food shortage. The persistence of baby fat as a normal and expected part of human infancy further hints that, even today, fat is functional.

Fat has other uses besides storing energy. Several vitamins, including A, D, E, and K, are not water-soluble, so they require fats in order to be absorbed as well as transported. Some fats are also necessary for the production of certain types of signaling molecules, such as prostaglandins, important in smooth muscle contractions such as those controlling blood pressure. The oily aspect of fats keeps skin smooth and soft, and fat deposits in several key areas protect and cushion different organs, such as the eyeballs and kidneys. Moreover, fat is a very good insulator, and fat deposits under the skin help shield people and other animals from extreme heat and cold. This may not seem very important now, for a species that can readily don a down jacket when the temperature drops, but consider how a whale or a polar bear would do without a thick layer of blubber to protect it.

Carbohydrates

Carbohydrates—"carbs," to their intimate friends—are composed, as their name suggests, of carbon, hydrogen, and oxygen. Made up of either single (monosaccharide) or multiple (polysaccharide) sugar units, carbohydrates are broken down in the stomach and small intestine into their component sugar molecules, which are then absorbed by the small intestine and distributed through the body. The monosaccharides are ingested in the form of simple sugars, such as honey, molasses, and white table sugar. The polysaccharides are consumed as starches (or, when they are being admired, complex carbohydrates), such as those found in potatoes, wheat, or rice. Polysaccharides can be made of hundreds or even thousands of linked monosaccharides.

The raison d'être—or raison d'eat—of both types of carbohydrates is fuel. We have seen how glucose (the most common monosaccharide) is broken down to produce ATP. The other monosaccharides meet a similar fate. Although per gram, carbohydrates store less than half the energy of fats, carbohydrate metabolism is necessary for fats to be used completely. If not enough carbohydrates are consumed, perhaps due to a diet that limits such consumption in order to promote the breakdown of stored fats, compounds called ketones can accumulate, resulting in a potentially fatal condition called ketosis. Mild ketosis actually reduces hunger, and as a result some popular eating fads such as the "zone diet" recommend restricting carbohydrate intake.

With some interesting exceptions, such as the Inuit or Eskimos, who used to subsist largely on fats and proteins, and certain peoples who relied very heavily on fish, carbohydrates represent by far the bulk of all human foods. Whether as wheat, corn, rice, barley, rye, or potatoes, they make up about three quarters of the human diet. They also comprise nearly all the dietary fiber, material that humans are unable to digest completely. Such fibers are found especially in plant material, mostly in such structures as cell walls (cellulose), skins of fruits and vegetables (lignin, the only fiber that is not a polysaccharide), and the membranes that divide sections of oranges or the strings in celery (cellulose and hemicellulose), in addition to the insoluble fibers such as pectins and other gums that are found between plant cells.

Don't confuse *indigestible* with *indigestion*, or *undigestible* with *unimportant*. Fiber is in fact important, because it helps give feces more bulk, making them easier to pass and less likely to cause problems such as hemorrhoids as a result of constipation. As feces pass through the large intestine, water is absorbed, passing into the body. The longer feces remain in the intestine, the more water they lose and the more likely they are to become hard and difficult to pass. Soluble fiber such as cellulose and lignin can hold a great deal of water, so a diet rich in fiber produces softer stools. The insoluble fibers, such as the pectins and gums, make the feces larger, causing the walls of the intestine to expand and increasing the force of peristalsis, making them move faster through the large intestine. In addition, because they speed the passage of waste, both types of fiber appear to reduce the risk of intestinal and colon cancer, since any carcinogens (cancer-causing agents) that are eaten or produced by intestinal bacteria will have less time to exert their harmful effects.

Proteins

Proteins are the third major class of food we eat. They are the most complex of all foodstuffs, made of amino acids joined in long, convoluted chains. Amino acids are the crucial building blocks of almost all the body's enzymes and some hormones, as well as an important component of structural features such as hair, the part of muscle that contracts, and the part of red blood cells that carries oxygen, as well as the basic framework of teeth and bones. The proteins that are found within a body need constant replenishment; as cells die and their proteins get used up, new ones must be made to take their place. Proteins can also be used for fuel, providing about the same amount of calories as carbohydrates, but this is not the body's preferred mechanism. Proteins are only metabolized for energy if a huge excess of protein is eaten, if not enough carbohydrates and fats are available, or in some unusual circumstances such as at extremely high altitude, where the available oxygen is dangerously low. (Incidentally, no one really understands why the body consumes its proteins in this situation, although it is well known that above 20,000 feet or so the human body actually loses muscle.)

The proteins that you eat are not terribly useful in the form they come in. What is useful is the amino acids that make them up. Digestion's job, therefore, is to break down proteins into ever smaller fragments, so that eventually individual amino acids can enter cells and be recombined in various ways to create new proteins, as needed. There are twenty amino acids, each one necessary to make the body's proteins. Human beings can synthesize twelve of these twenty amino acids from bits of carbohydrates and fats, plus nitrogen, but the other eight must be absorbed from food. These so-called essential amino acids need to be present in everyone's diet. Otherwise, our protein-building machinery is like a typewriter from which one or more of the keys has been removed.

"Part of the secret of success in life," wrote Mark Twain, "is to eat what you like, and let the food fight it out inside." For even greater success, however, the secret is to eat a balanced set of proteins, so that any deficiencies from one source are compensated for by another, without any fighting. Moreover, this balance should be achieved every day, since the body synthesizes new proteins all the time and does not store free-floating amino acids for very long. The easiest, although not necessarily the best, way for a human being to get all necessary amino acids is to eat meat or dairy products. Because all other vertebrates build their proteins from the same twenty amino acids that human beings use, eating them gives us what we, too, require. Beef, chicken, and fish, for example, all contain well-balanced proportions of amino acids, as do milk and other dairy products, including eggs.

Eating large amounts of meat in order to get enough protein, however, is unnecessary and in fact can be harmful. People appear to do best when about 15 percent of their calories come from protein, but most Americans eat about twice this. Moreover, animal protein is often associated with large amounts of saturated fat, which is certainly *not* needed. As much as 50 percent of a T-bone steak might be fat, much of it very difficult to remove, as it is interspersed between the muscle fibers. Even "extra lean" ground beef will typically have more than 10 percent fat, which means that about 30 percent of its calories are fat-based. If you do eat meat as your primary protein, it's better to eat either chicken, in which most of the fat is found right underneath the skin and is therefore easily removed, or fish, where the percentage of fat is fairly low and that which occurs is mostly in the form of fish oils, which are largely unsaturated.

All the essential amino acids can also be obtained without touching animal products. In contrast to those found in meats, the proteins found in various grains, nuts, and beans are not associated so much with fat as with carbohydrates, which are more benign. A vegetarian diet requires some planning, however, because unlike meats and dairy products, there is no single plant that contains enough of all eight essential amino acids. Critics of vegetarianism often trot out this argument, as though it is a showstopper, forgetting that people are no more likely to consume a diet of pure broccoli than one of pure T-bone.

Most grains, such as corn and wheat, are low in the amino acids isoleucine and lysine but high in the other essential amino acids. Beans and nuts, on the other hand, contain isoleucine and lysine but are low on tryptophan. A diet of only grains or only beans would therefore be deficient in some of these essential amino acids. By eating both corn and beans, however, a vegetarian would get not only enough protein but also a complete selection of the essential amino acids and much less fat. (Let's hear it for bean dip and tortilla chips!)

Perhaps the most common type of malnutrition in the world is caused by deficiencies in one or more of the essential amino acids. This usually happens in places where the food supply—especially meat and dairy products—is low in relation to the population size. In sub-Saharan Africa, this condition is called *kwashiorkor*, a word derived from the Ghanaian language meaning "the evil that afflicts the first child when the second child is born." *Kwashiorkor* occurs mainly among young children who are put on a diet of mostly starches after weaning, or who are given diluted baby formula instead of breast milk. In either case, the child's protein intake is inadequate. (Another disease, marasmus, occurs if a child's overall calorie level is too low; *kwashiorkor* is specifically a protein deficiency). *Kwashiorkor* results in retarded development, both mental and physical, and can even lead to death among especially young children.

Protein-deficiency diseases are rare among people in developed countries, although they are a possibility among fundamentalist vegetarians or those who cannot afford adequate protein. More frequent, however, is the inability to metabolize certain amino acids. The most common of these disorders is phenylketonuria, or PKU. It occurs when, because of a mutation, the body is unable to break down molecules of the amino acid phenylalanine and its derivative, phenylpyruvate. If children with PKU are given a diet that contains large amounts of phenylalanine, mental retardation will result, although the exact mechanism is not known.

Children suffering from disorders such as PKU experience something of a catch-22, because even though they cannot break down phenylalanine, and hence find it toxic in large amounts, it remains essential. PKU and its treatment provide a fine example of how genetic defects can be overcome: such children are put on low-protein diets, carefully monitored so they get enough protein, but not too much. PKU symptoms—including mental retardation—are easy to prevent if identified early, and testing newborns for PKU has become standard procedure in the United States, the reason for the well-known heel puncture to which infants are subjected.

Vitamins and Other Nutrients

The essential amino acids are "nutrients," substances that are necessary for a healthy body. Nutrients cannot be synthesized internally and must therefore be present in the diet. There are four types of nutrients: essential amino acids, essential fatty acids, vitamins, and minerals. The essential fatty acids are all unsaturated and are used to make certain structures in a cell's membranes. Fatty-acid deficiencies are extremely rare, especially in developed countries such as the United States, where most people eat too much fat. The essential fatty acids are also found in nearly all nonanimal fat, such as olive or canola oil, as well as some animal fats. Only a very small amount is needed. The vitamins and minerals, however, can have a much greater influence, even though the necessary quantities are tiny indeed.

Vitamins are organic (carbon-based) molecules that play a vital role in body functioning. They do not provide energy, as many people think; rather, they typically serve as "cofactors," which assist in the release of energy present in fats, carbohydrates, and proteins. Vitamins can be components of certain important molecules such as rhodopsin, which aids black-and-white vision (vitamin A); they can enhance certain processes, such as the growth of bone (vitamin D) or the healing of wounds (vitamin C); they can be necessary parts of enzymes (such as almost all the water-soluble vitamins, including thiamine, niacin, folic acid, and biotin). In fact, the different roles for different vitamins are so diverse that they are defined by what happens when they are *not* present: a vitamin is a molecule that, when absent from the diet, results in some ailment or disease that disappears as soon as that molecule is reintroduced. As a result, it is easiest to understand vitamins by looking at some of the diseases that arise in their absence.

Probably the first dietary-deficiency disease ever identified was scurvy. Its symptoms, including swelling and bleeding of the gums, under the skin, and in the muscles, were commonly experienced by blue-ocean sailors. When prolonged, death resulted. In 1498 Vasco da Gama sailed around the Cape of Good Hope, beginning with a crew of 160, of whom 100 died of scurvy. In 1754 a Scottish naval doctor named James Lind had a hunch that the disease was related to the fact that on long naval voyages the crew was deprived of fresh fruits and vegetables. He discovered that the symptoms could be quickly reversed and even prevented if the sailors were fed lemon juice. In 1865 the Royal Navy switched from lemons to limes, giving rise to the term "limey" for a British sailor. It was not until 1932 that the antiscurvy factor was isolated and identified as ascorbic acid, now known as vitamin C.

Pellagra is another vitamin-deficiency disease that was recognized and treated a long time ago. Common in Italy and the American South before the twentieth century, pellagra was found among poor people who subsisted mostly on corn. Although a few different factors contribute to the effects of pellagra, the most important is a deficiency in vitamin B_3, also known as niacin. Pellagra is characterized by the "3 D's": dermatosis (skin disease), diarrhea, and dementia (impaired mental functioning). It is rapidly reversed when the diet is improved, but it turns out that most of the B vitamins, not just niacin, have to be added.

The deficiency disease that first alerted doctors to the idea of vitamins, however, was beriberi. *Beriberi*, which means "extreme weakness," had been known for many years, especially in Asian countries where people subsisted on white, polished rice. Its symptoms include extreme fatigue, loss of appetite, weakness in the arms and legs, and digestive problems. In addition, beriberi can cause nerve degeneration, cardiac failure, and edema (fluid buildup). Researchers showed that birds could be induced to have beriberi-like symptoms if their diet was similarly restricted. Furthermore, these symptoms go away following treatment with concentrated rice polishings—that is, if they are fed the rice husks that are typically removed before human consumption! Many scientists and doctors in Japan, Indonesia, and Malaysia also reported that the symptoms of beriberi in humans could be reversed when people ate brown, unpolished rice or extra meat, fish, and vegetables.

These findings led Casimir Funk in 1912 to propose that there were some "vital-amines" (an amine is a type of molecule) whose absence was causing diseases such as beriberi. Later the term was shortened to *vitamin*, and the crucial substance missing in the diet of beriberi victims was isolated and found to be the vitamin B_1, now known as thiamine. Thiamine is present in most raw and unprepared foods, including brown rice, and the incidence of beriberi has decreased dramatically since its cause was determined.

Not all vitamin deficiencies have been eradicated, however. Rickets, a deficiency in vitamin D, is easily avoided but nonetheless afflicts significant numbers. Vitamin D is synthesized in the skin from a molecule that is either ingested or made from cholesterol. (Because human beings are capable of synthesizing it, vitamin D is not strictly "essential," but the dietary influence is large enough that deficiencies still happen.) A crucial step in the body's synthesis of vitamin D occurs when the reacting molecules are zapped by ultraviolet (UV) radiation from the sun, which is why the reaction occurs in the skin, and also why people who are kept from sunshine, either covered up when outside or housebound for a prolonged period, are susceptible to the disease.

Many foods, such as margarine and milk, are now laced with vitamin D to guard against rickets, but the disease still occurs in certain regions. Symptoms are all related to bone deformities: young infants may get soft skulls, children become either knock-kneed or bow-legged as a result of joint problems, and adults may develop contracted pelvises that can interfere with childbirth. These defects occur because vitamin D is an important player in facilitating the absorption of calcium and phosphorus, all of which are important for normal bone growth and formation.

In general, water-soluble vitamins (C and the B-vitamins) carry relatively little danger of overdose; excess quantities are usually peed away. On the other hand, water-soluble vitamins are liable to dissolve in boiling water and be lost, hence the wisdom of steaming vegetables. Fat-soluble vitamins (A, D, E, and K) are stored in the body and can be toxic in high doses. Sudden weight loss can also cause fat-soluble vitamin overdose, when accumulated vitamins are suddenly dumped into the body as fat cells shrink.

The final essential nutrients are the minerals. As with vitamins, minerals take on a wide range of roles. Many of them are important structural features of various proteins: magnesium, copper, selenium, zinc, molybdenum, and manganese are all parts of different enzymes; iron is in some ways the most important atom in a red blood cell because it forms hemoglobin, responsible for transporting oxygen around the body; phosphorus is an important ingredient in ATP (remember those phosphate groups?) and in DNA. Phosphorus is also significant in bone development and maintenance, along with calcium. In fact, calcium is needed in such quantities that dietary supplements are sometimes recommended, especially for older women who are prone to the bone-weakening disease osteoporosis. Most other essential minerals can be easily obtained in sufficient quantity by a well-balanced diet containing at least some whole, unprocessed foods.

Sodium, potassium, and chlorine are important for maintaining the correct fluid balance in the body, as well as in nerve transmission. These minerals are eaten in the form of salts, such as sodium chloride (table salt), and in certain fruits and vegetables, such as bananas for potassium. Because few salts are found in vegetables, a purely vegetarian diet can in theory result in a salt deficiency, which is why many herbivores such as deer and horses will congregate at either natural or manmade salt licks. People in developed countries such as the United States, on the other hand, tend to eat more salt than they need, largely from prepared foods as well as a widespread, almost reflexive tendency to add salt at every meal. For most Americans at this point, intentional salt consumption is recommended only in cases of dehydration, such as after strenuous exercise or prolonged diarrhea, or extremely low blood pressure. In these situations, the ingestion of salts (along with copious amounts of water) draws water into the cells while also increasing blood pressure.

Iodine is another essential mineral, and until recently iodine deficiencies were fairly common among inland peoples, especially in Asia, who did not eat many marine fish, shellfish, or dairy products. Iodine is an important constituent of thyroid hormones, which help regulate metabolic rate. When iodine intake is too low, the thyroid gland, located in the neck just underneath the chin, can swell dramatically, sometimes getting as big as a grapefruit. This condition is called goiter, and in addition to the unsightly lump, sufferers often feel listless and lethargic because of their thyroid deficiency. Only a *very* small amount of iodine is needed to correct the problem, however, and in countries such as Nepal, where it used to be fairly common, goiter has been almost completely wiped out by the introduction of trace amounts of iodine into the salt supply. Iodized salt has in fact made goiter rare in nearly every country.

In part because of the astounding improvements in public health that followed the discovery of vitamins and other dietary micronutrients, dietary supplements have become faddishly popular, especially in the wealthy West. Surprisingly little is known about the effectiveness of such home remedies as megadoses of vitamin C, various exotic plant and fish oils, antioxidants such as vitamin E, and so forth. A visit to a health-food store can be a dizzying experience, even without overdosing on any of the available products. Some supplements seem undeniably valuable for specific purposes, such as calcium to protect against osteoporosis in postmenopausal women, or additional iron for women with unusually heavy menstrual bleeding. Others may rely heavily on the placebo effect, by which a substance having no actual biochemical activity nonetheless can generate a perceived improvement. It may be, for example, that taking several grams of vitamin C at the onset of cold symptoms really does help "cure" the cold, if only by mobilizing the immune system of anyone gullible enough to believe that it will work. (*Placebo* comes from the Latin, meaning "I shall be pleasing.")

Don't sneeze at placebos. They are, in their own way, "real." After all, any therapy that averages about 33 percent efficacy, with no side effects, qualifies as nearly a medical miracle.

Food Intake and Body Fat

The Wisdom of the Body is the title of a now-classic textbook of human physiology. One of the best examples of such wisdom is the body's control of food intake. As anybody who has tried to lose a few pounds knows, once maturity has been reached, the chances are that one's body weight is going to remain stubbornly constant—if not indefinitely, then at least for a long while. It is difficult to lose weight and keep it off, and it tends to be difficult to gain weight, too, although few adults complain about that. Like it or not, the human body is reluctant to depart—at least, not for long—from a remarkably narrow weight range.

How is this accomplished? Why is it so common for people to lose weight on a carefully monitored diet, only to have the pounds come right back as soon as the dieter lets up? On the other hand, why are some people able to eat tons of ice cream and cheesecake and never gain a pound, while others just think about their favorite food and begin to feel their waists thicken? This is yet another mystery, but we are at least getting closer to understanding how people maintain their body-fat levels, and there is hope that the more we learn about such mechanisms, the more we may be able to control them. The magic diet pill may never become reality, but we are moving in that direction as we discover more about the body's regulation of its own weight.

Unlike such questions as brain function or the genetic code, it seems that the control of body fat should be straightforward, if only because fat is so much less esoteric and more immediately apparent than nerve transmission or the workings of DNA. After all, most of us can grab a handful of it, whenever we want.

Body-weight regulation turns out to be a remarkably complicated issue, however, with both environmental and genetic components. It is clear, for one thing, that people living in industrialized countries have been getting fatter, as measured by their body-mass index, or BMI. This is a modified ratio of a person's weight to his or her height: the higher the number, the more weight a person has for his or her height. You can figure out your own BMI using the formula $703 \times$ (weight in pounds)/(height in inches)^2. Thus, a woman who is 5 feet 4 inches and weighs 120 pounds has a BMI of 20.6, while one who is 5 feet 4 inches and weighs 150 pounds has a BMI of 27.5.

According to government guidelines released in 1998, anyone with a BMI of over 25 is mildly overweight, or "preobese," while those people with a BMI over 30 are genuinely obese. The lower your BMI, the less likely you are to develop weight-related health problems such as certain cancers (breast, uterus, cervix), heart disease, and strokes. These figures have to be taken with at least a small grain of salt, however, because they do not indicate anything about body composition—muscle and bone weigh more than fat and do not have the same harmful effects—or about cardiovascular fitness and activity levels.

In addition, it has not been proven conclusively that overweight people enjoy significantly improved health when they lose weight, although this seems likely. Most studies have compared skinny and fat people, rather than addressing the health effects on fat people when they lose weight. Weight-loss studies are also plagued with the problem of evaluating preexisting conditions as well as the difficulty overweight people have in losing and keeping off weight without either starvation or sickness.

Aside from the social disapproval, what's wrong with being plump? First, excess weight means excess strain on the body, be it the limbs and bones (which do not get bigger when body weight increases) or the heart (which enlarges and becomes a more efficient pump with exercise but not with weight gain). And regardless of whether being "circumferentially challenged," per se, is unhealthy is the fact that most people who are overweight are also not very fit, and there is abundant evidence that being unfit is unhealthy.

In one study of more than 21,000 men, unfit people with BMIs of less than 25 were twice as likely to die early as fit men with BMIs over 27.8. The reasons are unclear, but the major factor is probably the greater vulnerability of unfit men to heart attacks, combined with the blood pressure–lowering consequences of fitness. With increased blood pressure comes increased likelihood of strokes. Also, a fit person has a more efficient heart. It is undeniable, however, that simply being overweight influences health, too. Most dramatically, fatter people have elevated levels of blood lipids and usually of LDL cholesterol, both of which contribute to heart disease and possibly metabolic illnesses such as diabetes.

In its classic form, insulin-dependent diabetes mellitus (IDDM) strikes patients when they are children, whereas non-insulin-dependent diabetes mellitus (NIDDM) tends to strike older, obese patients and is much more common. In both cases, sufferers have elevated levels of blood glucose because their bodies are not responding enough to the hormone insulin. Insulin is released from a healthy pancreas when glucose levels are high; it stimulates muscles and fat cells to absorb more glucose. Sufferers of IDDM produce almost no insulin, which therefore must be provided artificially, either by injection or orally. People with NIDDM secrete some insulin but not enough, and moreover, they have become resistant to its effects. When this happens, they get dehydrated, which can eventually lead to a coma if glucose levels in the blood are not controlled. They also tend to have high levels of triglycerides and LDL cholesterol along with low levels of HDL cholesterol, all of which increase the likelihood of atherosclerosis and heart attacks.

The treatment for NIDDM consists primarily of diet control. Eat a low-fat, low-cholesterol diet and lose weight, and many of the symptoms of NIDDM disappear. A weight loss of only 5 to 10 kilograms (12 to 25 pounds) can completely cure it. In addition, it appears that exercise helps increase the sensitivity of the body to insulin. Although some drugs reduce the symptoms of NIDDM (mainly by causing patients to become mildly hypoglycemic, or glucose-deficient), none are as effective as simply losing weight and getting more exercise.

Clearly, controlling food intake and body weight can be very important for maintaining basic health. But as we have seen, this is easier said than done, because our bodies seem to have their own agendas, automatically controlling their own weight, often contrary to our intent. How are food intake and body composition controlled? Two separate physiological systems operate here, depending on whether the time frame is short—day-to-day or hour-to-hour—or long: over weeks, months, or years. Both systems are important, and they work together.

Every type of animal has its preferred meal pattern. Rats, for example, tend to eat in the dark, especially at dusk and dawn. This distinct feeding style can change dramatically if one or more stresses are encountered; for example, if the rat has a limited food supply, or a predator shows up, or the environment is new. In this circumstance, the rat—and all other animals, including people—regulates its food intake by changing the size of its meals. Chemical signals accumulate while the animal is eating, eventually reaching a level that tells the brain it is no longer hungry. These signals are called satiety factors, and several of them have been identified.

The first and most important is a peptide (small protein) called cholecystokinin (CCK). CCK is released from the gut in increasing amounts as the stomach fills up with food. Acting in concert with the stretching of the gut that occurs when a meal is eaten and some other satiety factors with names like bombesin, neuromedin B, and glucagon, CCK tells the brain to end the meal. In experiments, giving an animal excess CCK will reduce the size of its meal without causing nausea or any obvious distress, so it appears that the animal stops eating because it feels full, not sick. Similarly, blocking the effects of CCK increases meal size dramatically. So far, so good.

But surprisingly, dosing with CCK does not seem to cause an animal's weight to change! There are many ways for a rat to fill its belly: when a few large meals aren't in the cards, it substitutes a larger number of small ones. When a rat is given CCK frequently, the size of each meal decreases. But it eats more often, so the total caloric intake remains constant. This means that the animal's body is somehow aware of its overall metabolic needs, and that CCK and other satiety signals, while important for determining the size of individual meals, probably play a relatively small role in the control of overall body weight. Some other, long-term, general signals must be involved. These are thought to be directly related to the body's fat mass, keeping it relatively constant over time.

Two main signals appear to fit this bill: insulin and leptin. Insulin has been known for many years because of its role in diabetes. It is secreted from the pancreas depending on the amount of glucose in the blood. In someone not suffering from diabetes, circulating levels of insulin are also directly proportional to body-fat levels: more fat, more insulin. Insulin levels are detected by the brain, and if high enough, they cause the animal to lose weight. Adding insulin causes animals to keep their protein and carbohydrate intake constant while reducing both their fat consumption and overall body fat. It has to be administered directly into the brain, however. When given otherwise, insulin has almost the opposite effect, enhancing fat storage by encouraging glucose to leave the bloodstream and enter cells. It is not, therefore, a feasible "diet pill" or even diet injection.

Leptin was only discovered in the mid-1990s, and it may be *the* key to maintenance of normal body fat. Leptin came to light when it was discovered that certain genetically obese strains of mice, known as *ob/ob*, were missing a crucial chemical factor. Later, these mice were found to have a mutation in the *ob* gene that rendered the gene's product useless. This product was named leptin. As discussed briefly in Chapter 5, leptin is produced by fat cells so that its blood level is proportional to the amount of fat in the body; the more fat, the more leptin. Leptin's job, it seems, is to tell the brain how much fat is on board, so food intake and metabolic rate can be adjusted accordingly. If fat levels drop, so does the amount of leptin, which causes the brain to increase food intake and decrease metabolic rate until fat stores increase. With more fat, leptin levels increase, whereupon the brain reduces food intake and increases metabolic rate until the levels drop again. Acting like a thermostat, leptin has been designated a "lipostat"—we might call it a fatostat—a key player in the negative feedback that keeps fat stores (lipids) constant.*

If you give a normal mouse or rat excess leptin, it will eat less, exercise more, and have a higher metabolic rate as its brain is fooled into thinking that its body is obese. Moreover, the weight that the animal loses in response to leptin will be almost entirely fat, not muscle. This occurs without the animal exhibiting any signs of sickness or stress. The perfect diet pill? Unfortunately, the pathway that leptin uses to control body fat is long and convoluted, and many people (and other animals) who seem to be genetically obese are resistant to excess leptin administration. There is a type of mouse called *db/db* that looks and acts like *ob/ob* mice. The *db/db* mouse, however, has a genetic defect in its leptin receptor. This receptor is found in many different tissues, including important parts of the brain, and is essential to the body's response to leptin. You can give *db/db* mice as much leptin as you want (in fact, these animals produce very high levels of it naturally because they have so much fat), yet they will not lose weight because they have no way of recognizing that so much leptin is present.

What about obesity in humans? Do fat people produce insufficient leptin, so that their brains don't know how chubby they really are? Or is it that their receptors are insensitive to the message, as in *db/db* mice? Very likely, leptin is somehow involved in obesity, but at least so far, very few obese people have been found with defects in the gene for producing it or in the receptor sensing it, and moreover, giving obese people leptin does not cause weight loss. Leptin and insulin seem to act together, somehow: if the brain is unable to sense leptin, as is the case with *db/db* mice, then it will not react to insulin, either.

Let's look further at what happens inside the brain, where leptin, insulin, and other chemicals have their impact.

---

* *Thermostatlike systems of this sort should be familiar by this time, resembling those employed—also by the hypothalamus—to regulate reproductive hormone levels.*

There are two kinds of brain signals that influence body composition: those that stimulate the accumulation of fat stores and do so by lowering metabolic rate and increasing food consumption, and those that induce a loss of fat, raise metabolic rate, and decrease food consumption. Of the former, the most important seems to be neuropeptide Y (NPY), and of the latter, the so-called melanocortins. NPY is a ubiquitous neurotransmitter with many roles. The NPY that concerns us here is found in parts of the brain called the arcuate and paraventricular nuclei. These lie within the hypothalamus, that small but crucially important brain region just above the pituitary gland, implicated in many "vegetative" functions such as weight gain and reproduction. NPY increases fat storage, so much so that placing it directly into the brains of rats or mice causes obesity in just a few days.

NPY, in turn, seems to be strongly affected by leptin and insulin and is presumably at least one of the mechanisms whereby they exert their influence. *Ob/ob* mice, which lack leptin, have very high levels of NPY, as do *db/db* mice, which cannot detect leptin. Giving these animals more leptin decreases NPY levels in *ob/ob* but not *db/db* mice, indicating that when the brain detects leptin, less NPY is released. Insulin has similar effects.

The melanocortins act in the other direction. *Ob/ob* mice have low levels of melanocortins, and giving them leptin increases the melanocortin levels in their brain. Both NPY and the melanocortins seem to exert at least part of their influence by modifying the body's sensitivity to satiety factors such as CCK. Thus, an individual's fat level influences how hungry he is at any particular meal. It is likely that this effect is not precisely maintained during every meal; rather, food intake and energy output are integrated over several days to maintain a set weight.

What does all this mean for a healthy person? Let's follow the chain of chemicals activated when a normal animal has high levels of leptin, that is, when it is a bit on the fat side. High leptin works on the brain to cause low NPY and to increase the amount of melanocortins, enhancing sensitivity to CCK and increasing metabolic rate and activity level, resulting in weight loss. As fat stores drop, so do leptin and insulin levels; as a result, NPY increases and melanocortin decreases, causing weight gain and reduced sensitivity to satiety factors such as CCK.

This helps explain why it is so difficult to lose and gain weight: we each have our own set-point weight, which the body tries to keep constant. It does this through insulin and leptin, which in turn act on the brain. An obese person may have a defect in this regulatory system so that the body thinks that it is constantly too skinny and therefore needs to gain weight. This is clearly the case in *ob/ob* and *db/db* mice, but it is not known whether defects of this sort occur in people.

Persons who are slightly overweight may have a higher set point for fat content than those on the thinner side. A person's genetic constitution need not dictate his or her weight, however, since human beings are capable of self-moderation in food intake and also of increasing their exercise. Nonetheless, someone with a genetic predisposition toward a fatter body is going to have a harder time losing weight and keeping it off than someone whose set point is lower.

Diet Fads and the Yo-Yo Effect

People who diet aren't necessarily yo-yos, but their weight certainly is prone to going up and down. Oprah Winfrey has become a national symbol of what nutrition expert Dr. Jean Mayer refers to ruefully as "the rhythm method of girth control." In 1988 she proudly displayed 67 pounds of animal fat, piled on a red wagon, to illustrate the weight she had lost following a rigorous, near-starvation diet. By 1992 she had regained it all and then some, topping the scales at 237 pounds. There have been numerous additional ups and downs. As we have seen, Oprah's "fatostat" seems to explain much of her problem, one shared by tens of millions of Americans.

One source of difficulty is that the human body—like that of other mammals—is adapted to withstand periodic famines. So, when food intake suddenly drops, it can "helpfully" reduce its basal metabolic rate, by a whopping 40 to 45 percent. Another insidious problem: during a diet, people generally lose muscle along with fat, particularly if they are relying on diet alone, without significant exercise. Then, when they put weight back on, it is largely fat. Moreover, since fat has a lower metabolic rate than muscle, their internal engine is now set at an even slower idling speed, making it yet more difficult to lose weight in the future.

Notwithstanding these problems—more likely, *because* of them—diets abound. We have already referred to the so-called low-carbohydrate diets (the "zone diet"). Other contenders include the Scarsdale diet, the drinking man's diet, the Atkins supposed "diet revolution," and so forth. In addition to potential problems with ketosis, these present the possibility of kidney damage (because high-protein foods put extra strain on the kidneys to excrete additional nitrogenous waste) as well as osteoporosis. Liquid diets can be nutritionally balanced but are necessarily low in fiber. Diuretics may give the illusion of weight loss but only because water, which is heavy, is driven out. Eventually, it comes back, as does most if not all of the weight lost through various fad diets. As a general rule, any diet that promises to help you lose large amounts of weight quickly is probably too good to be true . . . or at least to last.

Which brings us to diet and antiobesity drugs.

Maybe people are overeager to solve their problems with pills. On the other hand, certain pills really do solve certain problems, and obesity, especially when severe, is a problem indeed. (There are also buckets of money to be made by catering to the widespread American obsession with slimness and the abhorrence of body fat.) In any event, the search for antiobesity drugs and diet pills has if anything heated up in recent years. If there is a genetic and molecular component to body composition, and this seems undeniable, then it would make sense that there should be some sort of chemical treatment as well.

When considering the possibilities, however, it is important to remember that there is a large environmental contribution to weight in addition to its molecular underpinnings. People who do not exercise enough, who regularly visit fast-food restaurants (where the cuisine is heavy on fats and salt), who drink large quantities of alcohol (which has lots of calories), or for whom eating satisfies psychological needs will always have a harder time losing and keeping off weight than those with a healthier lifestyle. All the same, new diet drugs emerge regularly. Indeed, part of the reason for their popularity is undoubtedly the false hope that the right pill will make it possible to lose weight and keep it off "painlessly," which is to say, without exercise and without eating less.

A number of strategies are being pursued by scientists looking for antiobesity drugs. The first and most obvious are the appetite suppressors, which limit the amount ingested, usually by making people feel full faster. Another strategy is to reduce the amount of energy actually absorbed, no matter how much someone eats. This type of antiobesity drug would allow "dieters" to eat as much as they want without gaining weight. A third type of drug would increase a person's metabolic rate, so that more energy is used up with less activity.

The final class would stimulate the use of fat or reduce the production of fat, again without any mandatory increase in physical activity.

All these strategies have the same goal: to promote negative energy balance until a desired weight is reached. Negative energy balance occurs when the amount of energy ingested and/or absorbed is less than the amount used. It should be clear that the optimum way to lose weight is also the most difficult: eat fewer calories and/or burn up more.

Not surprisingly, people try to have it both ways, seeking the satisfaction of fatty foods, without the calories, as with "fake fat" such as Olestra. Fat-free potato chips are appealing, but, it appears, less than satisfying. And herein lies a possible explanation for the contradictory finding that even though Americans have been consuming less fat in recent years, they have also been getting *more* obese. If a moderate amount of dietary fat acts as a natural appetite suppressant (and we suspect it does), then people who overzealously reduce their fat intake may actually wind up consuming more calories—via carbohydrates—and thus gaining weight! Although nearly all Americans would benefit from eating less fat, it is difficult to have much confidence in either the long-term efficacy or the healthfulness of any drugs or food substitutes that promise weight loss in conjunction with eating poorly and exercising less.

Currently, most of the approved weight-loss drugs in the United States are appetite suppressors. The most common of these act by affecting brain neurotransmitters that mediate appetite, specifically by influencing the release and reuptake of serotonin and norepinephrine. One common treatment, recently taken off the market because of potential side effects, was fen-phen, a combination of two older drugs: fenfluramine and phentermine. Doctors are concerned about fenfluramine, which increases the release of serotonin, which in turn affects peripheral systems such as the heart in addition to the brain. As a result, another, similar antiobesity drug, dexfenfluramine (Redux), has also been recalled.

Some weight-loss drugs remain available, however. Phentermine (the *phen* in fen-phen) is still considered safe and retains some efficacy by itself, although it only affects norepinephrine and therefore its effects are limited compared with fen-phen. There is another drug, called sibutramine or Meridia, that affects serotonin and norepinephrine. Unlike Redux or fen-phen, Meridia does not stimulate the release of serotonin from nerve endings and therefore does not seem to have the same harmful side effects. The current appetite-suppressor drugs do not always work, however, and they rarely provide more than about 10 percent weight loss. There is also the ever-present problem of regaining lost weight once the drugs are discontinued. (We do not mean celebrating a successful diet by consuming a gallon of ice cream, thus sabotaging one's accomplishment, but rather, the insidious consequences of the body's set-point weight-maintenance system that we have just discussed.)

Currently there is only one weight-loss drug approved by the Food and Drug Administration (FDA) that does not affect food intake: orlistat, or Xenical. It acts by decreasing the amount of fat that can be absorbed from food, by attacking and blocking the lipases released by the pancreas. These lipases break down fats into smaller fatty acids, which can be absorbed in the small intestine. When the action of the lipases is blocked, less fat can be absorbed. (Occasional side effects include diarrhea, gas, and an indelicate tendency for "anal leakage," as undigested fats seep through the intestines.)

The most promising new direction for weight-loss drugs involves leptin, which not only reduces food intake and hunger but also directly affects metabolism and the use of fat. Clinical trials are currently under way to evaluate leptin's effectiveness in people. Other possible weight-loss targets include our old friend NPY (when its action is blocked, animals are less hungry) and the melanocortins, which, when stimulated, reduce hunger.

A new direction for weight-loss drugs targets mitochondria themselves. As described earlier, ATP is produced in mitochondria by harnessing the power of an electrochemical gradient of hydrogen ions across the mitochondrion's inner membrane. This means that the space between the two mitochondrial membranes has lots of hydrogen ions, whereas the inner matrix doesn't. ATP synthase is dependent on this difference, using the movement of hydrogen ions across the membrane (into the matrix) to provide the energy to create ATP. For this to be effective, hydrogen ions must be unable to cross the membrane by themselves; their only exit passes through ATP synthase. If they leak through without driving the formation of ATP, the energy they contain is released, instead, as heat. It turns out that hibernating animals take advantage of this leakage across the mitochondrial membrane to do just this, producing heat, which they use when reviving from their prolonged "sleep." They achieve this through use of an uncoupling protein, or UCP. UCPs simply provide a "free" passage for hydrogen ions so the gradient is abolished and heat, instead of energy-rich ATP, is produced.

Mammalian fat, as already mentioned, is generally white or faintly yellow. Sometimes, however, it is brown, and it has long been thought that UCPs were only found in this brown fat typical of hibernating animals (as well as some infant mammals). It turns out, however, that UCPs, in inactivated form, exist in all fat cells. Researchers are therefore looking into the possibility that these UCPs could be activated in the fat cells of overweight, nonhibernating *Homo sapiens*. If so, human mitochondria would become less efficient, as the membranes are made more leaky, and the person's metabolic rate would, in effect, become higher, with more fuel "burned" and some resulting weight loss. As one researcher has said, it would be like jogging without jogging. Maybe it would also be like having an instant, internal Sterno can, available to keep Minnesota ice fishermen warm while they lose weight at the same time that they angle for more calories.

PART 3

IN PERSPECTIVE

In our final three chapters, we step back a bit and try to see human beings, and life in general, in context. Much of science involves analysis rather than synthesis, breaking things down into smaller and smaller parts rather than building them up into larger wholes and seeing how everything fits together. In the process much is gained, but much is lost, too; after all, the fact that human life is part of many different, larger systems is every bit as real as the different routines and subroutines that make up each life.

When we say that *Homo sapiens* is a perfectly good mammal, we mean not only that we share the fundamental wiring and architecture of other animals but also that we are part of the structure of life itself.

ECOLOGY:

The Cloud in the Paper

"If you are a poet," writes Zen master Thich Nhat Hanh, "you will see that there is a cloud in this sheet of paper. Without a cloud, there will be no rain; without rain, the trees cannot grow; and without trees, we cannot make paper." The beloved Vietnamese monk goes on to include the logger who cut the trees, the logger's mother, and so forth.

If you, too, can see the cloud in this sheet of paper, then maybe you also are a poet, a Zen master—or an ecologist. The cornerstone ecological concept is easy to grasp although often difficult to act upon. It is also remarkably similar to the fundamental insight of Eastern mysticism: the interconnectedness of all things. British poet Francis Thompson said it this way: "All things . . . near and far, hiddenly to each other, connected are, that thou canst not stir a flower without the troubling of a star."

Regardless of who sees it, there really *is* a cloud in this sheet of paper, as well as a bark beetle, a handful of soil, a bit of bird poop, even the gasoline that powered the logger's chain saw. It is even possible that a few of the carbon atoms you are now holding were once part of Alexander the Great, a woolly mastodon, or (and!) a redheaded woodpecker before they found themselves incorporated into the loblolly pine that eventually became this page of this book.

We human animals have been stirring flowers and troubling stars for a long time, and the word for this connection is *ecology*. It derives from the Greek *oikos*, meaning "home," and it is a realization no less than a technical discipline: that all things, often hiddenly to each other, are connected. A house becomes a home when it is lived in; ecology is the study and acknowledgment of that process. For all its technical expertise and sophisticated analysis, ecological wisdom comes back, time and again, to this unitary, unifying message of interconnectedness: the mutual dependence of all things, living and nonliving, the fact that our skin doesn't separate us from the rest of the environment but joins us to it, and so there is no "out there," neither for human beings nor anything else.

A seemingly simple message, this, but in such simplicity can be great power and truth. In a famous essay titled "The Hedgehog and the Fox," the late British philosopher Isaiah Berlin used these animals and an observation by the ancient Greek poet Archilochus as a metaphor for various human styles of knowing: "The fox knows many things," Archilochus wrote more than two thousand years ago, "but the hedgehog knows one big thing."* Ecological wisdom is also derived from just one thing, but it, too, is a very good one. "Good" in two ways: for one, it is scientifically valid, and for another, it points toward something that doesn't often appear in scientific discourse, namely, that which is ethically correct as well.

In his essay "The Uses of Natural History," Ralph Waldo Emerson suggested that above all others, natural history (what would subsequently come to be known as ecology) holds lessons for humanity, and that right consisted of "conformity to the laws of nature." A century later, Aldo Leopold—founder of wildlife management and one of the gurus of what subsequently became the environmental movement—wrote that "a thing is right when it tends to preserve the integrity, stability, and beauty of the biotic community. It is wrong when it tends otherwise."

Ecology is many things: a science, a worldview, a cautionary tale. It can be nearly incomprehensible in its mathematical modeling, downright tedious in its verbal pomposity, verging on the poetic and

---

* *In its original usage, the fox referred to Athens, a multitalented city, and the hedgehog to Sparta. Like a hedgehog, Sparta knew but one way of defending itself . . . but did so very effectively.*

mystical in its insights, theoretically abstruse yet dirty-under-the-fingernails practical. Although it is often ignored and derided, it is desperately needed as a voice for basic planetary hygiene and a corrective to human hubris.

Interconnectedness: A Case Study

At one point in *The Origin of Species*, Darwin speculated playfully that by keeping cats, English spinsters made London a more pleasant and flower-ful place. Here's how he explained it: Cats, as everyone knows, eat mice. Mice, as fewer people realize, occasionally destroy the nests of bumblebees, which are typically dug into the ground. And bumblebees, of course, pollinate flowers. So: More cats, fewer mice. Fewer mice, more bumblebees. More bumblebees, more flowers. Therefore: more cat lovers = more flowers.

To our knowledge, no one has ever tested Darwin's proposal. But many other such connections have been elucidated. Here is just one, notable for, among other things, being carefully worked out over a period of three years, involving Herculean efforts at ecological field experimentation, and having unanticipated implications for human health.

The project was originally conceived to investigate the causes of periodic infestations of gypsy moths, an introduced pest from Europe that, about one year in ten, causes great damage to forests in eastern North America. Researchers suspected there might be a connection between these gypsy-moth outbreaks and the abundance of acorns, because of the intervention of another species: white-footed deer mice, common rodents of eastern forests, whose populations, like those of the gypsy moths, tend to fluctuate. The abundance of deer mice seems to be influenced by the availability of acorns; in fact, deer-mouse numbers skyrocket following a good acorn crop, which happens about every two to five years. And white-footed deer mice don't only eat acorns. They are also major predators on the pupae of gypsy moths. So, it seemed reasonable that in the immediate aftermath of a heavy acorn production there would be a large mouse population, which, in turn, would keep the gypsy moths in check. Similarly, following poor acorn years, there should be relatively few mice and therefore relatively more gypsy moths.

During the summer of 1995, a year after the most recent heavy acorn crop, mice were abundant in an upstate New York forest. The ecologists removed most of the mice from three patches of forest, each measuring about 2.7 hectares* (no small task!). Then they compared the numbers of gypsy-moth pupae in these experimental areas with comparable forest plots from which mice had not been removed. Sure enough, fewer mice yielded more moths; forty-five times more, in fact. The researchers were even able to identify mouse tooth marks on freeze-dried pupae left as bait.

The next step was to simulate a banner acorn year. With the help of a local Girl Scout troop, they scattered nearly four tons of acorns over the experimental, mouse-depleted plots. Mice quickly repopulated the area; in fact, mouse populations went through the roof, demonstrating that more acorns indeed means more mice. But the story isn't over yet.

White-footed deer mice not only eat acorns and gypsy-moth larvae, they are also a major reservoir for the parasitic organism that causes Lyme disease (an increasingly troublesome illness that afflicts people). Infected mice are bitten by tick larvae; these larvae find their way to deer, whereupon they develop into reproducing adult ticks, which are then capable of transmitting Lyme disease to people, among other mammals. The ecologists found that following a good acorn crop, there was, in conjunction with the increased mouse population, an eightfold increase in tick larvae. There were also heightened numbers of deer, attracted by the acorns, who brought along their burden of adult ticks, which bred and bestowed their larvae upon the flourishing mice. Mice occupying acorn-enriched plots had about 40 percent more tick larvae than did mice in normal, control forests.

Are there any practical implications to be derived from all this? First, a warning to the medical community: it seems likely—but is not yet proven—that Lyme disease outbreaks will be especially likely in so-called mast years, when acorn production is unusually high. (Remember, more acorns means more mice, more deer, and more ticks.) Second, foresters might be tempted to try inhibiting gypsy-moth outbreaks, thus improving lumber yields, by distributing additional acorns or some other mouse food. But this might bring about Lyme disease epidemics! Alternatively, public-health officials, wanting to reduce Lyme disease, might look into various ways of chemically suppressing mast production. But this might bring about gypsy-moth infestations!

A hectare is equivalent to a square 100 meters on a side.

Probably the most valuable take-home lesson from this saga is the most practical of all: "When we try to pick out anything by itself," wrote John Muir, "we find it hitched to everything else in the universe." Therefore, maybe we shouldn't pick too hard. Or, as an old television advertisement used to claim, "It's not nice to fool Mother Nature."*

Food Webs and Trophic Levels

Among the most important images of the late twentieth century—perhaps of all time—are those renowned photographs of the planet Earth, taken by the Apollo astronauts from space. The picture of that small, precious blue sphere emphasized the uniqueness of Earth and the fact that good planets are hard to find. It also highlighted that for all practical purposes, the earth is a closed system. There is the occasional meteorite and a controversial theory that comets regularly splatter the atmosphere with water in the form of celestial snowballs, but when it comes to matter generally, we are pretty much stuck with what we have.

Two important ecological concepts can be derived from this. One is the reality of recycling, not just the contrived reuse of plastics, glass, aluminum, or paper, designed to reduce the human wastage of natural resources, but rather the much larger and deeper reality of how the materials of planet Earth are naturally shuffled around in vast systems of biogeochemical cycling. We'll look at this shortly. The other take-home lesson is that whereas stuff is constantly recycled, energy—unlike matter—is not intrinsic to our shared planet. Regarding energy, the earth is an open system, with consequences that are not only theoretically fascinating but practical and immediate as well.

Some additions to Earth's energy budget come from internal sources of heat (volcanoes, hot springs, ocean vents, radioactive decay), but for practical purposes, the sun powers our planet. The solar energy striking the earth is immense, equivalent to 100 million Hiroshima bombs every day. Most is absorbed or reflected by clouds, water, snow and ice, or bare ground, with only a feeble 1 percent even reaching Earth's green things. And chlorophyll captures only about 2 percent of this! On balance, then, less than .02 percent

---

* Or to fool around with.

(2 percent of 1 percent) of the energy reaching the earth from the sun is converted to chemical—that is, usable—energy. But it is a crucial percentage, responsible for everything that lives.

The key process by which life grabs energy and turns it into more living matter is photosynthesis: specifically, taking carbon dioxide and water, adding light (hence *photo*), and making the sugar, glucose (hence *synthesis*). All organisms ultimately owe their existence to the green mantle of plant life, whether terrestrial, marine, or aquatic, that performs this crucial capture; if a living thing is not a plant itself, then it eats plants, or eats the eaters of plants, and so forth. Thus, we are all vegetarians at heart.

Plants are the primary ecological producers, known to ecologists as the first trophic level (from the Greek *trophos*, meaning "feeder," and in a sense, plants "feed" on photons). Then come the animals, initially the herbivores: grasshoppers, ants, bugs, seed-eating birds, grass-eating sheep, tree-eating elephants, algae-straining whales; these so-called primary consumers occupy the second trophic level. On the next, tertiary trophic level are the carnivores, who eat the flesh of the herbivores; these include songbird and snake, bobcat and wolf, lion and tiger. In many complex systems, there can even be quaternary consumers. For example, algae in a Georgia pond are consumed by daphnia (water fleas), which are eaten by small sunfish, which are eaten by striped bass, which are eaten by green herons, whose eggs are eaten by cottonmouth moccasins (snakes), which in turn are consumed by osprey, and on it goes.

Add to all this the nontrivial existence of frogs, ducks, snails, and insects, and what had started as a food chain quickly becomes a food web and then something more like a dense network of interwoven connections, with the same living thing often occupying different levels at the same time. Human beings, for example, are sometimes secondary consumers, as when we eat chicken or steak. But we are also tertiary consumers or higher, when eating, for example, swordfish (which eat smaller fish, such as mackerel, which in turn live on yet smaller fish). When a snake eats a mouse, it is being a secondary consumer; when it eats a shrew, it is a tertiary consumer. And what is a Venus flytrap, when it both photosynthesizes and eats a fly? "Oh what a tangled web we weave, when first we practice to deceive!" noted Sir Walter Scott in *Lochinvar*. More tangled yet is the web that nature weaves, deception or no. It may be tedious to memorize the warp and woof of such relationships, but it is glorious—and unavoidable—to be part of them.

Food webs are also cycles. Unlike energy—which comes to us from on high, specifically as photons from the sun, and is thus constantly replenished from beyond this planet—when it comes to the physical, material stuff of life, there is no manna from heaven. Accordingly, there is also the less-than-elegant side of things, those slugs, beetles, fungi, bacteria, and their hardworking colleagues among the decomposers, who make sure that what goes around comes around . . . literally. They don't get much respect, but they keep the equation balanced. And when, in the closing image of Gary Larson's hilarious eco-parable *There's a Hair in My Dirt*, a friendly family of worms looks smilingly up at the reader as though to say, "We'll be seeing you soon," they are factually and ecologically right.

But this isn't the only practical consequence that ecological food webs hold for *Homo sapiens*. There is what we might call "pyramid power," not the feel-good delusion of New Age crackpot mystics but the genuine ecological physics of how energy is transmitted through different trophic levels.

Let's go back to Newton, specifically his second law of thermodynamics, the one that explains how the world tends toward increased entropy or disorder. According to this law, inefficiency is rampant in the real world, which is why, for example, it is impossible to construct a perpetual-motion machine: no matter how you start it going and how smoothly it functions, eventually such a machine will run down, because of that ever-present inefficiency known as friction. A similar kind of ecological friction operates in the biological world, between trophic levels in particular.

The efficiency of energy transfer between trophic levels, although higher than the 1 to 2 percent from sunlight to green plants, is nonetheless appallingly low, about 10 percent. In other words, about 90 percent of usable energy is lost in the step from, say, primary producers to primary consumers, then again from primary consumers to secondary consumers, and so on. Why this high rate of energy loss? There are many reasons. For one, not all potential food gets eaten. Most leaves remain unconsumed, which, after all, is why the world looks green. Similarly for most caterpillars, fish, mice, and moose—indeed, all those creatures that have not yet gone extinct. For another, animals have other things to do than just sit around, eat, and transform what they have eaten into more of themselves; they walk, trot, run, fly, and swim; they court, mate, and rear offspring; they often must exert themselves to get food, avoid predators, keep out of the sun or the rain, migrate, and scratch when they itch, not to mention the basic metabolic costs of keeping their motors idling, which is to say, remaining alive, which itself is a constant and ultimately losing battle against the ubiquitous second law. Moreover, only energy stored as growth is actually available for the next level of consumers. This is why, despite the extraordinary complexity of food webs, most natural communities do not have more than four or five trophic levels: if you keep decreasing by a factor of ten, you very quickly run out of available "stuff."

It is also why, for most ecosystems,* the number of individuals as well as the amount of calories they represent decreases dramatically from primary producers to primary consumers, secondary consumers, and so on. And it is why this ecological pyramid is generally steep, each level being about one tenth that of the one below it. The Alaskan tundra, for example, will have vast amounts and numbers of grasses, lichens, sedges, and dwarf willows. Just above them are lemmings, arctic hares, and caribou, lots of them, but not so many (in numbers) or so much (in biomass) as the plants on which they—literally and figuratively—stand. Then there are the snowy owls, lynxes, arctic foxes, and wolves that consume the consumers, and perhaps a very few wolverines and others that occasionally eat these predators. Not surprisingly, there aren't any critters that make a living by eating eagles, orcas, or wolverines; there simply isn't enough eagle meat, for example, to balance yet another layer of the ecological pyramid. (The only exceptions, interestingly, are parasites and other small, indwelling creatures: for them, a single eagle can be a universe of potential food.)

Ecologist Paul Colinvaux wrote a useful book titled *Why Big Fierce Predators Are Rare*, and ecological pyramid power—the tenfold diminution in energy transfer—is why. It is why if you walk in any natural environment, you are almost certain to see green stuff and may well see at least a few herbivores but unless you are very patient, skilled, or lucky, you will probably not see a single carnivore. It is why there are many more caribou and moose than wolves (even though a wolf is smaller than its prey), why it takes huge herds of wildebeest and zebra to support comparatively meager numbers of lions and leopards, why tuna is so much more expensive than mackerel. It also helps pinpoint one of the major issues in wildlife conservation: namely, the fact that so-called charismatic mega-fauna, the large, flashy animals such as tigers, cheetahs, or orcas, tend to be few and far between.

---

* Pronounced "eeko," not "echo"

It should be clear that trophic ecology is not something to ignore, even though—or perhaps because—human beings are not easily classifiable in the world's multistep, intersecting biological pyramids. Specifically, we need to visit two major trophic consequences, both of them especially relevant to *Homo sapiens*: biological intensification and hunger. For biological intensification, sometimes called biological magnification, the 10 percent rule is especially relevant. Although organic chemicals are constantly being broken down and often excreted as they pass from body to body, some are very long-lived and slow-moving, notably the chlorinated hydrocarbon pesticides (DDT and its relatives, including PCBs), heavy metals such as mercury or lead, and certain radioactive isotopes such as strontium-90 and iodine-131.

What does this have to do with trophic levels? Consider DDT sprayed on plants. These plants are then eaten by mice; because of the inefficiency of energy transfer, one mouse, in the course of its lifetime, will likely consume many times its weight in plant stuff. And while the mouse's digestive system goes about rearranging 100 grams of plant matter into 10 grams of mouse meat, DDT is broken down so slowly that it accumulates in the mouse's fat and liver, concentrated about tenfold in the life of each mouse.

Enter a hawk. Just as a mouse eats many plants, this hawk will likely eat many mice, and for the same reason that DDT accumulated in mice, it will accumulate in the hawk. Let's say we started with 1 gram of DDT per 1 million grams of plant matter; this concentration was magnified to perhaps 1 gram per 100,000 grams of mouse meat, which in turn became about 1 gram per 10,000 grams of hawk meat. By this point, a seemingly innocuous amount of DDT has become intensified to a dangerous level, enough to disrupt the hawk's ability to reproduce.

We humans are similarly vulnerable, especially insofar as we exist high on the food chain: the higher a creature feeds, the greater the intensification. For example, mercury that was dumped into Minamata Bay in Japan was in an insoluble, inorganic form. However, it accumulated in the mud at the bottom of the bay, where it was eventually converted into an organic compound, methyl mercury, by the action of bacteria living there. Thus enabled to enter living organisms, the methyl mercury didn't simply pass up the food chain, it was *magnified* along its way: absorbed by plankton and diatoms, then concentrated in small invertebrates, and then more so among small fish, to larger fish, eventually to human beings who ate those upper-food-chain-dwelling predators and got a whopping dose of mercury as an unintended bonus. The result was mercury poisoning, with brain damage and often death. Significantly, those creatures lower in the food chain, although victimized by the predators above them, were not victimized by the poison.

One lesson to be learned from all this is to avoid broadcasting poisons into the environment, if only because every environment is ultimately a human environment. Another is to eat lower on the food chain. It's a way to ingest less of someone else's compacted trash.

Hunger, too, is intimately connected to the matter—and energy—of ecological trophic levels.

An average human being consists of about 150 pounds of meat. Going back to our 10 percent rule, this requires consuming something on the order of 1,500 pounds of cow meat, which would in turn need to have eaten about 15,000 pounds of grain. If, instead, that person consumed the grain directly, the middleman (or middle cow) would be cut out, thereby permitting ten times as many people to be fed. This is not an argument for increasing the human population but for taking better care of those we already have. In a nutshell: you can feed ten times more people, or the same number of people ten times as well, if they eat rice, wheat, corn, and barley instead of fried chicken, T-bone steaks, lamb chops, and bacon. And because of the diminished biological intensification, not to mention the reduced fat consumption, they would also be healthier.

Pyramid power, in this sense, also explains why the agricultural revolution—which transformed the human diet from that of a hunter-gatherer to one based on crops and domesticated animals—was so important, and why it was responsible for a massive jump in human population around 4000 BCE. This greatest of all technological advances enabled *Homo sapiens* to eat *lower* on the food chain, although ironically, many of our efforts ever since have been directed (rather, misdirected) toward eating higher.

Cycles

Einstein pointed out that matter and energy are interconvertible, and we are not going to argue with him. Relativity works great when it comes to the realm of incomprehensibly large things or those traveling at remarkable speeds. At the level of ecological processes, however, matter and energy are effectively quite separate. Moreover, they follow different pathways. The sun's energy, transformed into plant stuff, may be changed into animal (including human) stuff, until death and decomposition. Thanks to the ever-present sun, energy is always arriving, some of it corralled into creating and maintaining organic complexity. But matter is, well, a different matter.

The basic chemical building blocks present on and in the earth are pretty much what there is, was, and will be. New hydrogen, oxygen, nitrogen, and carbon are not created. Unlike energy, fundamental substances are not continually replenished, at least not from the "outside." Rather, they are recycled, either on a vast global scale (in the case of water, carbon, and oxygen) or as part of more local systems (as with phosphorus, potassium, sulfur, etc.).

Biogeochemical cycles have been studied and worked out in great detail; they are different for each substance and crucial for everything that lives. Like body processes such as digestion, respiration, or nerve conduction, global ecological cycles mostly receive attention when they are disrupted; when in good working order, they are largely taken for granted. In historical time at least, disruptions of this sort have been mostly human-caused; the technical term is *anthropogenic.*

Of all the global biogeochemical cycles, carbon is the one most disturbed by human activity. In its nonliving incarnation, carbon is stored especially as carbonate minerals in rocks and as dissolved bicarbonate and carbonate ions in the oceans. It is also intimately connected with the great cycle of life, in that organic compounds are all carbon-based: organic chemistry is the study of carbon, upon which all life depends. During photosynthesis, carbon dioxide from the atmosphere is converted into organically usable carbon in the form of glucose, then into other carbohydrates, which, when eaten, become incorporated into animals, in the grand trophic trophy hunt.

The carbon loop has a return part, too. Thus, all living things, plants included, cycle carbon back into the atmosphere as a result of respiration. Essentially, the carbon present in energy-containing foods, notably the simple sugar glucose, is oxidized to carbon dioxide when that energy is released during cellular respiration. Decomposition of dead animals and especially plants adds yet more carbon dioxide. Inorganic processes also contribute: the gradual erosion of limestone returns comparatively large amounts of carbon back into carbon dioxide gas. But it is the fourth component of the carbon-return cycle that is especially troublesome, because it is especially recent: combustion.

In the prehistoric past, huge amounts of organic carbon were locked away when plants died in a low-oxygen environment and were converted into fossil fuels: natural gas, coal, and oil. A new balance in the carbon cycle was established while this material was literally out of circulation. Then, since the industrial revolution, vast quantities of these fuels have been burned, adding greatly and suddenly, as ecological time is reckoned, to the carbon dioxide (CO_2) burden of the environment. Something like 5 to 6 billion metric tons of CO_2 are released annually via the combustion of fossil fuels. (An automobile coughs out roughly its weight in carbon dioxide during one year.) In absolute amount, carbon dioxide could actually be considered a rare gas, since it comprises only about .035 percent of the total atmosphere, which is 79 percent nitrogen, 20 percent oxygen, and less than 1 percent everything else. There is nearly thirty times more argon in the atmosphere than carbon dioxide. But argon is not increasing, nor does it have much effect on anything.

On the other hand, the carbon dioxide load has been rising, from about 265 ppm (parts per million) around 1840 to 350 ppm today. At the present rate of increase, it will reach nearly 600 ppm by the middle of the twenty-first century. And it is clear that carbon dioxide has lots of effects on practically everything.

Eventually, carbon dioxide levels will reach a new equilibrium, particularly as the gas is transferred to the oceans and deposited in sediments as calcium carbonate. But the rate of such carbon lockup is very slow compared to the rate at which carbon dioxide is being added, anthropogenically, to the world ecosystem. Aside from the combustion of fossil fuels, yet more CO_2 comes from the burning of tropical forests, especially in the Amazon and Indonesia; this is doubly pernicious, because in addition to contributing carbon dioxide, such destruction affects the lungs of the earth, since trees play a major role in removing atmospheric CO_2.

The primary downside of increased atmospheric carbon dioxide is global warming, a connection that is not intuitively obvious. The phrase "greenhouse effect" helps clarify the problem. Carbon dioxide in the atmosphere works much like glass in a greenhouse, which creates a warm environment for the plants inside because it is essentially a one-way street for heat energy. Of the sun's energy that reaches the earth, some is absorbed as heat and some is used by plants to build carbon-based organic compounds, but a lot is reflected back into the atmosphere. Like the glass of a greenhouse, atmospheric CO_2 is permeable to the sun's incoming energy (which enters as light) but comparatively resistant to the passage of infrared radiation (heat), and it is as infrared that most of the reflected energy bounces back from the earth.

The glass of a greenhouse produces its effect by permitting sunlight to enter, then refusing to let the reflected infrared get back out.

Carbon dioxide has a similar effect in the normal atmosphere. In fact, if it weren't for the atmosphere's carbon dioxide, reabsorbing infrared energy that would otherwise be lost in space, our planet would be plunged to a frigid and life-denying average temperature of −18 degrees Celsius!

Cloud cover, like carbon dioxide, also retains heat. At high latitudes in particular, the coldest winter nights are associated with clear skies, because without clouds to reabsorb it, heat is reflected away from the earth. For another common occurrence very much like the greenhouse effect, consider what happens if you leave an automobile with the windows rolled up, even on a moderately warm day: the car gets remarkably hot, because the glass allows energy to enter (as light) but keeps it from leaving (as heat).

Carbon dioxide isn't the only greenhouse gas. Nitrous oxide ("laughing gas," sometimes used by dentists, also produced when atmospheric nitrogen is oxidized), chlorofluorocarbons (CFCs), and ozone have a similar effect. So does methane, produced in remarkably large quantities by farting cows. But CO_2 is clearly the major culprit. And even small amounts of "greenhouse warming," on the order of a few degrees, is guaranteed to have disastrous consequences for life on earth. For example, there would be major shifts in agricultural productivity, almost certainly for the worse, although perhaps Canada and Russia would become good places for growing pineapples and oranges. As the polar ice caps melt, the oceans would rise, dooming low-lying islands and wreaking havoc on the world's great cities, most of which are at sea level. Maybe the oceans would obligingly absorb some of the excess CO_2 more quickly than they do at present, and maybe some forests would pitch in as well, and grow more rapidly. But on balance, it is distressingly likely that greenhouse warming is leading toward potential global catastrophe.*

Many of the major pollution problems confronting the world are associated with other disruptions to biogeochemical cycles, often on a more local scale than that of CO_2. For example, the nitrogen cycle,

---

* We must acknowledge that there are some scientists who insist that greenhouse warming is nothing to worry about. But they are very much in the minority. And our review of the situation suggests that they fall into one of two predictable categories: (1) the tiny number of cantankerous contrarians who can be found in any walk of life, on any issue, and (2) those otherwise mainstream professionals who receive financial backing from the fossil-fuel industry, automakers, manufacturing associations, chambers of commerce, and other compromised sources.

long a bane of beginning biology students because of the various kinds of bacteria involved, which all sound the same yet do different things (nitrifying, denitrifying, nitrogen-fixing), is a key global process often in regional disequilibrium. Atmospheric nitrogen (N_2) is abundant, making up about three fourths of the atmosphere, but it is also chemically sluggish, requiring specialized nitrogen-fixing bacteria to convert it into usable form. At the same time, nitrogen is an essential component of proteins, thus something that living things desperately need. The abundance of "biostuff," technically "biomass," is therefore often limited by nitrogen availability. Enter chemical fertilizers. Unknown 150 years ago, they add nearly 100 million metric tons of nitrogen to the planet each year. Natural denitrification (which returns nitrogen to its less active, molecular, atmospheric form) can't keep up with this vast amount of so-called reactive nitrogen added to the environment.

The result, in areas of excessive fertilizer use, is eutrophication (from the Greek, meaning "good food"), whereby lakes and rivers choke on too much of a good thing. Another significant source of local excess in reactive nitrogen is hog and poultry farms, which deposit huge amounts of nitrogen-rich pig and chicken manure into ecosystems unprepared for such an influx. Algae, cyanobacteria, and shallow-water weeds bloom, then die, killing many fish as well, all of which produces a tremendous blast of organic material that feeds decomposers, which deplete the water of remaining oxygen, causing further death and deterioration. Instead of trout, which require clear, oxygen-rich water, we get—if we're lucky—carp and catfish. Instead of mayflies, blood worms. Instead of clean water, pea soup. (Everything is connected to everything else, remember?)

Problems with the nitrogen cycle are exaggerated by further additions of phosphorus, which normally has its own rather ponderously slow cycle but which is given a troublesome boost via fertilizers and detergents. The phosphorus cycle does not have an atmospheric component. Instead, in terrestrial ecosystems, phosphorus—necessary for making such molecules as ATP and DNA—is recycled through the various food webs. It begins as phosphates in rock, which are then released by weathering and incorporated into plants and eventually animals, ultimately traveling downstream to the oceans where they may gradually be recycled via vertebrate skeletons and mollusk shells. For many years, before the advent of chemical fertilizers, the major way phosphorus was returned from ocean to land was as fertilizer via the guano of tens of millions of fish-eating seabirds, especially concentrated off the west coast of South America, notably Peru.

At first blush, it might seem that adding to organic productivity would generally be a good thing—making the deserts bloom and all that. But there is a remarkable delicacy to ecological systems, if only because everything is tied to things as they currently are. Ozone is similarly confusing, a chemical version of Dr. Jekyll and Mr. Hyde. Thus, in its Mr. Hyde form, when it hovers at low elevations—such as in the Los Angeles basin—it is a major component of photochemical smog. But in the upper atmosphere, ozone is extraordinarily important in absorbing much of the sun's ultraviolet light and keeping the earth livable: Dr. Jekyll.

Whereas molecular oxygen is O_2, made up of two atoms of oxygen, ozone is O_3. Ozone can switch to molecular oxygen, and vice versa, under complex conditions of atmospheric chemistry. Basically, when ultraviolet radiation strikes an ozone molecule, the energy is absorbed and the ozone briefly transformed into molecular O_2 and an unpaired atom of oxygen. This oxygen "free radical" is highly unstable; within seconds, it recombines with O_2 to produce more ozone, which is ready to absorb yet more ultraviolet energy.

Here an unanticipated anomaly of chemistry rears its head: it turns out that certain chemicals, chlorine in particular, interfere with this high-speed, high-altitude ozone cycle. The particular culprits are chlorofluorocarbons (CFCs), notably freon, which—like DDT in its heyday—was originally seen as a wonder chemical. CFCs are highly stable, nontoxic to human beings, relatively cheap, and immensely useful in refrigeration systems and as aerosol propellants. Not realized, however, was that these compounds would rise into the upper atmosphere, break down into chlorine, and eventually wreak havoc with the ozone layer, creating the now-notorious hole over Antarctica and a counterpart in the Northern Hemisphere.*

The threatened results? Heightened rates of skin cancer, cataracts, possible blinding of pollinating insects (which are especially sensitive to ultraviolet), maybe even incapacitation of photosynthetic plankton.

---

* In its harmful, Mr. Hyde mode, ozone is created from auto emissions in particular by two chemical reactions. First, nitrous oxide (NO_2) reacts with molecular oxygen (O_2) to produce nitric oxide (NO) and free-radical oxygen (O). The latter then reacts with O_2 to yield ozone (O_3), just as in its Jekyll-like stratospheric incarnation. The problem, of course, is that ozone causes respiratory distress when inhaled. There might be some poetic justice if heightened levels of ultraviolet radiation—reaching the earth because the high-altitude ozone layer is being depleted—in turn facilitate the breakup of low-altitude photochemical-smog ozone back into O_2 and free-radical oxygen, thereby re-creating the previous balance. But don't count on it.

The details aren't clear, but the likelihood is that by randomly disrupting a system to which much of the planet's life is already adapted, such anthropogenic intrusions will make things worse, not better. The industrialized nations have agreed to phase out CFC production; this has caused the *rate* of CFC increase to diminish. But since CFCs are very stable and rise into the upper atmosphere quite slowly, the actual amount of atmospheric CFC is *increasing* and will continue doing so for at least several decades.

Then there is acid rain, produced when the burning of fossil fuels releases nitrogen and sulfur oxides into the air; these in turn are converted into sulfuric and nitric acids, which sterilize lakes and stunt forests. The problem is especially severe downwind of major sources of industrial pollution: in New England, for example, which is atmospherically downstream of the Ohio Valley. Some regions are blessed with calcium carbonate in their rocks; this chemical essentially absorbs excess acidity in fresh water, buffering local water systems against the worst effects of acid rain. But in New England and upstate New York, the rocks tend to be granite and basalt rather than limestone, and so the lakes are unable to neutralize the low pH. The forests and lakes of Scandinavia, affected by industry in Germany and Great Britain, have been experiencing a similar fate.

Sadly, there are many more environmental horror stories. Happily, they are at least becoming understood, and more information as well as some possible solutions are available. Our point, for now, is that effective citizenship requires informed bioliteracy, not just about the internal workings of our own mammalian bodies but also about the interconnecting natural world of which we are an important and increasingly disruptive part.

The Tragedy of the Commons

There is an episode in English history—pointed out by ecologist Garrett Hardin—that sheds light on a major factor underlying human-caused environmental problems. The tradition in Great Britain had long been that some grazing lands were privately owned, while the remainder were common property of the community at large. Citizens were free, for example, to graze sheep on their own private lands or on the public "commons." It was well known that overgrazing was harmful to the productivity of any grassland, so sheep owners generally avoided overgrazing their own property. But the commons were treated differently: all recognized that healthy commons were of benefit to everyone, but each also reasoned that even if he refrained from grazing sheep on the commons, there was nothing to inhibit the others from fattening their sheep on public land.

Those sheep owners who tried to preserve the commons suffered degradation of their own holdings, while those who were less socially and ecologically responsible profited by overgrazing the commons instead. As a result, tendencies to be prudent and ecologically minded were suppressed because individual sheep owners reasoned that as long as someone was going to take advantage of the commons, it may as well be them! The result was deterioration of the commons until they were no longer fit to support sheep or shepherds.

The tragedy of the commons, then, is that individuals—or corporations, or governments—each seeking "private" benefit, find themselves behaving in ways ultimately disadvantageous to everyone. The principle can be generalized to cover a wide range of circumstances whenever short-term, selfish benefit conflicts with long-term public good. For example, there may be a selfish reason for a factory owner to use the atmosphere as a public sewer; after all, if his effluents pollute the air, the cost will be borne more or less equally by everyone who breathes, whereas the factory owner personally is saved the expense of having to install pollution-control devices.

The same is true when we overuse scarce resources: it may be inconvenient to recycle or to use only our share of scarce commodities. After all, we may derive some immediate personal gain or comfort by doing so, while the cost—in overcrowded dumps or worldwide resource shortages—is by contrast diffuse and general. Besides, if we don't abuse the environment, surely someone else will!

The tragedy of the commons has global dimensions, too. As just described, Scandinavian forests and lakes suffer from acid rain because of English smokestacks, while England gets the immediate economic benefit. Japan and Norway have defied international outcry while hunting the world's great whales to the verge of extinction (the great whales are, in a sense, "commons," vulnerable to selfish exploitation). Brazil seeks to benefit economically from the Amazon rain forest, even though such "benefit" requires that it be destroyed, to the ultimate detriment of everyone.

There are two apparent solutions to the tragedy of the commons, and interestingly, they point in different directions. One is enhanced privatization: if, the argument goes, people have ownership of a resource, they treat it less like a commons and thus more respectfully.

The alternative is the exact opposite: since people are tempted to degrade a publicly owned asset and rarely think beyond short-term maximization of profits and other gains, it is necessary for governments to impose obligations and restrictions, rules of conduct designed to assure that everyone behaves responsibly, because left to themselves, most *Homo sapiens* are less than truly sapient, only rarely acknowledging their connection to everything else.

Community Interactions

We hear a lot about communities these days, whether the community of scholars, the gay community, the African-American community, even—something of an oxymoron—the "world community." In our earlier example of acorns, mice, gypsy moths, and Lyme disease, there seemed little doubt that Everything Touches Everything Else (call it the ETEE principle). But it is still possible that every component of this system is basically independent, connected only fortuitously to the others, each of which is also behaving individualistically. The players in any particular ETEE drama undeniably influence one another, but it is not certain that they have evolved together to constitute a biological community.

Now we turn briefly to a major scientific enterprise known as community ecology, which focuses on the degree to which different species are precisely sculpted by their interactions with one another, resulting in the kind of lock-and-key interacting systems that ecologists mean by a "community." In such cases, natural selection appears to be acting on individuals, but the individuals of other species comprise an important part of the selective environment experienced by each; the resulting "coevolution" can only be understood by looking at their mutual, interacting impacts. In other words, ETEE.

Take this example. Passionflower vines produce toxic chemicals and are therefore avoided by most herbivorous insects. There is, however, one notable exception: *Heliconius* butterflies, which have evolved the ability to tolerate the chemicals. But *Heliconius* caterpillars are ravenous eaters, and each one can survive better if there aren't too many others competing with it on a given patch of passionflower leaves. So, the adult butterflies have evolved a nifty way of minimizing competition. Females avoid depositing their eggs—conspicuous yellow beanlike structures—on leaves that are already occupied. So far, so good . . . for the butterflies.

But passionflower plants aren't simply passive victims of *Heliconius* butterflies (which, by the way, seem to have avoided their own tragedy of the commons). In some species, passionflower leaves are regularly outfitted with conspicuous yellow beanlike structures, which resemble *Heliconius* eggs but aren't; the mimicry is good enough, however, to induce some of the credulous butterflies to lay their eggs elsewhere! Not only that, but these egg mimics are also provided with a nectar source that is attractive to certain ants and wasps that prey on *Heliconius* eggs and caterpillars. It appears that the mere presence of such predatory insects is enough to divert a *Heliconius* butterfly from her egg-laying mission.

Complex community interactions also have practical environmental consequences. Here is a true story that harkens back to the mid-1950s, before biological magnification was understood and when DDT was being hailed as a potential savior of humankind. It tells why the World Health Organization (WHO) found itself parachuting cats into remote villages in Borneo. The saga begins when the WHO, hoping to control malaria (which is carried by mosquitoes), sprayed large amounts of DDT in the Borneo backcountry. It happens that a species of caterpillar fed on the thatched huts of rural villages; these insects were hardly affected by the poison, but a predatory wasp that fed on the caterpillars was quite vulnerable. (Don't forget biological magnification.) The wasps died. With their primary predators removed, the caterpillar population soared and village huts began falling down. At the same time, gecko lizards, which also ate insects and thus concentrated their DDT levels, were poisoned as well, along with the village cats that ate the lizards. The rat population skyrocketed, introducing danger of plague. The chastened WHO sought to make amends by airlifting uncontaminated cats.

Ecological wisdom encompasses an immense range of interspecies interactions. Predators are closely adapted to their prey, and prey are equipped with an array of counteradaptations, from camouflage to deceptive coloration that makes them resemble anything from a bird turd to their predators' predators. Of course, species interact in a huge variety of other ways, too. They cooperate, compete, act as part of one another's "selective environment" in a complex composite of ways that ecologists are just beginning to unravel.

A crucial concept, as relevant to *Homo sapiens* as to every other species, is the niche. The word may evoke an image of a small, cozy cubbyhole, but the concept of ecological niche is quite different. If habitat is an organism's address, its niche is its profession. But *niche* implies more than simply how a living thing earns its keep (what it eats, for example, and how). It also includes what eats it, what interacts with it in any way and at any time, and well as the sum of all its nonliving environmental requirements (temperature, pH, moisture, etc.). One now-classic treatment described each species' niche as an "*n*-dimensional hypervolume," meaning that it has as many different components as one can measure or even imagine, and then at least a few more!

Niche theory has implications for the way natural communities are put together, for extinction, and for species diversity. Human beings, too, occupy a niche, although it is one whose boundaries are huge, shifting, and difficult to define.

As a general ecological principle, *intra*species competition (within a species) is more immediately apparent and significant than *inter*species competition (between different species). Thus, males of the same species often compete for females; females of the same species may compete for access to nesting sites; and males and females of the same species will often compete for food during the nonbreeding season (consider chickadees milling about a birdfeeder during the winter). All of this helps determine the makeup of each species. At the same time, however, there is *inter*species competition, especially important in determining the makeup of communities.

Archimedes is said to have been the first to put forth the basic principle that two different objects cannot occupy the same place at the same time (connected, incidentally, to his insights into why things float). Its ecological equivalent was first postulated by two mathematically inclined biologists, A. J. Lotka and V. Volterra, who predicted that for two or more species to coexist, their ecological requirements would have to be at least slightly different. Otherwise, reasoned Lotka and Volterra, even a tiny advantage, over time, would result in the superior species replacing the inferior.

This was later tested by the Russian microbiologist G. F. Gauss. He found that two different species of paramecium, each capable of thriving when kept by itself, could not coexist in the same simple laboratory system, even if supplied with abundant food. If the two species competed significantly—that is, if there was a high degree of "niche overlap"—then eventually one eliminated the other. The loser was "competitively excluded." Hence, the competitive exclusion principle. When Gauss introduced two paramecium species whose niches were different, in that one fed at the bottom of the flask while the other lived in midsuspension, they were able to coexist.

But what happens in nature? Often the contenders divide up the environment, each extending the tendrils of its "niche space" to places that are not entirely overlapped by another. Via "resource partitioning" of this sort, several different species with needs that are similar but not identical are able to coexist, resulting in the species diversity that makes the living world so rich and interesting. This speaks to the importance of habitats being heterogeneous rather than homogeneous; the greater the patchiness, the greater the number of different species that can be accommodated, without excluding one another. Simple habitats are, in this sense, comparatively boring not only in how they look but in the biological diversity they can support, via resource partitioning. As far as we know, resources are not partitioned as a result of general agreement among the participating partitioners but because natural selection acts on individuals, favoring those that experience less competition, giving a reproductive edge to any that are a little bit different from their competitors.

Imagine two species of North American warblers both occupying spruce trees. Like Gauss's paramecia, they would presumably compete, and perhaps over time only one species would persevere. But spruce trees, unlike laboratory flasks filled with bacteria, are pretty complicated places, heterogeneous and readily partitioned. Consider now that members of one warbler species might be a bit more adroit at eating insects on spruce trunks, whereas the others might do slightly better on the tips of branches. In each case, such individuals would have an advantage over others that are less distinct and that therefore encounter more competition. Those that partition the spruce-tree niche would be more fit than those that do not, and evolution would produce divergence, or "character displacement," between the two species.

In fact, one classic ecological study revealed that spruce trees of northern North America are occupied by at least five different yet closely related birds, each of which nest and hunt for insects in different places, often in the same trees. The cast of characters includes myrtle warblers, black-throated green warblers, Cape May warblers, bay-breasted warblers, and blackburnian warblers. They minimize niche overlap, thereby reducing competition, by partitioning their resource use: each one exploits its environment in a slightly different way. Another way of looking at it: when species with similar needs coexist, their niches are likely to be narrower than if either occupies its niche alone, in which case it tends to spread out ecologically.

In the case of spruce-occupying warblers, as in many others, different species "compete" without necessarily even encountering one another directly. The point is that, in a sense, they do *not* compete, or rather, they minimize competition. As ecologist Joseph Connell once put it, what biologists observe is the "ghost of competition past." Connell himself performed a now-famous piece of ecological research, investigating a situation in which direct competition appears to have determined resource partitioning.

The habitat is the rocky marine coast of Scotland, occupied in particular by two species of barnacles, *Balanus* and *Chthalamus*. *Balanus* dominates the lower reaches, which stay wet most of the time and are exposed to air only rarely, during the lowest tides; *Chthalamus* occupies the upper and thus drier rocks. At first glance, this might seem to be a straightforward case of specialists partitioning their environments and occupying slightly different niches. As it happens, *Chthalamus* is more desiccation-tolerant, which correlates with its occupying the higher, drier rocks. But Connell noticed that *Balanus* is also larger and literally capable of growing underneath *Chthalamus* and pushing them off their rock hold. When he scraped away the *Balanus*, Connell found that *Chthalamus* extended its range, living lower and wetter, where otherwise only *Balanus* was found. On the other hand, scraping away *Chthalamus* did not have a reciprocal effect: because *Balanus* is not able to survive regular drying, it did not spread into *Chthalamus* habitat, even if the latter were removed, giving them a clear field.

An important distinction here emerges, between a species' "theoretical" niche and its "realized" niche. The former reflects the range of conditions under which a species could survive if it were not for its competitors; the latter is its actual situation. In the barnacle case, the realized niche of *Chthalamus* is normally smaller than its theoretical niche. *Chthalamus* occupies the high, dry zone not because it likes it better but because *Balanus* is a bit of a bully, pushing it away from where it would otherwise thrive. By contrast, the theoretical and realized niches of *Balanus* are pretty much the same, since it is normally able to exclude its competitor. *Balanus* distribution, therefore, is limited not by the other species but by its own physiological limitations.

It's a long way from mollusk to mankind, yet the barnacle research is loaded with implications for *Homo sapiens*, a species whose niche is exceptionally broad, so broad in fact that it impinges on every other species on earth. In paleohistory, human beings may well have competitively excluded closely related species (notably *Homo neanderthalensis*) whose niches overlapped. But rather than accepting its share of resource partitioning, *Homo sapiens* has aggressively pushed itself, *Balanus*-like, onto terrain previously occupied by other species. Our theoretical and realized niches overlap, leaving many of the world's living things either extinct or with realized niches that are but a tiny fragment of their theoretical possibilities.

In short, the human impact on natural environments stems not only from such obvious perturbations as introducing pollutants, using up nonrenewable resources, killing off certain species, and destroying natural habitats with abandon, but also by disrupting the delicate ecological patterning of communities worldwide. We have become ecological bullies.

The introduction of "exotic" species provides many examples of such disruption. In these cases, introduced species may lack native predators, while also finding themselves facing local species that aren't equipped with defensive adaptations for dealing with them. For example, in the mid-1980s a cargo ship released ballast water into the St. Lawrence Seaway. This water, originally taken up in the Caspian Sea, contained zebra mussel larvae. In just a few years, these introduced mollusks have spread through the Great Lakes and down the Mississippi to New Orleans. Numbering in the hundreds of thousands per square foot, zebra mussels have been decimating native species, clogging water-intake pipes, and doing literally billions of dollars' worth of damage.

Other introduced pests are starlings and English sparrows, now the most common birds in much of the United States: starlings, for example, seriously threaten native bluebirds. Nonnative fire ants and "killer bees," introduced into the southern United States, have been making their way north.

The Japanese kudzu vine, originally introduced in the American South as a decorative plant, has become a scourge, overgrowing native vegetation and even swallowing up houses left untended for a few years. Another southern plant, the water hyacinth, was introduced as an ornamental and now clogs waterways by the millions of tons. And lest you conclude that only the South is confronted with introduced plant pests, consider well the dandelion.

Around the globe, introduced species—especially rats, pigs, goats, and mongooses—have chaotically disrupted natural ecosystems, often leading to extinction of entire species, especially on islands. (Hawaii has suffered immensely in this regard.) Nile perch were introduced by Europeans into Lake Victoria, a huge, isolated East African lake with no large predators that had evolved a unique array of fishes, found nowhere else. The introduced perch have annihilated many of the locals, destroying the native fishery, and yet are insufficient to provide substantial food for humans.

There is no meaningful way to compare the ecological devastation wrought by such introductions with the other, better-known environmental catastrophes brought about by, for example, depletion of water sources, overhunting, overgrazing, air and water pollution, soil erosion, or increased salinization and resulting desertification. All involve disruptions in natural systems, and all, in one way or another, highlight the interconnectedness of life. Each case is specific, unique, fascinating, extraordinary, and often tragic, an ecological tale unto itself. The general rule connecting them all: ETEE, or the cloud in the piece of paper.

We repeat that no one can demarcate the biologically given boundaries of the human niche. This is especially true since technological innovation enables *Homo sapiens* to expand beyond the merely biological, or, if you wish, to expand the meaning of biological. Ecology's warning is that by extending themselves in this way, human beings disrupt the otherwise precise adjustments by which numerous species, including ourselves, coexist. In the opening lines of Goethe's *Faust*, Mephistopheles complains about "man" as follows: "Would he still lay among the grass he grows in! Each bit of dung he seeks, to stick his nose in." By sticking our noses everywhere, we cause trouble not only for other species but for ourselves as well. AIDS, for example, has been dubbed the "revenge of the rain forest," since this disease apparently lurked in the Old World rain forest, after which it seems to have made the leap to *Homo sapiens* as logging expanded, highways were opened, and sexual practices combined with worldwide transportation permitted HIV to spread rapidly.

When it comes to ecosystems, those dominated by human beings tend to differ from their natural counterparts in a variety of ways. For example, although natural ecosystems are sun-powered, human ecosystems rely largely on nonrenewable fossil fuels. Also, natural ecosystems recycle their nutrients, whereas human ones lose nutrients, which are then replaced artificially. Garbage and trash are collected and taken "away," sewage is flushed "away," while erosion carries about 3 billion tons of topsoil "away" in the United States alone, every year. (Then we seek to make up the loss via chemical fertilizers.) In natural ecosystems, water is generally stored and cleansed, whereas it is polluted and passed rapidly through human-dominated ones. Deforestation and the paving over of natural vegetation greatly decrease the human ecosystem's ability to absorb water, clean it, and release it slowly, increasing the frequency and severity of floods. Finally, human ecosystems tend to be much simpler biologically, less diverse and less stable than their natural counterparts. And so we turn to biodiversity.

Biodiversity

Alarms have been sounding about loss of species richness, or biodiversity, and for good reason. We don't even know how many species there are on earth or how rapidly they are going extinct; ironically, we are losing biodiversity before even having the opportunity to catalog it. Reasonable estimates place the number of species at about 10 to 12 million, of which only about one tenth have been described and named. Perhaps as many as 3 species go extinct per hour, or about 30,000 lost each year. Most of these are small, indeed tiny: bacteria and minuscule invertebrates. Some are the size of dodos or passenger pigeons (both already gone); others are the size of Indonesian rhinos, giant pandas, or Siberian tigers (endangered). None of them, once gone, can be called back.*

Although extinctions happened long before human beings arrived on the biological scene, the rate of species loss is now extraordinarily high, much of it due to human activity. There have been five previous periods of mass extinctions, associated, it is now believed, with planetary catastrophes such as an asteroid that may have killed off the dinosaurs 70 million years ago. And it may be that even without another such global impact, we are heading for a sixth. The rate of species extinction in modern times appears to be from 100 to 1,000 times higher than the usual "background" rate. Let's examine some ecological wisdom concerning natural communities, especially what determines biodiversity and its decline.

We have already described problems of competitive exclusion and the disruptive effects of introducing exotic species, destroying and otherwise messing around with previous systems of mutual involvement and resource partitioning. We have also alluded to the consequences of disrupting the great natural cycles such as carbon, nitrogen, water, and so forth. There is more.

"DNA banks" have been established for many endangered species, in the hope that some day they might be clonally resuscitated if they go extinct. A good idea, but cold comfort.

When it comes to wildlife, one doesn't need a refresher course in bioliteracy to realize that overhunting eliminates living things or that habitat destruction (including but not limited to overgrazing, clear-cutting, drowning behind dams, or covering with asphalt) does the same thing, only a bit less directly: no species can survive if its niche is done away with. This is ecology, of course, but it is also just common sense to understand that giant pandas are doomed without their bamboo forests, and grizzlies without huge tracts of northern temperate wilderness. To be sure, there are sophisticated details to be teased out by careful ecological insights and research: unraveling, for example, such otherwise hidden niche requirements as nesting sites, migratory pathways, precise habitat needs (e.g., old-growth forests for northern spotted owls), and critical minimum population size for a species to have a reasonable hope of weathering unpredictable population crashes as well as the dangers of insufficient genetic variety.

An important recent trend in ecological research has been to conduct studies on a relatively large scale, measured in acres instead of square meters, and in nature instead of laboratories. For example, pathbreaking research in the Hubbard Brook valley in New Hampshire's White Mountains involved experimentally clear-cutting several small watersheds and directing literally all of the water outflow so that its nutrient levels could be measured. The results were dramatic: compared with intact forests, deforested systems lost six times more calcium and other necessary minerals.

In most related cases, however, we suspect that science takes a backseat to ethical, social, and political judgments. Some Ph.D. ecologists find it infuriating, but it is altogether appropriate that in common usage, many people call themselves ecologists when they mean environmentalists. Although nearly every ecologist is also in fact an environmentalist, there is at the same time a realm of scientific ecology that is independent—but not very—of its policy implications.

In any event, despite our lack of knowledge—or perhaps because of it—species diversity is being lost in many different ways. For one, because of the economics of energy flow through trophic levels, higher-level consumers and predators tend to be comparatively rare even in pristine environments. Because they are especially eye-catching and thus provocative, they are among the first to be extirpated from natural systems. Animals occupying the highest trophic levels also require the largest areas of pristine habitat. And the fewer individuals, the more vulnerable they are to extinction. Such "upper-level" creatures are also more prone to magnifying toxic materials, as we have seen. And they are more likely to be seen as competitors with *Homo sapiens*, who competitively exclude them by hunting, trapping, poisoning, or selectively destroying their habitat.

When it comes to species diversity, which is both of practical concern to environmentalists and of theoretical concern to ecologists, it is important to distinguish abundance (numbers) or biomass (amount of living stuff) from diversity as such. A field of corn is rich in numbers of corn plants and in biomass, but impoverished when it comes to diversity. A plantation of pine trees, all of the same species, is a virtual monoculture compared with a natural forest ecosystem that may contain several dozen species, as well as additional structural diversity that comes from having individuals of different ages, shapes, and sizes, each offering different niche opportunities. The ecological principle is that strength comes not in unity but in diversity.

Think about a field of corn, lacking diversity in two ways: it consists not only of just one species, *Zea mays*, but likely of one variety as well. All these plants are therefore vulnerable to a parasite or disease organism adapted to its particular genetic stock. Compare the fate of this monoculture with a cornfield planted with diverse genetic strains. Or a natural grassland composed of many different species, even different genera and different families.

The correlation of species diversity and ecosystem stability is an important general principle, but like everything else in ecology, it is complicated and filled with exceptions. For example, a long-term ecological study has recently monitored total biomass, ecosystem stability, and species diversity in more than two hundred experimental plots of Minnesota grasslands. By varying the amount of nitrogen, the researchers were able to generate plots with different degrees of species diversity. They found that most of the time, species-poor areas produced about the same biomass as those that were species-rich. Accordingly, it is an oversimplification (at least in this case) to conclude that species number alone determines ecosystem productivity. On the other hand, the Minnesota ecologists found that after a severe drought, plots containing the fewest species lost fully seven eighths of their biomass, whereas those that were richest in species lost only half theirs. Species diversity may thus be especially important in times of environmental stress, when it can act as an ecological shock absorber.

It also seems likely that species diversity as such is less important than the presence of a relatively small number of players that perform key ecological roles, sometimes called "keystone species." At the current stage of sophistication in ecological research, even experts are hard-pressed to identify these species and their precise roles, which leaves ecologists with the fallback proposition, "When in doubt, save whatever you can."

We do know, however, that removal of certain predators can have surprising and devastating effects on species diversity, as revealed in a classic bit of ecological research, this one emphasizing that although all species are important (and, for the conservation-minded, precious), some are more important than others.

Ecologist Robert Paine studied the rich intertidal habitat found in rocky tidepools on the Washington coast. It is a balanced and complex community, with starfish, crabs, anemones, mussels, snails, barnacles, and many other species. When Paine experimentally removed all the starfish of one particular sort, the effect was dramatic: one type of mussel became extraordinarily abundant, crowding out other mussels as well as urchins and seaweeds, greatly reducing the overall diversity of the tidepool community. Removing other species did not have a comparable effect. Starfish evidently constitute a keystone in this particular community. By consuming their favorite prey—the one type of mussel—they normally keep it from overwhelming the system and add greatly to its diversity.

Some keystone species are comparatively easy to spot, although their exact significance is only now becoming clear: North American bison, for example, not only served as a resource base for prairie wolves and Native Americans, but they also had a major role in maintaining the great tallgrass-prairie ecosystem that at one time stretched from Texas to southern Canada. Experiments on large patches of tallgrass prairie in northeastern Kansas, for example, have shown that bison grazing—not unlike starfish predation—greatly increases grassland biodiversity, in this case by reducing the preponderance of heavy, taller grasses, thereby permitting an array of smaller species to survive.

Beavers have been similarly important in modifying freshwater habitats in the Northern Hemisphere, as are termites in sculpting arid environments in the Southern Hemisphere. Elephants are comparably consequential in much of Africa. Other keystone species are less obvious, although perhaps no less key: the red-naped sapsucker, for example, drills holes in willow trees throughout the Rocky Mountains. The sticky, nutritious sap that oozes out is then available as food for literally dozens of other species, from butterflies and wasps to hummingbirds and chipmunks, most of which are unable to access it by themselves.

Moreover, sapsuckers typically dig nest holes in hardwood trees, especially aspens, and they obligingly excavate a new one each year, creating nest sites for other species, notably two kinds of swallows, which, unable to create their own nests, are only found in regions occupied by sapsuckers. Human beings have eliminated keystone species with abandon, and a dramatic reduction in species diversity has ensued.

Any discussion of species diversity, even one so abbreviated as this, must grapple with the prime dilemma of biodiversity: the tropics. It is a two-pronged issue, important for both theory and practice. Underlying both prongs is this not-so-simple fact: the tropics are immensely richer in species than the temperate zones, which in turn are more biodiverse than the arctic. If you travel from the equator to higher latitudes (either north or south), you will notice a continuum of change, with species diversity declining the farther from the equator you go. This applies to plants, insects, fish, reptiles, birds, and mammals. There are more different species of birds, for example, to be found on a single thousand-acre farm in Colombia than in all of Alaska! (Although Alaska has more birds in sheer number.)

Insects aren't rare in the Alaskan summer. There are billions of them . . . nearly all mosquitoes. By contrast, tropical insects are not especially abundant, but there is a remarkable array of kinds: gaudy butterflies, beetles and bugs of all shapes, parasitic wasps beyond counting, and so forth. The arctic oceans are home to great schools of herring, mackerel, or haddock; a tropical reef, by contrast, offers scores of species but not very many of any one kind.

In North America, forests are consistently of one type or another, such as the Douglas fir/western hemlock forests of the Pacific Northwest, or the oak/hickory woodlands of Pennsylvania. By contrast, in the Amazon forest or the rain forests of Indonesia or central Africa, any tree is likely to be quite different from its neighbors, and the nearest member of the same species may be a mile or more away! A single hectare (about 2.5 acres) of tropical rain forest can easily contain a hundred different kinds of trees; by contrast, a hectare of temperate forest may very well hold only ten, of which two or three may comprise 90 percent of the individuals.

Here is another of life's mysteries, although ecologists have proposed many possible explanations. Perhaps tropical environments have been relatively stable over long geological periods, allowing greater opportunity for more species to evolve. In addition, their communities, too, would be comparatively old, setting the stage for the evolution of complex interactions and increased diversity via resource partitioning. Regardless of their stability over hundreds of millions of years, tropical environments clearly are more stable than their higher-latitude counterparts over short time spans, notably the course of a single day, or even one year. (The difference between the highest and lowest temperature in a single day is quite a bit less in the tropics than at higher latitudes, and this difference is greatly accentuated if we compare the range between winter and summer in the tropics with that between winter and summer in New York, never mind Alaska.)

So, perhaps the greater stability and predictability of the tropics have permitted greater specialization of its inhabitants, which are under less pressure to adapt to severe weather shifts and thus able to partition their environment more precisely, which means more species. Arctic and temperate environments are, in a sense, "boom and bust" ecosystems, with booms in the spring and summer, followed by busts in the fall and winter. During boom periods, emphasis is on reproducing rapidly, to fill the available niche. By contrast, tropical communities, if undisturbed, are always booming, which means that tropical organisms compete with one another instead of with a comparatively empty environment.

Or perhaps there is greater environmental heterogeneity—patchiness—in the tropics, which in turn results in more different kinds of species exploiting this greater niche diversity. Tropical rain forests, after all, feature complex microhabitats from the canopy down to the ground; a Kansas prairie, the Siberian boreal forest, and the Yukon tundra, by contrast, offer less structural variety and thus fewer opportunities for different kinds of species.

Whatever the reasons, biodiversity is a tropical specialty. And the tropics are under ecological siege. This is the practical horn of the tropical dilemma. Because of their extraordinary biodiversity and species richness, tropical habitats— and rain forests in particular— contain a remarkable proportion of the world's species. Although covering only about 6 percent of the planet's land area, they are home to between 50 and 70 percent of the total number of species! They are being destroyed at an equally remarkable rate, mostly for agriculture: about 40 million acres per year, or 80 acres per minute. In the process, biodiversity unparalleled on the planet is being transformed into ecological wasteland.

Ironically, given their extraordinary variety and richness, tropical forests turn out to be very unproductive as food-growing regions. Tropical soils, by and large, are exceedingly poor in nutrients. This surprises many people, since intact tropical rain forests are so magnificently productive. But in fact, nearly all the richness of tropical rain forests is in the vegetation itself, not the soil. Under natural conditions, when tropical vegetation dies, its nutrients do not remain long in the soil; instead, they are quickly recycled into new growth, leaving virtually no reservoir in the ground. After these forests are cleared for agriculture or harvested for their valuable hardwoods such as mahogany, they are typically burned before being planted. This is supposed to release nutrients into the soil, but is largely a myth. Heavy tropical rains leach out whatever nutrients might be left, and after just a few years, the soil "gives out," forming a hard, water-resistant crust.

What remains at this point is not only useless for agriculture but very resistant to returning to its natural state. This is because tropical soils tend to be "lateritic," high in oxidized iron and aluminum compounds, which, when exposed to sun and rain and deprived of their normal vegetative cover, turn into what is essentially reddish concrete. (The famed thousand-year-old temples of Angkor Wat in Cambodia were constructed of laterite.)

The biodiversity crisis is not limited to the tropics, however, although it is most pronounced there, as measured in species extinctions. Species loss is even greater on islands: by some estimates, fully 80 percent of all species to go extinct in the past few hundred years were insular. To a large extent, this is because islands are particularly likely to harbor distinct, endemic forms, since their inhabitants are generally isolated from other populations. This enables island populations to go their own way, genetically as well as ecologically. And because islands are necessarily smaller than mainlands, they usually harbor smaller populations, which are more vulnerable to extinction both because their numbers are lower and because they have less genetic diversity to draw upon.

In the late 1960s Robert MacArthur* and Edward O. Wilson developed the first modern, comprehensive statement of "the theory of island biogeography." It described how islands eventually achieve a kind of equilibrium in species number, based on many intersecting factors. These include distance of the island from the mainland (nearer islands receive more immigrants); number of species already there (more species results in more extinction, both because more exist to get into trouble and because with more already present, more immigrants are competitively excluded); and size of the island (larger

---

* MacArthur also conducted the pioneering study of resource partitioning among North American warblers, described earlier.

islands are not only more likely to be contacted by dispersers from the mainland but are also more likely to provide the necessary diversity of niche opportunities for species to establish themselves and thrive).

One renowned field test of the ecological theory of island biogeography involved killing all the resident arthropods on a series of very small mangrove islands (averaging about 10 meters in diameter) off the southern tip of Florida. As expected, size of the island and distance from the mainland strongly influenced recolonization; interestingly, although the community composition and overall species diversity eventually reached an equilibrium very much like that before the experiment, the actual species that arrived and settled in were rather different. This suggests that chance effects—literally, who gets where first—may also be crucially important.

The relationship between habitat area and species composition, often referred to as the "species-area curve," is especially consequential. John Donne was more correct than he could have known when he wrote that "no man is an island." In fact, not only islands are islands! For ecological biogeographers—and really, there can be no other kind—mountaintops are islands (isolated patches of trees, often amid a "sea" of prairie), and so are lakes (islands of water amid a "sea" of land), rivers, and patches of forest or natural prairie surrounded by agricultural land. As wild habitats are increasingly fragmented, the earth is becoming a series of islands whose overall diversity is sorely threatened.

Much ink has been shed by ecologists attempting to use principles of island biogeography to design the optimum pattern of nature reserves. Called the SLOSS debate—Single Large or Several Small—it involves arcane calculations over minimum population sizes, niche requirements, and the details of the species-area relationship. In the SLOSS spirit, we suggest that the best solution is simply "AMA-PEALAP": As Many As Possible, Each As Large As Possible.

There is no way to measure precisely the effects of reducing the world's biodiversity. Maybe the impact will be "only" aesthetic and philosophical, a diminution of the excitement, interest, and variety of our planet. Maybe what remains will go on, indefinitely, as it has in the past, with human ingenuity somehow compensating for the various disruptions that *Homo sapiens* manages to create. Maybe we'll lose access to potential wonder drugs. Or maybe we'll actually destabilize the planet, as suggested by Stanford University biologists Paul and Anne Ehrlich, when they described the loss of biodiversity as equivalent to removing rivets from an airplane: confident that there are more rivets than needed, yet knowing little or nothing about the role played by each one that they eliminate, the rivet removers keep at it.

Maybe the airplane won't crash.

Population

When the definitive history of the late twentieth century is written, one of the great underreported stories will be overpopulation. In a sense, the nightmare numbers are already known, although rarely appreciated as they deserve. Here is one way to think about it. If you read at an average rate, you began this paragraph about ten seconds ago; since then, approximately 35 people have been added to the earth's population. By tomorrow, those numbers will have increased by more than 250,000; and a year from now, by nearly 100 million, about the population of Nigeria. In India alone, 100,000 new job seekers enter the already overcrowded market every *week*, while at the same time, 100 acres of potential croplands are withdrawn from agriculture every *day* because of salinization (intensive irrigation adding more salts to the land than the generally sparse rainfall can wash away).

But numbers are numbing. Some important realities remain strangely distant. Americans, for example, generally know what it means to be *hungry*, but have very little experience of *hunger*. Overpopulation is a political, social, and economic problem, but also a biological and ecological one. We emphasize, by the way, that the population dilemma is worldwide, not unique to the developing world. It is actually two-sided. Because people in the developed—or, perhaps, overdeveloped—world use a disproportionate share of the planet's resources, it can be argued that per capita, population growth in Europe and especially the United States is more environmentally destructive than anywhere else. (Since the United States comprises about 6 percent of the world's people but uses about 30 percent of the world's energy, the average American uses roughly five times his "share.") But whereas overpopulation in the rich countries is more burdensome to the ecology of planet Earth, overpopulation in the poor countries is more burdensome to their own people, since in the short term it is those already overburdened economies, ecosystems, and individuals that are especially likely to suffer.

There is little doubt that we are overgrazing our planet. The world fish harvest, for example, peaked in 1989 and has been dropping ever since. Soil is eroding, deserts are advancing, biodiversity is declining, poisons are building up, water tables are drawing down, tropical forests are disappearing, and natural, unspoiled ecosystems of any sort are becoming ever more rare. Some reliable estimates state that for the world's people to enjoy the standard of living now characteristic of the United States, the total population would have to be about 500 million, less than one tenth its current numbers. Or: we would need ten Earths to satisfy those currently alive.

Fortunately, the ecology of population is increasingly understood. (In this case, the biological science is easy compared with the political science required to overcome indifference as well as outright hostility to population control, especially from fundamentalist religious traditions.)* Although the science of population biology quickly tends toward mathematical opacity, its key concepts are not only worth understanding, but capable of being understood, and stated in plain English, too. Here are a few. They tend to occur in pairs.

First, exponential growth. Like money left at compound interest, population accumulates. A fundamental point is that even if the *rate* of growth remains the same, the *amount* of growth increases exponentially, or geometrically. Start with $1,000 or 1,000 people and a seemingly modest rate of increase; the result is a huge amount of money, or of people, in a remarkably short time: 3 percent of 1,000 people is 30. Three percent of 1 million people is 30,000. The growth rate remains the same, but as the numbers grow an increasing number of individuals is added to the population during each succeeding period.

The human population around ten thousand years ago was approximately 5 million, about the current population of El Salvador. The total impact of so few people on the world's ecology was probably quite low. That is no longer the case, at least in part because our numbers have been increasing, more rapidly than most people think. Thus, it is commonly believed that, as the gloomy English clergyman Thomas Malthus famously pointed out, the human population has been growing exponentially. This would mean that it doubles in a constant time period, such as every hundred years (the interval would depend on the actual rate of growth). But in fact, human numbers have increased at a rate that exceeds exponential growth.

About 2,000 years ago, there were perhaps 250 million of us; it took fifteen to sixteen centuries for this number to double, to 500 million.

---

* *Einstein once remarked—when turning down the presidency of Israel—that politics is much more difficult than physics!*

We reached our first billion (another doubling) in just two or three centuries, between 1825 and 1845. Only one century was required for the next doubling, totaling 2 billion persons by around the year 1930. From here, the jump to 4 billion took only 44 years. The human-population "doubling time," in other words, decreased from 1,600 years to just 44. That is exponential growth with a vengeance! As population biologist Joel Cohen put it, "Never before the second half of the twentieth century had any human being lived through a doubling of the human population, and now everyone who is 40 years old or older has seen the Earth's population double."

Although the rate has changed over time, human population has increased almost continuously throughout our evolutionary history, producing what population ecologists sometimes call a J-shaped curve, because it first rises very slowly as it progresses toward the right (like the bottom of an uppercase J), then turns alarmingly upward, until it is going nearly straight up, like the rest of the J. The J-shaped curve of exponential growth can be described mathematically and is also observed in practice, either in laboratory populations—typically, of one-celled organisms or small invertebrates—or in natural environments when an area is newly colonized by species with relatively few competitors or predators. At first, life is Edenic, as reflected by the increasing numbers. But eventually things go sour.

This leads to the J-shaped curve's paired alternative: logistic growth, or the S-shaped curve. Waste products from the growing population may poison the environment; scarcities may develop, so there isn't enough food to go around, or enough nesting sites, or predators may appear, or diseases may take advantage of the increased population density combined with declining physical condition, and so forth.

"There is no exception to the rule," wrote Charles Darwin, "that every organic being naturally increases at so high a rate that, if not destroyed, the earth would soon be covered by the progeny of a single pair." *The point is that no species has ever increased its population indefinitely.* And a moment's reflection will show that no species ever will. Not even us.

Eventually, a limit must be reached. This limit is described as the species' carrying capacity. It varies with the species and its circumstance. Carrying capacity sets a ceiling on the height of our J-shaped curve, changing it from a vertical line to one that is essentially horizontal. Assuming there is progressively greater environmental resistance as numbers increase, we can expect a gradual transition from exponential growth to some sort of steady state as the carrying capacity is reached. Accordingly, the upward sweep along the vertical part of the J eventually bends around and levels off. (Mathematically, this environmental resistance is reflected by adding a simple term to the exponential-growth equation; it is self-correcting, so that the rate of increase becomes smaller as the population becomes larger.)

The resulting "S-shaped" curve is only somewhat S-shaped, actually more like a flexible S that is held firmly at the bottom left while being gently stretched and pulled toward the upper right. The result, a stretched-out S, begins with the usual J shape, which accelerates briefly before reaching its maximum rate of increase; then it slows down and becomes more horizontal as it approaches carrying capacity. There is some reason to think this is beginning to happen: in 1965, the overall growth rate of the human population peaked and began to decline. Only once before in recorded history has this happened: during the fourteenth century, because of wars, famine, and especially the epidemic of bubonic plague. The twentieth-century drop in birthrate, by contrast, occurred because of conscious decisions: millions of people choosing to have fewer babies.

It is something to cheer, because the alternative, which has often been observed in animal populations that become too large for their own good, is a precipitous decline to carrying capacity and often below. That cheer should be at best halfhearted, however, because a reduction in the *rate* of increase still means an increase in the overall population. Imagine a bus headed for a cliff and going ever faster as it approaches. Just because it stops accelerating does not mean that it isn't heading for trouble!

Another important concept: sometimes population size contains within itself the seeds of its own restriction, as when increased density makes disease transmission more likely, which in turn reduces such density. In other cases, reproductive rate declines when population rises. Ecologists call these self-limiting effects "density-dependent." (The flattened, horizontal component of the S-shaped, logistic population curve is due to density dependence in one form or another.) There are also "density-independent" effects, which act to reduce population size regardless of how large (or small) it may be: a sudden flood, volcanic eruption, tornado, or equivalent, something not brought about by the numbers of its victims but that influences those numbers just the same.

For humans, density dependence and density independence are typically intertwined. For example, flooding in Bangladesh recently caused tens of thousands of deaths. Although floods are usually described as density-independent, these were due in large part to deforestation on the slopes of the Himalayas, itself due to local population pressure. In addition, overflowing rivers would not have caused such loss of life if there had not been so many people living downstream, and especially if they were not occupying vulnerable floodplains, to which they were pushed, in large part, by sheer force of numbers.

When populations are in a phase of exponential growth, along the steeply rising vertical of the J, they are characterized by a combination (a "suite") of reproductive factors that are quite different from those of populations occupying the relatively flat, horizontal stages along the upper right-hand corner of the S. This leads to our last set of paired concepts: r-selection versus K-selection.

The *r* in *r-selection* comes from the ecological term for maximum reproductive rate, whereas the *K* in *K-selection* derives from ecological shorthand for carrying capacity. The idea is simple; its implications, complex. For r-selected populations, the emphasis is on rapid increase in numbers. Among animals, this means early and frequent reproduction, small body size, large litter size, and (counterintuitively, perhaps) comparatively high mortality, if only because an emphasis on quantity rather than quality results in significant wastage. Populations are likely to be r-selected if population control is density-independent, so that they occupy "boom or bust" environments that are often relatively empty and that therefore reward sheer increase in numbers.

By contrast, K-selected populations are already at or near carrying capacity. For them, population control is likely to be density-dependent, and competition is mostly among individuals of the same species (for food, mates, living space, etc.) rather than between an individual and its environment. Therefore, K-selected species emphasize quality over quantity; they are likely to have larger bodies, breed later in life, produce smaller litters, and enjoy lower infant mortality. Elephants, for example, are K-selected compared with mice. (Notice that r- and K-selection are relative terms; whereas mice are r-selected compared with elephants, they are K-selected compared with bacteria.)

What about *Homo sapiens*? Human demography, although quite precise when it comes to describing population levels and rates of change, has been notoriously unreliable as a predictor of future trends. One pattern, however, has been consistent: the so-called demographic transition. As mammals go, we are pretty much K-selected: large-bodied, late-breeding, with small "litters" and generally low mortality. And yet, we continue to occupy the vertically ascending part of a skyrocketing J-shaped curve. Our numbers have been increasing like an r-selected species. But herein lies some hope.

The basic observation is that along with industrialization and improved living standards there comes a transition to lower birthrates, increased age at first reproduction, and reduced death rates . . . in short, a shift from r-selected characters to their K-selected counterparts. Although such changes do not reflect a genetic or evolutionary shift, since they take place very quickly—often within the space of a single generation—they are remarkably similar to the adaptive changes ecologists have long noted in animal populations living under natural conditions.

The reasons for the demographic transition are many-faceted. With improved public health and thus lower mortality comes less reason for large families. Diminished emphasis on nontechnological agriculture makes children less important to rural families as cheap sources of farm labor. More emphasis on education, which is costly, increases the payoff to having fewer children so as to be able to invest more in each one. Whatever the precise causes, there is much to be said for transitioning from a mouse to an elephant style of reproduction.

Another insight from population ecology seems especially relevant to *Homo sapiens*, although it is more descriptive and less hope-inspiring than the demographic transition. It is simply this: every population has its characteristic "age structure," which, in turn, predicts short-term population growth. By age structure, we mean the relative numbers of individuals who are, say, less than one year old, between age one and two, and so on. The usual situation is something of a pyramid, wider at the base (more younger individuals) and becoming more slender at the top (fewer older), simply because the longer people live, the more likely they are to die.

Populations whose age structure is relatively top-heavy, such as in Germany or Scandinavia, are unlikely to grow in the near future and may even decline, because they will have fewer people entering their reproductive years than are currently there. Others, whose age structure is more moderate, such as the United States at present, are likely to grow very slowly or achieve an equilibrium. Still others, such as Nigeria and Saudi Arabia, where the youngest individuals predominate, can be expected to increase rapidly as a demographic bulge reaches childbearing age. Largely because of this difference in age structure, it will take approximately 1,000 years for the population of Europe to double, 100 years for North America, 35 years for Asia and Central America, and fewer than 25 years for Africa.

In rapidly growing countries, nearly all of which are drawn from the ranks of the developing and impoverished (which used to be called the Third World), medical advances have reduced the death rates, but birthrates have not yet followed suit. Ironically, a vicious circle often develops, sometimes called the "DEE trap," for "demographic, economic, environmental." In such cases, rapid population growth exacerbates the poverty, undereducation, and environmental deterioration that frequently lead to high birthrates in the first place. Which in turn leads to more poverty, undereducation, and environmental deterioration, and so on.

This leads to another observation, one that can be seen as either a glass half full or half empty: there is a huge unmet demand for birth control and family-planning services worldwide, so that many women currently have more children than they want (the glass half empty) and would have significantly fewer if given the opportunity (the glass half full). Moreover, one surefire way of diminishing family size—thus ultimately reducing the rate of population increase—is to improve the education and socioeconomic status of women.

"Every child a wanted child" is the slogan of Planned Parenthood. When "woman's work" includes more than baby making and women come to feel valued and validated in other ways, too, then perhaps the wanting of children will correspond more closely to their actual production, and the population of our shared planet might even come into harmony with its biological limits.

We began this chapter with the single great truth of ecology: all things are interconnected. Awestruck by this fact, we dealt lightly with many of the specifics of ecological science, tending more, we fear, toward preaching than teaching. But we do not really apologize: if you understand ecology's single great truth—better yet, if you act upon it—you are already bioliterate beyond measure.

EVOLUTION:

The Road Stretches Out

All of us have evolved. "Out of the dreaming past," writes Don Fabun in *The Dynamics of Change*,

> with its legends of steaming sea and gleaming glaciers, mountains that moved and suns that glared, emerges this creature, man—the latest phase in a continuing process that stretches back to the beginnings of life. His is the heritage of all that has lived; he still carries the vestiges of snout and fangs and claws of species long since vanished; he is the ancestor of all that is yet to come. Do not regard him lightly—he is you.

Not everyone would agree, but then, this book is not for everyone.

Biologist Edward O. Wilson coined the term *biophilia* to describe the universal human fondness for nature and living things. We suggest the simultaneous existence of biophilia's antithesis: call it *biophobia*, a fear, sometimes even a loathing, of nature and living things. People may be biophobic for perfectly good reasons, such as the sting of a scorpion, the bite of a cat, the toxin of *Botulinus*, or the great stomping feet of an enraged elephant. But there is more—or less. Aside from presenting real or potential threats to physical well-being, the living world in general and animals in particular pose a subtle psychic threat to our self-image.

Beneath any fear of animals, we suspect, lies the deeper anxiety of acknowledging our kinship with "lower" life-forms. "In an aversion to animals," wrote Walter Benjamin, "the predominant feeling is fear of being recognized by them through contact. The horror that stirs deep in man is an obscure awareness that in him something lives so akin to the animal that it might be recognized." That recognition does not need elaborate scientific backing; it is usually enough to look into a dog's eyes (which, interestingly, does not usually evoke horror). Anyone inclined to look further, into evolutionary biology— or indeed any area of biology—cannot help being overwhelmed by the truth and the beauty of human connectedness to the rest of life.

As a species and as individuals, human beings have done remarkable things, while also frequently declaring, loudly and insistently, their independence from biology. We compose symphonies, write laws, build rockets and space stations, worship gods, worry about death, explore the atom, the universe, and our own psyche. But such declarations ring just a bit hollow. Attempting to separate ourselves from biology, we people doth protest too much.

In *Civilization and Its Discontents*, Sigmund Freud refers approvingly to a nineteenth-century German playwright, Christian Garber, who gave this advice to a would-be suicide: "We cannot fall out of this world. We are in it once and for all." This caution applies to everyone: we'll eventually die, but aside from that, the world is irretrievably with us. We are stuck in the muck and glory of it all, living creatures among many, created by our biology no less than is a dandelion or a rhinoceros, and fully connected.

This connectedness is at the heart of evolution. It is also a cure for humanity's overwhelming hubris. Here is George Bernard Shaw, in *Back to Methuselah*:

> The sense of the kinship of all forms of life is all that is needed to make Evolution not only a conceivable theory, but an inspiring one. St. Anthony was ripe for the Evolution theory when he preached to the fishes, and St. Francis when he called the birds his little brothers. Our vanity, and our snobbish conception of Godhead as being, like earthly kingship, a supreme class distinction instead of the rock on which Equality is built, had led us to insist on God offering us special terms by placing us apart from and above all the rest of his creatures. Evolution took the

conceit out of us, and now, though we may kill a flea without the smallest remorse, we at all events know that we are killing our cousin. No doubt it shocks the flea when the creature that an almighty Celestial Flea created expressly for the food of fleas, destroys the jumping lord of creation with his sharp and enormous thumbnail . . . .

And no doubt it (evolution) equally shocks those who persist in seeing their species, whether flea or *Homo sapiens,* as the apple of God's eye, possessing a divine spark that sets it apart from all others. But at the same time, it is why being a perfectly good mammal—or, for that matter, a perfectly good flea— is not in any way a put-down. For Darwin, there was grandeur in being linked to all living things both by historical continuity (that is, common ancestry) and as the products of the same fundamental process: organic evolution.

Messy Misconceptions

Noting the role of the Royal Air Force in saving his country during the Battle of Britain, Winston Churchill observed that never had so many owed so much to so few. We owe much—indeed, literally everything—to evolution, yet never have so many said and written so much about something that they misunderstand so greatly. Not that evolution is so difficult to understand. It appears, rather, that many people have such strong feelings about it (both pro and con), often connected to so many regrettable stock phrases ("survival of the fittest," "good of the species," "only a theory"), that clear thought has often been obscured.

Let us review some common misconceptions surrounding evolution.

"Its only a theory." Admittedly, biologists often speak of "the theory of evolution," but not because evolution is a guess or mere speculation. Our *Random House Dictionary* provides, among its definitions of *theory*, the following: "a proposed explanation whose status is still conjectural, in contrast to well-established propositions that are regarded as reporting matters of actual fact." Someone might express a "theory" that Elvis Presley still lives, or that trailer parks cause tornadoes. The same person might also say, "Biologists have a theory that human beings evolved," in which case, knowingly or not, a very different use of the word has been employed. Indeed, our dictionary also gives an alternative definition of *theory* as a "more or less verified

established explanation accounting for known facts or phenomena," with examples that include number theory, the theory of relativity, atomic theory, and so forth. In this sense, and this only, is evolution a theory. It is, in fact, as close to truth as science is ever likely to get.

"Evolutionary logic is circular: the fittest are those that survive, and those that survive are the fittest. So it doesn't say anything." First, natural selection is not about survival; it's about reproduction: specifically, individuals and genes reproducing themselves. Survival is evolutionarily important because—and only because—it contributes to reproduction. Depending on their characteristics, some living things survive only a few hours, others for decades; living is only a means to the evolutionary end of projecting genes into the future. Second, "fitness" does not determine natural selection; rather, natural selection reveals how "fit" something is (or, equivalently, it defines something's fitness) by its success in promoting its genes. As such, fitness leads to the important prediction that natural selection favoring a particular type should result in a larger proportion of that type in future populations. As we shall see, this prediction has been tested and confirmed many times.

"Natural selection is just a negative process; it cannot create anything new." Natural selection is testimony to the simple fact that living things are not equally successful in living and, most important, in reproducing; that is, projecting copies of their genes into the future. It is only "negative" in that certain individuals and their genes fall by the evolutionary wayside in preference to others, which prosper. But evolution is not merely a question of deleting those organisms that are less fit. Because of mutation (which provides genetic novelty) and sexual reproduction (which combines DNA in unique ways), new genetic material is constantly being produced.

Much depends on this regular generation of genetic diversity, on the world being, as the poet Louis MacNeice put it, "incorrigibly plural." In his poem "Snow," MacNeice went on to feel "the drunkedness of things being various," a variousness that is essential as the building blocks from which evolution constructs those things that we identify as highly adapted organisms, including ourselves.

But although the production of diversity is fundamentally random, the beauty and power of natural selection is that, contrary to what many people think, it is not simply at the mercy of haphazard events, which eliminate the unfit but do not "create" anything new. Nearly everyone has heard the old story about the infinite number of monkeys banging on typewriters: given infinite time, they will have written out all of Shakespeare's plays, and indeed, everything else that has ever been written. Although this argument is technically correct, it isn't very compelling, if only because infinity is a very long time. But in fact, the ability of evolution to generate biological novelty is only a tiny bit like those mindless, industrious monkeys. The big difference is that natural selection adds a crucial process: a mechanism for "selective retention."

Imagine that instead of creating all of Shakespeare, we just wanted a single phrase, "to be or not to be." This phrase consists of 18 characters, including spaces. Given an alphabet of 26 letters, plus a blank space between words, we have a total of 27 possibilities for each slot. The chance that one of our hypothetical monkeys might randomly get the initial t is thus 1/27. The chance that it would simultaneously and randomly place an o in the second slot is $1/27 \times 1/27 = 1/729$. The chance of getting all 18 characters correct, by chance alone, is thus 1/27 times itself 18 times, which is inconceivably small.*

But all this presumes that each attempt is treated the same as every other. What if, instead of tossing out every meaningless word combination and starting afresh each time, those patterns that were promising, even by a little bit, were somehow retained, then randomly modified again, once more retaining (that is, selecting, either naturally or unnaturally) those that were more "fit"? In this case, it is possible to produce the desired pattern in a very short time. Let's say that you start with 18 random strikes on a computer keyboard, and you program the machine to make a few small changes—introduce some new letters, at random—every "generation." These changes would be equivalent to mutation and sexual recombination, providing regular sources of random variation on an existing theme. Next, add the simple requirement that you selectively retain whatever most closely approximates "to be or not to be" (that is, whatever is "more fit"). You discard the others, after which the selected string of letters and spaces proceeds to vary randomly again, with the outcome screened once more for resemblance, however slight, to "to be or not to be." After a very small number of generations—usually on the order of 30—the desired outcome will be obtained.

In one test, for example, we started with "fuwl sazgh ekm fje." Discouraging, perhaps, but after several runs it had become the dimly recognizable "tubl hot noq ioby." And by run 22, it was "tu bep ok not ts e." By 29, "to be ok not to bo," which even the most cynical

---

* *The following approach, with some modification, is borrowed from Richard Dawkins's* The Blind Watchmaker *(New York: W. W. Norton, 1987).*

monkey is likely to acknowledge as having just about arrived, and a whole lot more quickly than 1/27 to the 18th power would suggest.

Starting with gobbledygook and using only random variation and selective retention, something new has been created, something so meaningful, in fact, that it is perhaps the most famous phrase in the English language! One might object, of course, that there is a crucial added factor: human intelligence was injected into the process. "To be or not to be" is composed of very short, simple words (not unlike the DNA code for amino acids, incidentally), but it was arrived at only retrospectively, *after* it was first created by a renowned human intelligence. Only then did a seemingly blind process obtain it via our computer trials, by selecting certain versions over others. For example, "to pee or not to pee" is also syntactically correct—and appropriate to certain occasions—but is unlikely to have echoed down the corridors of literature for hundreds of years.

Shakespeare presumably considered various possible alternatives to Hamlet's renowned dilemma, although he assuredly didn't puzzle through every option. Like a human chess master as opposed to IBM's chess-playing program, Deep Blue, a creative intellect takes numerous shortcuts.

But this, too, is very much what natural selection does. Living things offer it only a limited subset of what is possible. Instead of a creative human intelligence rejecting certain verbal combinations over others, the environment faced by every living thing rejects certain genetic combinations over others. What it rejects depends on the environment. In arid conditions it rejects combinations that waste water; in cold conditions it rejects combinations that waste heat. Among predators it rejects combinations that are clumsy at stalking their prey, while among prey species it rejects combinations that are incautious or inept when it comes to avoiding their predators. As Robinson Jeffers put it, in "The Bloody Sire,"

> What but the wolf's tooth whittled so fine
>
> The fleet limbs of the antelope?
>
> What but fear winged the birds, and hunger
>
> Jewelled with such eyes the great goshawk's head?

The fleet limbs of the antelope, the wings of birds, and the eyes of goshawks are all marvels of natural design, in no way inferior to the human design of Hamlet's melancholy question. And to understand how all of these—including the magnificent cerebral hemispheres of *Homo sapiens*—were created, we need only understand how natural selection, proceeding by a random process, can nonetheless generate highly nonrandom results.

"Evolution is no longer going on, especially in the case of human beings." Evolution happens any time there are changes in a population's genetic makeup. The most powerful mechanism of evolutionary change is natural selection, which operates whenever some individuals leave more genetic representatives than others. So the only way for evolution to cease would be if everyone reproduced equally—more precisely, if genes continued to replace themselves in exactly the same proportion as they currently exist. Just a moment's reflection should convince anyone that evolution is very alive, for human beings as for everything else, so long as "differential reproduction" is going on.

This doesn't mean, however, that the conditions of evolution are the same as they have been in the past. The "selective environment" for human beings, for example, has changed dramatically from the Pleistocene. Certain traits that almost certainly were strongly selected against, such as myopia or diabetes, are now neutral or, at worst, only mildly negative. Human ingenuity has come up with eye-glasses and insulin, and this only scratches the surface of the ways modern *Homo sapiens* has been modifying natural selection, hence its own evolution. Whereas it might have been selectively advantageous to be a good hunter, gatherer, or mastodon-avoider, now it is selectively advantageous to be able to reproduce despite strontium-90 in our bones and DDT in our fat and, perhaps, to be positively attracted to ideologies that are unsympathetic to birth control. In any event, we have changed our own evolution but not ended it.

A related misconception is that human evolution is moving toward a future in which people will have large heads and small bodies (because with modern technology, we are supposed to use our heads more and our bodies less) and in which the appendix and tonsils will have disappeared because we don't use them anymore. The only way for large-headed, small-bodied people to evolve would be if such specimens had more offspring than their smaller-headed, larger-bodied counterparts, and this seems unlikely. Similarly for the purported demise of the appendix and tonsils, although this is more plausible, if even a small proportion of people with larger appendices and tonsils die young because of complications from tonsillitis or appendicitis.

If use and disuse led to such evolutionary change, a theory known as Lamarckian evolution, we would be unusual indeed among living things. It is tempting to slip into Lamarckian thinking, however, if only because using part of our body often tends to make it larger; think of the effect of exercise on human muscles. In Longfellow's poem "The Village Blacksmith," we learn that

> Under a spreading chestnut-tree
> The village smithy stands;
> The smith, a mighty man is he,
> With large and sinewy hands;
> And the muscles of his brawny arms
> Are strong as iron bands.

We are not told, however, that because of the blacksmith's daily labor, his children will inherit those brawny muscles. Which is just as well, because "use it or lose it" does not apply to genetics. Biological information flows from DNA, via RNA, to proteins, not vice versa. What happens to our bodies may indeed influence our evolution but only if it influences our body's success in passing along its genes, not because such experiences reach into our gonads and directly modify those genes.

There is, however, one sense in which human beings *do* experience Lamarckian evolution, via a kind of inheritance of acquired characteristics. It may also be unique to *Homo sapiens*. But even this special kind of Lamarckism does not negate or even supersede Darwinian evolution. It operates on a parallel track, although the two evolutionary systems occasionally collide. It is cultural evolution.

Cultural evolution involves change based on cultural practice: language, technology, fashion, lifestyle, traditions including culinary, religious, military, intellectual, social, sexual, and just about everything else that human beings do outside the anatomy and physiology of their own bodies. Like biological evolution, cultural evolution requires variation, but instead of genetic mutations, its raw material comes from ideas and concepts, innovations of mind and matter that may ultimately be traceable—however indirectly—to humanity's DNA, but which are not a simple matter of biochemical alterations in genetic material. Cultural evolution, like biological evolution, proceeds by selective retention of whatever works or is favored for any other reason (such as obvious efficiency, the vagaries of fashion, or even the dictates of the powerful). Most important, cultural evolution, because it is Lamarckian and can be "inherited" nongenetically and passed on to others within a single generation, is much faster than biological evolution.

On October 19, 1879, the incandescent light bulb was invented by Thomas Edison. Within a few decades, millions of people had "inherited" Edison's cultural mutation, and today literally billions make use of it. Had Mr. Edison's invention been genetic instead of cultural, it is unlikely that more than a hundred people—his biological descendants—would have light bulbs in their lives. Agriculture, domesticated animals, written language, fire, the wheel, Coca-Cola, computers, nuclear weapons, rap music: all these things and of course many more have been acquired, nongenetically and nonbiologically, by people who did not create them but who employed them in one way or another, many of whom passed them on to others, regardless of their genes, their reproductive success, or their fitness.

Of all the trials and tribulations faced by *Homo sapiens*, perhaps the most daunting are those that derive from its anomalous position: subject to two masters, biological and cultural evolution. These two are not necessarily incompatible, and in fact, a case can be made that cultural innovations are attractive if and when they accord with biological needs. Those with staying power (like agriculture, language, domesticated animals) are those that contribute, however circuitously, to biological fitness. Personal satisfaction from a particular source may well be closely tied to whether that source offers a deeper, often unconscious, biological satisfaction. Nonetheless, there are some genuine perils in the fact that cultural and biological evolution proceed at such different rates.

During the 150,000 or so years that *Homo sapiens* has existed as a biological entity, the gene pool of our species has probably not changed very much. During the past 10,000 years or so, it has remained almost constant; certainly, it has been slow-moving at best, a mere tortoise compared with the hare of cultural evolution that has been racing ahead, nearly oblivious to its plodding, Darwinian counterpart.

And so, unlike all other living things, whose rudimentary "cultures" are inevitably in synch with their biology, the human species finds itself confronted with all sorts of cultural innovations for which its biology is not prepared. For example, our ability to kill, at long range and in vast numbers, has outstripped any possible biological inhibitions against doing so. The same is true of our ability to pollute and destroy natural environments, as well as to reproduce. We surround ourselves with strangers, especially in our cities, and then resent the impersonality, alienation, and confusion that ensue. (Primitive *Homo sapiens* probably spent their entire lives interacting with only a few dozen fellow members of their own band.) We separate ourselves, often by thousands of miles, from biological kin, then lament the strain of raising children without a network of caring relatives. We invent labor-saving devices and sedentary lifestyles, then suffer from lack of exercise. We consume diets high in fats and sugars—rare treats for hunter-gatherers—and then develop obesity, heart disease, stroke, and cancer because our culture and technology enable us to tempt our primitive sweet tooth without "burning off" the consequences.

The "noble savage" probably wasn't all that noble, and in any event, we cannot turn back the cultural clock or stuff the genie of extrabiologic innovation back into its bottle. Yet human biology remains immutably shackled to genetics even while its cultural creations soar, or plummet, into the future.

Not for the Good of the Species

All sorts of things evolve: galaxies, stars, a person's thinking, a government's policies. But none of these qualify as evolution as biologists understand the term. Wolves have evolved, as have elm trees and bacteria, but an individual wolf does not evolve, nor does a single elm tree or a bacterium. People have evolved; a person does not. Evolution involves change, but not the change that takes place in growth, learning, aging, changing one's clothes or one's mind. To qualify as biological evolution, there must be change in a population's gene pool over time. This is why only populations evolve, not individuals: because each of us is stuck with our private genetic endowment. Yet individuals—both individual bodies and individual genes—are crucial to evolution. In fact, one of the most important and useful realizations of recent years is that evolution operates most strongly at the lowest possible levels: that of individuals and genes, not groups or species.

"Good of the species" arguments are the bane of modern biology. Asked about the evolutionary rationale for living things doing whatever they do, most people blithely reply, "For the good of the species." In fact, natural selection virtually never operates that way: neither for the good of the species nor even for the good of the group.

Natural selection works by differential reproduction, with some individuals and their genes more successful than others. A species is the sum of its individuals and their genes. It has no metaphysical existence of its own and, as far as can be determined, no one is looking out for the good of the species as a whole. Insofar as its component parts have been selected to do a good job gathering food, avoiding drought, migrating, courting, or competing (or, for that matter, breathing, pumping blood, resting when tired, eating when hungry, and so forth), a species may be "fit," "adapted," or just plain "healthy." But in fact, the species isn't doing anything at all: its subparts are.

Analogously, in a free-market society, individuals and corporations seek to maximize their profits; any larger-order benefit derived by the nation is simply the unintentional summed effects of all these private, enterprising activities going on at a lower level. In fact, this is probably the greatest single weakness of capitalism, and why private enterprise functions best when leavened by government efforts to look out for the greater good: pursuit of private gain can occur *at the expense* of overall public benefit.

In the world of living things, there is no equivalent to Medicare, the FBI, the Department of Education, or the Centers for Disease Control. And indeed, with individuals and genes selected for their own benefit and none of them looking out for the interest of the larger group, the overwhelming majority of species that have ever lived are now extinct! Furthermore, when a species is endangered and thus at risk of going extinct, there is no indication that its constituent individuals are especially inclined to deprive themselves for the good of the larger whole; indeed, there is every sign that living things do whatever it takes to promote their own success, not that of the species.

In many cases, the separation of individual from species benefit is a distinction without a difference, because the two generally point in the same direction. When selection works on individuals, favoring those bodies—and the genes that created them—that are most successful, the result is a species that is also well adapted, if only because it is composed of successful individuals and their genes. Things get interesting, however, when benefits to individuals point in one direction and to their species in another.

For example, it has been known for some time that many birds are capable of laying more eggs than they usually do. Although swallows may occasionally fill their nests with as many as eight eggs at a time, the average clutch size is typically four. Seeking to explain why these animals usually reproduce less rapidly than they appear capable, biologists used to naively point to some need to rein in their reproduction "for the good of the species," so as to prevent overpopulation.

Overlooked in this facile explanation was the simple but troubling fact that any parent who sought to place the good of the species above his or her own evolutionary best interests, and accordingly who reproduced at less than his or her possible maximum, would be replaced by others who selfishly followed their own best interests, damn the consequences. The species might be better off if its members were more self-restraining, but those individuals who followed such a policy would be replaced by selfish types who had no "species conscience." (The result is analogous to what economists call the "free-rider problem," in which individuals are tempted to avoid paying their way, especially if they can profit from the sacrifices or expenditures of others.)

In recent decades, research has looked carefully at this question of what appeared to be a species-aiding, altruistically restrained breeding rate. And it has been shown conclusively that when animals such as swallows produce unusually large numbers of eggs, they almost always end up actually rearing *fewer* successful offspring than do others who lay fewer eggs but are able to take better care of those that hatch. The conclusion is unavoidable: animals that breed at a slower rate are not doing so out of concern for their species; rather, they are reproducing as rapidly and efficiently as they can, as individuals.

Natural selection is not likely to smile on behavior that reduces one's own reproductive success; conversely, it will reward self-serving acts regardless of whether such acts benefit the larger group. The essence of this pattern was captured in a cartoon that showed a bunch of lemmings jumping off a cliff. Among the crowd was one individual, with a sheepish grin, wearing a life preserver! (Incidentally, lemmings do *not* really jump off cliffs in episodes of mass suicide.)

There is an important and far-reaching exception to the generalization that natural selection promotes biological selfishness: namely, the question of altruism and "kin selection," which we will explore in Chapter 9. For now, we note that even this is not really an exception, since it highlights the general rule that selection operates most effectively at the lowest possible level. Hence, it can favor self-sacrificial behavior so long as individual genes profit even though their bodies lose out. Once again, the fate of the larger unit is no more than an incidental consequence of selection operating on the smaller one: genes favored over individuals, just as individuals are favored over species.

There are many cases of natural selection ignoring species benefit in favor of individual payoff; in fact, it was only after biologists achieved this insight that sociobiology was born, and with it a slew of revelations about how evolution fine-tunes behavior. Here is another example. Among many species, including the majority of mammals, polygyny or harem-keeping is the rule. For elk, most seals, and gorillas (or human beings, if our biology alone were calling the tune), an especially dominant male will strive to control the reproduction of numerous females. In the case of elk, for example, this means that one bull mates with perhaps twenty females. As a result, there are nineteen unsuccessful bulls for every one that hits the jackpot. But this, of course, presumes that the sex ratio among elk is 50:50, which it is, not only for elk but for pretty much all animals that reproduce sexually. The question is, why?

If evolution were operating for the good of the species, then it should produce, on average, one bull elk for every twenty cows. This would maximize the amount of elk flesh around at any time, by minimizing the wastage of growing nineteen nonbreeding bulls for every one who passes his genes along. Moreover, it would also minimize the disruption that elk—and every other polygynous animal—experience. Excluded bachelor bulls don't accept their celibate situation lying down, graciously forgoing reproduction in favor of their more fortunate colleague. Instead, they mill about, making all sorts of trouble, resentfully fighting with one another and periodically attempting to overthrow the dominant harem keeper. (Incidentally, when such an overthrow succeeds, in many species the new harem keeper typically kills the infants sired by his predecessor.) Things would be nicer all around if these troublesome bachelors didn't exist, and the species as a whole would certainly be better off.

But evolution doesn't look toward the species as a whole. Because of natural selection's relentless focus on individual and gene benefit, it stubbornly keeps producing equal numbers of each sex. Here's why.

With our hypothetical elk, every female is likely to breed every year, producing one calf and hence enjoying an annual reproductive success of one. The harem keeper has a reproductive success of twenty. Imagine there are no bachelor bulls, so our elk are enjoying a species-beneficial Shangri-la. Under these conditions, parent elk who produce female offspring would get an evolutionary payoff of one per cow, via her eventual reproductive success. But parents who produced a male child instead would get a payoff of ten, since there would now be two males competing for the favors of twenty females. (Assuming that each male is equally likely to succeed in monopolizing the harem, each gets either twenty offspring or none, for an average of ten.) As a result, selection would favor the production of male offspring, so long as there were fewer males than females in the breeding population, and regardless of the size of the typical harem. Another way of looking at it: the rarer sex conveys a higher evolutionary payoff than the more abundant one. As a result, selection favors producing whichever sex is rarest, automatically righting the imbalance.

At what point would the system reach equilibrium? When there are equal numbers of males and females. At this point, in our example, every female elk would have a reproductive success of one and every male would also have a reproductive success of one (actually, twenty for the harem master and zero for the nineteen bachelors, for an average of one per male). The same argument works the other way: if an excess of females should develop, selection would favor producing additional males, not to restore the balance for the good of the species, but because selection working on individual elk rewards those that produce offspring who are most likely to be successful and thus to convey success on those parents.

Don't miss the forest for the trees. If evolution worked on species, it would yield a sex ratio that approximates the breeding ratio. It is only because it works on *individuals* that there are equal numbers of males and females, regardless of the breeding ratio.

Speaking of forests and trees, they provide another argument for individuals being favored over species. There is no good reason, after all, for forests to be so tall, wasting all that metabolic energy constructing huge trunks, when all they really need is enough leaves or needles to photosynthesize and make seeds. If the issue were simply species benefit, "forests" would consist of dwarfed bushlike plants. But instead, they are absurdly, if grandly, vertical, shading out their competitors and in the process "wasting" vast amounts of matter and energy creating something we call wood. All this happens because each individual tree does better if it gets more light, and to do so, it must be at least as high if not higher than its neighbors, resulting in competition that is useless if not hurtful at the species level but is driven by natural selection's partiality for the individual.

Links, Missing and Found

Mark Twain once said that it was easy to stop smoking: he had done it hundreds of times. In evolution, there is no missing link: there are hundreds of them—or thousands, or millions. Consider two points, representing different species, one of which gave rise to the other.

Think of them as connected by a line, representing the evolutionary continuity between them. Now, add a third point, more or less midway, and call it "the missing link." Having located this missing link, have you finished your task, and bridged the gap between the two points? Not at all. In fact, you have just produced two new "missing links." Fill in both of them, and you are faced with four missing links. Like the horizon, which constantly recedes if pursued, the discovery of transitional forms merely adds to the transitional forms not yet identified. Mathematicians say there are an infinite number of points between any two identified points on a line; presumably there are fewer than an infinite number of missing links, but the more we find, the more there are.

Nonetheless, some people are truly bothered by what they see as the paucity of transitional forms in the fossil record. They might just as well be impressed by how many have been found. This applies to forms ancestral to *Homo sapiens* just as to other species. Probably the closest thing to a "missing link" in the human evolutionary lineage is the famous fossil Lucy, so named because the Beatles song "Lucy in the Sky with Diamonds" was playing in the background while paleontologists were celebrating her discovery. Lucy was a female Australopithecine ("southern ape") of the species *Australopithecus afarensis*, who stood about 3 feet tall and weighed around 66 pounds. Lucy is as much an intermediate between apes and people as can be imagined.* But she isn't alone.

There are several other species of *Australopithecus*, some relatively slender and most if not all of them on the line that gave rise to *Homo sapiens*. Others are heavy-bodied and—with the 20/20 vision of hindsight—identifiable as evolutionary dead ends whose descendants eventually went extinct. There is also a growing list of species belonging to the genus *Homo*, including *Homo habilis*, which is pretty much a link between Lucy and us, just as Lucy is a link between ancient apes and modern human beings. Other "found links" include *Homo erectus*, remains of which are known from Asia as well as Europe. Our purpose here is not to provide a detailed list of fossil prehumans, their dates,

---

* *The foolishness of the concept "missing link" becomes clear when you consider that for there to be some sort of midpoint, one must specify the two ends. Granted that one is modern* Homo sapiens. *But what ancestral form, precisely, holds down the other end of the linked chain: an anthropoid ape, a monkey, a primate, a mammal, a primitive reptile, an early amphibian, a primordial vertebrate, a pre-Cambrian worm? The "missing link" between modern human beings and the Devonian fishes, for example, might be a predinosaurian reptile.*

cranial capacities, or precise relationships, but to note the existence of numerous forms variously intermediate between *Homo sapiens* and earlier, ancestral animals.

Neanderthal fossils were first found in the Neander Valley, near Dusseldorf, in the mid-nineteenth century. Most specialists now agree that they were not a separate species but rather a subspecies of *Homo sapiens, Homo sapiens neanderthalensis*. It seems unlikely that Neanderthalers were transitional. They may have interbred with early Cro-Magnon man (*Homo sapiens sapiens*) or have been wiped out or simply outcompeted by them. Neanderthalers have also been calumniated in the popular imagination. As it happened, the first fossil specimen unearthed suffered from severe arthritis, and thus seemed unusually stooped and deformed. Although they were more heavily built than modern humans, Neanderthalers were a far cry from the crude and dull-witted stereotype. Archaeological evidence shows, for instance, that they buried their dead with elaborate ceremony, using flowers. Careful ritual positioning of bear skulls suggests a form of religious sensibility as well.

Another comment on transitions and common misconceptions: the caricature of people evolving from monkeys or apes is just that, an exaggerated fabrication. Evolution does *not* teach that modern human beings evolved "from" chimpanzees, even though chimps are undoubtedly our closest living relatives. Rather, human beings and chimpanzees share one or more common ancestors (as do human beings and hippopotamuses, hummingbirds and horseflies). Chimps and people are equally modern; we are cousins. You will never be a monkey's uncle, and a monkey has never been yours.

While it is alive, there is no way to identify a transitional form as such. Maybe its descendants will remain largely unchanged for millions of years, so that it is not transiting to anything else but rather is just something that evolved early and persisted late. Or maybe its descendants will go extinct, in which case it is transitional . . . to a dead end. Perhaps its descendants will be somehow recognizable in the present day, in which case it is transitional in the usual sense of the term. In any event, such transitional forms are actually quite abundant.

There is, for example, *Archaeopteryx* ("early feather" or "early bird"), a clumsy, toothed bird that appears to be transitional between reptiles and modern birds. More recently, we have the discovery of *Protoarchaeopteryx*, whose name implies its position, as prior to *Archaeopteryx* and more of a reptile. And now it appears that there was also *Caudipteryx* ("tail feather"), a flightless dinosaur, amply endowed with a plumed behind. Both *Protoarchaeopteryx* and *Caudipteryx* were flightless; their feathers must have evolved because of other advantages they conveyed, perhaps in courtship or heat regulation. This exemplifies an important evolutionary principle: sometimes traits evolve for one purpose, then switch over to a different function later.

Here's another case of transitions. It had long been known that modern whales, being mammals, must have evolved from some sort of mammal that returned to the ocean and became a kind of marine mammalian amphibian ("amphibian," that is, in habits only, not technically similar to frogs or salamanders). A creature known as *Ambulocetus* ("walking whale") fits the bill; it is an elongated mammal with heavy leg bones that might well have been about equally comfortable (or uncomfortable) on land or in the ocean. Of course, this left a new missing link, between *Ambulocetus* and the totally water-dwelling modern whales. Recently, that gap, too, has been filled, with the discovery of *Basilosaurus*, an extinct whale that sported definite hind limbs, more pronounced than on today's Moby Dick but less than on *Ambulocetus*. And so it goes.

Transitions don't only exist as fossils. There are "living fossils" whose ancestors may or may not have been transitions in their day. For example, the simple, immobile marine invertebrates known as tunicates or sea squirts produce larvae that are remarkably similar to the simplest chordates (close relatives of modern vertebrates). And a rare denizen of tropical rain forests known as *Peripatus* sports a wormlike body plus fleshy, unjointed, yet claw-bearing legs and a thin body cuticle containing small amounts of chitin. Thus it possesses many of the traits one might expect of the transition creature(s) between early annelids and arthropods. (Like chimpanzees, tunicates, and the deep-sea, lobe-finned *Coelacanth*, *Peripatus* is nonetheless a modern creature, despite its seemingly primitive traits.)

Students of evolution can even trace transitional aspects in the physical structure of species that are currently quite distinct in other ways. Here we encounter an interesting paradox: evolution's handiwork speaks loudly as a natural, biological process—as opposed to supernatural and theological—precisely because it is so *imperfect*. Time after time, close examination of living things shows that they have been cobbled together when natural selection modified some ancestral pattern, whereas creating an animal (or human being) from scratch would have provided a cleaner design and a better outcome.

For example, the basic vertebrate blueprint for front limbs consists of a single long bone, the humerus (upper arm in human beings), then two bones, radius and ulna (forearm), followed by smaller carpals (wrist bones) and phalanges (fingers). This pattern persists and can be identified among reptiles, birds, and mammals as diverse as dogs and cats, human beings, and bats. It is inconceivable that the same design features serve equally well in generating arms used for such different purposes as running, leaping, grasping and throwing, and flying. Yet it is easy to see how evolution, faced with a basic ancestral condition, modified the available raw material, creating an array of transitions en route, none of which are perfect, but each of which works.

Similar examples abound in the natural world, cases in which living things, carrying the historical baggage of their ancestors, are good enough to get by but less efficient or effective than if they had been designed specifically for the problem at hand. Rather than argue against evolution by natural selection, such cases do just the opposite: they argue against an omnipotent creator and for a mindless, blundering, tinkering process that is not so much masterful as make-do. Here are a few, drawn entirely from *Homo sapiens*.

The human eye, for all its effectiveness, has a major design flaw. The optic nerve, after accumulating information from our rods and cones, does not travel directly inward from the retina toward the brain as any minimally competent engineer would demand. Rather, for a variety of reasons related to the accidents of evolutionary history plus the vagaries of embryonic development, optic-nerve fibers first head away from the brain, into the eye cavity, before coalescing and finally turning 180 degrees, exiting at last through a hole in the retina and going to the brain's optic regions. The result is a small blind spot in each eye where the optic nerve leaves the cavity of the eye itself, where it should never have strayed in the first place. (By contrast, the eyes of an octopus, lacking the troublesome historical constraints of human eye-volution, gather appropriately on the far side of the retina, from which they travel directly to the brain. There is no blind spot.)

Then there is the awkwardly close association of the human excretory and reproductive systems, which almost seems like a celestial effort to embarrass our species or any other inclined to be squeamish or self-conscious. This, too, is the result of evolutionary happenstance rather than grand, perfect design. Hundreds of millions of years ago excretory systems didn't even exist, since animal bodies were so small that simple diffusion to the surrounding water would have been sufficient to eliminate waste products. Then, as animals became larger, the evolving excretory organs simply appropriated the nearby reproductive sluiceways, rather than laying down brand-new plumbing for their own purposes. Poor design? Yes, except that it wasn't really design at all, so much as making do with something already available and serviceable, even if indelicate.

Neither the human knee nor the lower back, moreover, is likely to win any prizes for structural engineering, as most people over age thirty can testify. (Not that they don't work; indeed, they are wonderfully effective, but as examples of making the best of a bad situation—the transition from quadruped to biped—rather than wise structural foresight.) Perhaps the best example of transitional awkwardness is found in that primordial human labor: giving birth. Most four-legged mammals have an easy time with parturition. But human beings, mammalian to the core, have nonetheless insisted on walking upright, which required a peculiar reorientation of the connection between pelvis, leg bones, and spine. The result is a distressingly narrow skeletal channel between the pubis in front and the left and right ischium in back. Add to this the absurdly large head of a newborn baby, and the result is that getting a fetus from inside to outside can be painful and dangerous.

By the way, it hasn't escaped the watchful eye of anatomists and physicians that there is lots of room elsewhere in the human skeleton for a baby to exit its mother's uterus: a huge gap between the pelvis and the ribs, for example. But the female reproductive system just isn't oriented in that direction. It stubbornly reflects its quadruped, mammalian ancestry, something upon which upright posture was rudely foisted. (That inviting passageway between pelvis and ribs is occasionally made use of: during Cesarean section, when evolution's bequeathal is potentially life-threatening and obstetricians take the obvious shortcut.)

Finally, any discussion of evolutionary transitions should note the basic biological transitions that all of us experience in utero and also carry around inside our cells. Every human embryo develops gill arches, a tail bud, and other stigmata of our ancestral history, which form and then recede, retained as developmental stages only because evolution has once again built upon preexisting, primitive patterns rather than creating something altogether special or new. (In the case of human beings, embryonic gill arches do not become gills; rather, they contribute to parts of the adult's lower jaw, larynx, tongue, and middle ear.) But perhaps the most impressive evidence of the underlying unity of all life—a unity due to our common evolutionary history—are those patterns we *don't* outgrow.

Despite the vast range of living forms, from fungi and oak trees to salamanders and vampire bats, with all their detailed and unique specializations, certain underlying biochemical mechanisms remain shared, part of our biologically based conjoint heritage. Take the case of cytochrome c, a protein that is used in mitochondrial electron transport chains, when cells store energy as ATP. In one form or another, cytochrome c is found in virtually every living thing that is aerobic, including bacteria, apple trees, elephant seals, and human beings—and just about everywhere else. Reflecting its basic functionality—and the connectedness of life—the gene that codes for cytochrome c is nearly unchanged from one species to the next, with the degree of difference correlating neatly with "evolutionary distance" as measured independently by other evidence from embryology, comparative anatomy, the fossil record, and just plain common sense.

For example, the same sequence of 104 amino acids comprises cytochrome c in both human beings and chimpanzees. (This doesn't mean that humans and chimps are identical, although the fact that they *are* identical with respect to cytochrome c confirms that they are similar indeed.) Comparing cytochrome c in people and rhesus monkeys, one amino acid is different; for people and dogs, the number is 13; for chickens, 18; for rattlesnakes, 20; for tuna, 31; and for yeasts, 56. It is always possible that God arranged such similarities to fool human beings into believing in false evolutionary transitions, or because of a sublime unwillingness or inability to come up with new polypeptide chains in each case. But it is probably less blasphemous to consider that God arranged for evolution and then stepped back to let it do its thing.

For a last example of molecular biology revealing evolutionary transitions, there is DNA hybridization. This important technique relies on the fact that DNA is double-stranded, with the complementary strands bonding to each other in a precise way: A to T, G to C. It is possible to take DNA from two different species, treat it enzymatically to unzip the molecules, then stir them together so as to allow new, hybrid DNA to form, consisting of one strand from each species. The more similar the DNA between the two "parent" species, the more the strands will match and the more firmly the new hybrid DNA will bind together. Strength of attachment can readily be assessed by measuring how much heat is needed to rupture the hybrid DNA and cause it to separate into single strands.

This technique provides yet another independent measure of evolutionary distance and closeness: as expected, it takes more energy, for example, to disrupt DNA hybrids between chimps and gorillas than between chimps and macaque monkeys, and it is quite easy to disrupt DNA hybrids formed between primates and rodents. DNA hybridization was used to resolve the long-standing question of whether the giant panda is more closely related to bears or raccoons. (The answer is bears. But its seeming relative, the red panda, turns out to be a raccoon.)

A Revolution in Evolution?

Creationists have been energized of late by signs of disagreement among biologists, interpreting a well-publicized dispute as indication that the whole Darwinian framework is wobbly. They couldn't be more wrong.

Look at it this way: there is *the theory* of evolution, and then there are various *theories* of evolution. The former is not in doubt; it is the bedrock upon which all of modern biology is based, the grand unifying theory of life, confirmed time and again in nearly every biological fact that is uncovered. By contrast, the different theories of evolution refer to various ideas regarding exactly how the evolutionary process unfolds. Advocates of "creation science" (an oxymoron) grasp eagerly at any indication of internal debates among biologists as evidence that the science itself is shaky. They do not understand that debate and creative ferment are the very stuff of science. Unlike theology, science is founded on ideas, discovery, testing, and refinement rather than on presumably unerring doctrines of faith. Disputes about evolutionary fine tuning, far from undermining the validity of evolution, testify to the vitality of the whole enterprise, since any worthwhile science raises more questions than it answers. (Analogously, there are disagreements among physicists about the details surrounding subatomic particles but no dispute about whether such particles exist.)

The most commonly cited recent "disproof" of evolution is the claim by paleontologists Stephen Jay Gould and Niles Eldredge that evolutionary change sometimes occurs more rapidly than traditional Darwinian teaching had previously held. This is a challenge only to one of the "theories of evolution," namely, that evolutionary change occurs by the steady, gradual accumulation of minuscule genetic changes (the traditional view, sometimes called "phyletic gradualism"). Gould and Eldredge offer instead the alternative that steady equilibrium states are punctuated by occasional bouts of rapid evolutionary change; this is sometimes called "punctuated equilibrium."

As paleontologists, Gould and Eldredge derive their data for the study of evolution from fossils; accordingly, they have access to only a small sample of the living things that once flourished. Indeed, it is remarkable that *any* fossils have survived the hundreds of millions of years since ancient times. Moreover, evolutionary change is generally so slow that it is virtually invisible even from millennium to millennium. It is not unlike continental drift: movement of a few millimeters per year does not register on the human psyche, although when continued for hundreds of millions of years, it smashed India into Asia (in the process crumpling up the Himalayas), separated Africa from South America, and literally rearranged the face of the planet.

In our judgment, punctuated equilibrium isn't nearly the revolutionary approach that Gould and Eldredge—or the creationists—believe it to be. Regarding the pace of evolutionary change, even those biologists most committed to a gradualist perspective never seriously considered that evolutionary change occurs at a precisely steady rate, reflecting the slow, monotonous grinding of some great cosmic mill. *Of course* there have been periods of relative stasis, punctuated by spurts of comparatively rapid change. Environments change irregularly and unevenly; hence, evolution is bound to be irregular and uneven. The point is that even times of galloping evolutionary change involved tens of thousands, even hundreds of thousands of years, and when they occurred, perfectly normal Darwinian processes—notably natural selection—were responsible.

Zoologist Richard Dawkins, a master of evolutionary metaphor, provided the best analogy for the gradualist-punctuationist debate. Dawkins notes that according to the book of Exodus, the children of Israel wandered in the wilderness for forty years, traversing approximately two hundred miles, before coming to their promised land. This means that their average speed was about five miles per year, or twenty-four yards per day, or one yard per hour. This is much slower than the average snail moves and is in fact a pace that is nearly impossible for humans to maintain.

Let us now imagine that two biblical scholars, troubled by the one-yard-per-hour dogma, offer a new insight: the Israelites didn't really move this way at all! Instead—wonder of wonders—they traveled in fits and starts, likely camping in the same location for days, weeks, perhaps months, with such "equilibria" punctuated by sudden spurts of many miles of travel in just a day or so. The prevailing "gradualist" analysis, they proclaim, is wrong! Next, legions of atheists, eager to highlight dissension in the ranks of Old Testament historians, publicize the new view as proof that the legitimacy of biblical scholarship and thus, somehow, the Bible itself, is in doubt. (Actually, we can empathize with Gould and Eldredge: if we had come up with something that might appear to be a dramatic reworking of the most important theory in science, evolution by natural selection, we too would probably have been tempted to trumpet it as a major modification, even if—or rather, especially if—it seemed to give Darwin a small black eye and thereby elevate our own contribution.)

Witnessing Evolutionary Change

Some creationists go so far as to claim that biologists have never actually witnessed natural selection causing an evolutionary change and that the whole enterprise is therefore conjectural. Wrong again. Nonetheless, there is understandable frustration, shared by many biologists, at the difficulty of actually observing evolution in action.

Evolution is slow, usually taking many thousands of generations. This is not surprising, since it takes time to accumulate an observable effect when, for example, a certain gene may enjoy an advantage of only one in one thousand over its fellows. Given enough time, such a selection differential will make a genuine difference, but no biologist has ever lived long enough to detect significant evolutionary change in sequoia trees, for instance, or blue whales, which take many years to produce even a single generation. This is not an excuse for evolutionary change being inferred rather than observed (as we shall see, such changes have indeed been observed), but it does explain why it is less common than might be expected.

There is another reason why it is difficult to catch evolution in flagrante delicto: namely, natural selection generally opposes change rather than promoting it. This seeming paradox makes sense when we realize that since life has been around for several billion years, most living things are very well adapted to their circumstances. And so, selection is likely to discriminate against departures from the norm, rather than reward such deviations with enhanced reproductive success. The technical term is "stabilizing selection."

Perhaps the best example of this comes from a study of house sparrows that had bumped into buildings during a hurricane. In this work, the researcher—appropriately named Bumpus—obtained wing size, body weight, and other physical measurements of the many birds he found littering the University of Rhode Island campus. As it turned out, about half the animals survived, having been merely stunned. The rest died, in a dramatic example of stabilizing selection in action. Thus, Bumpus found that the surviving birds were those whose measurements were close to the norm; those that died were likely to be exceptionally heavy or exceptionally light, to have unusually large wings or unusually small ones. The pre-hurricane population of sparrows had been pretty much optimum. Those with unduly small wings, for example, lacked the power needed to overcome dangerous gusts of wind, whereas those sporting very large wings were in trouble for a different reason, comparable to a boat carrying too much sail during a storm.

This is not to say that natural selection is always stabilizing. When the environment changes—and eventually, it does—then the advantage will lie with individuals who are somewhat different, and this is when an evolutionary change takes place. Which leads us to an important point: lots of evolutionary change has in fact been witnessed and studied.

Most dramatic has been the evolution of antibiotic resistance by microorganisms, especially bacteria. Since the generation time for bacteria can be as brief as twenty minutes, for a population of bacteria a few days can be equivalent to many centuries for a slower-breeding creature. Take a human being infected with, say, streptococci, and give her some penicillin. It kills the bacteria, but not necessarily all of them. Rather, it destroys (selects against) those sensitive to penicillin. Any that may be resistant—perhaps because of a mutation that alters their metabolism—will survive, reproduce, and in a short while become abundant. This is evolution: a change in the genetic makeup of a population over time, due to the differential reproductive success of some individuals and their genes over others. (As mentioned in Chapter 2, it is also a looming public-health menace, as careless use of antibiotics has been selecting for a number of drug-resistant disease organisms.)

The evolution of drug resistance involves human intervention, but that doesn't mean it isn't evolution. The same applies to innumerable cases of "artificial selection," in which people instead of natural forces do the selecting. Domesticated plants and animals differ substantially from their wild forebears because certain individuals and their genes were given a selective advantage by plant and animal breeders. For example, starting with ancestral wolves about ten thousand years ago, people began choosing the more docile and human-friendly individuals as companions, mobile garbage-disposal units, helpers in the hunt and in herding. Then particular breeds were created: large ferocious dogs to hunt bears, small perseverative ones to chase rats, others blessed with remarkable smell detection to track missing persons, or with extraordinary speed for racing, or with a fantastic variety of peculiar appearances just because they strike the human fancy. The basic procedure has been the same: choose those individuals that most closely approximate what you are looking for, and breed them with each other. Then choose those among their off-spring that are closest to your goal, and keep at it.

This process, which began millennia before anyone understood the details of evolution or the mechanisms of gene action, gave us not only the myriad breeds of dogs that now grace our world, from St. Bernards to Chihuahuas, but also the many different kinds of cats, horses, cows, pigs, pigeons, chickens, wheat, corn, rice, barley, apples, roses, carnations—indeed, all the extraordinary variety of domesticated plants and animals we know today.

But what about evolution among free-living populations? That, too, has been documented. The most famous involves so-called indus-trial melanism among English peppered moths. For centuries—prob-ably millennia—these animals were speckled. We know this because of the long-standing English penchant for collecting moths and but-terflies: inspection of centuries-old collections revealed the persis-tence of a light salt-and-pepper pattern among these moths, but with some notable exceptions. About 1 to 5 percent of speckled moths were not so much lightly peppered as darkly charcoaled; these black or "melanistic" forms owed their color to the pigment melanin. (Because of their rarity, collectors paid large sums for them.) In any event, such forms, when they arose, were presumably selected against in nature, because their numbers remained consistently low.

Then, with the industrial revolution in Great Britain, an evolu-tionary change occurred. The lightly peppered moths declined in number, and the melanistic form skyrocketed, until by the mid-1950s the situation had reversed, and about 95 percent of peppered moths were melanistic. Why? A clue comes from the persistence of the light form in rural areas, such as northern Wales, while the dark form was becoming especially abundant near factory cities such as Liverpool and Manchester.

In a preindustrial world, peppered moths had been beautifully camouflaged when they rested on light-colored, lichen-covered tree bark. The occasional melanistic mutant, by contrast, was lethally conspicuous against such a visually contrasting background. Then soot from English smokestacks killed the lichens and darkened the trees. Because of this dramatic change in their environment, the tables were turned for the lightly speckled moths and their melanistic counterparts. Previously inconspicuous, the light forms now stood out on the newly darkened arboreal background, whereas the melanics—which earlier had been sitting ducks—became hard to find; hence, their frequency increased.

Direct experiments confirmed that selection was favoring the dark form in polluted woods and the light form in unpolluted environments. When equal numbers of light and melanistic forms were released in an unpolluted woodland (light background), more of the light forms were subsequently recaptured, the melanistic ones having perished. Conversely, when both types of moth were released in a polluted woodland (dark background), more of the dark forms were subsequently recaptured, because the light forms had suffered a higher mortality. Films from hidden cameras showed birds preferentially eating dark moths in unpolluted woodlands and light-colored moths resting against soot-covered trees. (There is also a coup de grace: as a result of strengthened antipollution laws that restrict the burning of low-grade coal, airborne soot has diminished, the natural lichens have returned to English trees, and sure enough, the light-colored form of the peppered moth is once again preponderant.)

Here are two more cases in which biologists have been able to witness evolution in action. One involves a simple experimental modification by researchers; the other is entirely natural.

For the field experiment, we must travel to the island of Trinidad. Here, in the Aripo River system, live wild populations of guppies, the same small, attractive fish that are common aquarium pets in North America. Free-living Aripo River guppies occupy small pools, in which different populations are pretty much isolated from one another, often because the pools are separated by waterfalls. Biologists David Reznick and John Endler noticed that some of these guppy populations differed from one another in a variety of so-called life-history characteristics, notably the size of their broods as well as the size and age at which the adults reproduced. Moreover, these traits varied as a package deal: in some guppy populations, large brood size, early reproduction, and small adult size were all found together, whereas in others, guppies tended to produce smaller broods, to breed later, and to grow into larger adults. Not only that, but the latter suite of characteristics was found among guppies who had to cope with predators known as killifish, relatively small animals that are especially fond of eating small, young guppies; we'll call them small-guppy eaters. The other combination of guppy traits (larger broods, smaller, early breeding adults) was found in environments lacking killifish but containing pike-cichlids, which specialize in eating larger, adult guppies. So, guppies in small-guppy-eater pools are larger and slower-breeding; those in large-guppy-eater pools are smaller and faster-breeding.

Reznick and Endler hypothesized that the difference in guppy populations is an evolutionary one, due to differences in predation pressure in the two environments. Specifically, in a world filled with large-guppy eaters, especially likely to pick off bigger individuals, there should be an advantage to guppies who "live hard, love fast, and die young." On the other hand, when small-guppy eaters are present, guppies who survive beyond a minimum, vulnerable size can afford to grow larger and reproduce at a more leisurely pace.

To test this, the two researchers took guppies that had occupied large-guppy-eater ponds (and which therefore followed the "hard, fast, and young" strategy) and introduced them into pools that had small-guppy eaters but no guppies. Then they waited, keeping track of the guppies as they adapted to their new environment. In fact, they waited for eleven years, during which the transplanted guppies underwent from thirty to sixty generations, a long time for human patience but a drop in the evolutionary bucket in terms of natural selection. The results were persuasive. The average guppy weight at maturity increased by nearly 14 percent compared with others who had been left in large-guppy-eater ponds. Age at sexual maturity similarly increased, while average brood size decreased. This research confirmed that predation had selected for differences in guppy life-history characteristics. Even more striking, however, is the fact that an observable evolutionary change took place among free-living animals in a mere eleven years.

The other example of evolution in action entails a visit to the Galapagos finches, a group of birds that strongly influenced the young Charles Darwin when he traveled around the world as naturalist on H.M.S. *Beagle*. Darwin had been struck by the observation that on the Galapagos Islands a single family of birds had become specialized to occupy a variety of different ecological niches. The thirteen species of Galapagos finches have since become icons of evolutionary biology, appropriately lauded in textbooks as poster-child examples of "adaptive radiation," whereby a single group can expand to fill diverse biological roles, occupied in other places by entirely different species. But evolution isn't done with them. At least one species of Galapagos finch has joined the Trinidad guppies in showing how quickly evolution can operate, in this case with no human intervention.

For more than twenty years, Peter and Rosemary Grant, a husband-and-wife biologist team from Princeton University, have been studying one species of finch on a single very small island in the Galapagos chain. The Galapagos finches are usually identified by the culinary specializations revealed by their bills: great, heavy bills for crushing large seeds, medium bills for medium seeds, small bills for small seeds, narrow and pointy bills for insects, and so forth. The "medium ground finches" studied by the Grants are especially fond of medium to small seeds; they can also eat large ones but find them difficult to crack, resorting to them only in unusually dry years, when smaller seeds are scarce.

Assiduously charting the dimension of finch beaks, the Grants found that beak height varied with rainfall: wet years correlate with smaller beaks, more efficient for eating the smaller seeds that are abundant at such times. Dry years correlate with larger, heavier beaks, more suitable for large seeds, which plants are more likely to make when moisture is in short supply. During wet years, smaller beaks are more efficient and are selected for, which simply means that parents with smaller beaks leave more offspring at such times. During drought years, larger beaks are favored. (To clarify: dry years don't *produce* large finch beaks, nor do wet years *produce* small ones, just as guppy predators of one sort or another do not *produce* a particular constellation of reproductive tactics among their prey. Rather, genetic variation is constantly available for all these traits, and the weather, like guppy predators, simply provides a selective environment that favors certain variations and disfavors others. That's evolution.)

The key point for our purposes is that once again natural selection has been shown to give rise to evolutionary change during a very short time. Of course, guppy breeding and finch beaks may appear to be of little import, cases of evolution laboring mightily to bring forth trivialities. Such examples might seem a far cry from horses evolving from terrier-sized *Hyracotherium* to modern Budweiser behemoths, velociraptors arising from the swampy slime, or the human brain expanding from shrewlike insignificance to the crowning, cerebral glory of modern sapient humanity. All these have also happened, powered by natural selection, but such major transitions can only be witnessed in the fossil record. Nonetheless, all evolutionary journeys (including the big ones, so-called macroevolution) begin with small steps (microevolution) and are nothing but their accumulated consequences, over time.

The key is that natural selection allows biological voyagers to place one foot in front of the other, and we are all along for the ride.

The ABCs of ESSs

"May the best man win"—or the best woman, or chimpanzee, or Tyrannosaurus. This is the image most people have of natural selection. Alas, it isn't always accurate. For some evolutionary problems, there is no single best solution, leading to a situation that has recently has been making quite a splash among theoretically inclined evolutionary biologists. It is closely allied to so-called game theory in mathematics. Fortunately, however, one needn't be immersed in arcane numerology to understand the fundamentals of this phenomenon, known as an evolutionarily stable strategy, or ESS.

We can illuminate ESS theory by looking to the childhood game Paper, Rock, and Scissors. Two contestants stand with their hands behind their backs and at each round bring their hands forward, simultaneously, in one of three "moves": flat open for paper, clenched for rock, and extended second and third fingers for scissors. There are clear rules for who wins: Paper covers rock, rock breaks scissors, and scissors cuts paper. What is the winning move? Every child knows that there is no "best" way to play this game, since for every move there is a countermove. (This is a crucial part of game theory: a player's score—or, in evolutionary terms, his fitness—is determined not simply by what he does, but by what the other player does at the same time. As a result, there is no way to determine a best score without taking others into account.) If someone playing Paper, Rock, and Scissors begins to repeat any move, the opponent can specialize in whatever defeats it. So the winning strategy is to choose paper, rock, and scissors unpredictably, but with equal frequency—a probability of one third each time—which is why this game is effective at resolving disputes by chance alone.

Now turn to the world of nonhuman living things, and imagine an exaggeratedly simple situation that offers even fewer options. Let's say there are only two players and two possible moves. Is there some circumstance that might result in both alternatives persisting in the population, without one driving the other to extinction? The answer is yes. (Don't lose track of why this answer is important: it means that evolution can maintain two or more types, without either one "winning" or being in any sense "best.")

To see how this might work, let's look at the Hawk-Dove game. It seeks to explain how natural selection might keep both aggressive and unaggressive individuals at the same time within the same species, with neither type supplanting the other. Imagine there are two kinds, hawks and doves. Hawks fight; doves don't. So whenever they squabble, hawks defeat doves. One might think that hawks would therefore increase in frequency at the expense of doves, so that eventually there would be only hawks. But this need not be the case.

Start with a population of doves. When dove meets dove, competing, for example, over a food source, they don't fight, but instead divide the food 50/50. Introduce a hawk into this amicable world. When hawk meets dove at a food source, the hawk threatens to fight, but the dove backs down. As a result, the hawk gets the food and the dove gets nothing. Repeated many times in this way, such interactions should result in doves going extinct, since all things being equal, food equates to fitness. But here is the tricky part: as hawks become more abundant, they begin to encounter one another. And when they do, they fight, since hawks don't back down. Afterward, each hawk may well get half the food, just as doves do when encountering other doves. But hawks may also suffer damage from their fights with other hawks. If the food is sufficiently valuable and if the wounds that result from hawk fights are sufficiently mild, it may still pay to be a hawk. On the other hand, if food is not in especially short supply and/or if hawks are liable to inflict serious injury or even death on one another, it is entirely possible that hawks begin to do poorly at this point, so their numbers decline, whereupon the number of doves will correspondingly increase. In such a case, the numbers of hawks and doves eventually equilibrate at a certain proportion of each, at which point a condition of evolutionary stability will have been reached. At this equilibrium, the average hawk is as successful as the average dove, so both persist.

A key concept in such interactions is known as "frequency dependence," which simply means that the success of a particular type depends on its frequency. In the hawk-dove example, hawks do well when they are rare (because they only encounter doves, and when they do, they win). But as their frequency increases, they begin to do worse (because they start meeting other hawks, and suffering as a result).

Generally, if there are types A and B, and A does well when rare, it should increase at the expense of B. If it continues to do well as it becomes increasingly abundant, then eventually A will replace B altogether. But if something about A carries within its increasing frequency the seeds of its own destruction (or rather, restriction), then A might do well when rare but poorly when abundant. And if—as in the hypothetical case of hawks and doves—the same thing happens to the alternative form, then both types can be maintained.

Among evolutionary biologists, appreciation and application of ESS theory is still in its infancy. The future may well reveal its relevance in a number of biological situations, whenever variation is stepwise or discontinuous as opposed to gradual and smooth (for example, hawk versus dove instead of differing degrees of hawkishness/dovishness). Something very much like an ESS seems to explain the maintenance of equal male-female sex ratios, as already described.* Another ESS-like situation occurs regularly in supermarket checkout lines or commuter traffic, in which any line that is shorter or has fewer cars attracts more shopping carts or automobiles, until the speed of all "players" tends to reach equilibrium. In any event, it should be clear that evolution leaves room for outcomes that are sometimes more complex than simply one winner, one loser.

Obstacles to Optimality

Faced with evolutionary adaptations of great precision and intricacy, it is easy to conclude that evolution is perfect, or close to it, that the living world is like a gigantic, elaborate game of pool, in which the momentum, carom, and trajectory of every shot are perfectly calculated, with accuracy that would make Minnesota Fats green with envy. But the real world is rarely so obliging. Instead of resembling a competition among hotshot pool sharks, the evolutionary reality is more like the punishment decreed for such people in Gilbert and Sullivan's light opera *The Mikado*: forced to play "extravagant matches" on a "cloth untrue, with a twisted cue, and elliptical billiard balls." As we have seen, evolution can yield results that are remarkable, sometimes even breathtaking, but far from perfect. Here are some reasons why.

First, a slew of genetic factors. Mutation, although necessary for evolution in the long run, nearly always retards adaptive perfection in the short term. This is because mutations are errors, and random errors at that. If you take any well-functioning machine and fiddle

---

Note, incidentally, that an ESS does not necessarily imply equal numbers of contestants. Depending on the payoffs, evolutionary stability could be achieved with any ratio, so long as the individuals of each type receive the same average payoff.

with it randomly, the likelihood is that you will make things worse, not better. But occasionally, perhaps one time in a million, your mindless monkey business may help instead of hurt. So it is, roughly, with mutations. (The special strength of natural selection, as we've seen, is that it eliminates the failures and conserves the rare improvements.)

Early in the history of genetics, it was thought that every gene had just one effect. Today, we know this isn't true: nearly every gene does double, triple, quadruple duty. For example, a gene for a particular eye color in fruit flies also influences their courtship behavior and the way they metabolize certain chemicals. Let's say that the courtship pattern is especially advantageous, maybe enough to override its disadvantageous eye-color effect and its neutral metabolic effect. In this case, researchers looking only at fruit fly metabolism may well notice that it is below the possible optimum, while others, looking at eye color, may also be surprised to see such a clearly maladaptive trait. The answer is that such traits, taken by themselves, may in fact be nonadaptive or even downright maladaptive, although the gene as a whole is selected *for* because it conveys a sufficient advantage in a different arena, in this case, courtship.

Sexual reproduction introduces yet another genetic obstacle to optimality. Imagine a perfectly adapted individual. Indeed, imagine two such superior creatures, who have the good fortune to mate with each other. Although their offspring will probably be more highly adapted than most others randomly chosen from the population, there is no guarantee that the parents' exceptional quality will be passed on to their children. In fact, it is certain that, if nothing else, the offspring of such supercreatures will be different—and randomly so—from either parent. It is also likely that random changes from a high level of perfection will make things worse.

Such is the nature of sexual reproduction: it is a mindless, unplanned process by which the genes of each parent are arbitrarily chosen and then haphazardly recombined to produce new individuals. Renowned evolutionary biologist George C. Williams has referred to sex as a "Sisyphean" process, after the character in Greek mythology who was doomed to spend eternity rolling a large rock up a steep hill, only to have it inevitably roll back down. Selection labors to produce highly adapted individuals, who, when they reproduce, yield random outcomes that are often less "elevated" than the parents. As W. B. Yeats put it, "Things fall apart, the center cannot hold. . . ."

Finally, let us consider one more genetically based obstacle to optimality: the phenomenon sometimes called genetic drift. This is the statistical fact that small population size can sometimes override the effect of natural selection. If there were, say, 1 million orangutans in Borneo, and a sudden event (such as the massive forest fires that began in 1997) wiped out 20 percent, there would still be 800,000 left. Not only would sheer numbers buffer the species from extinction, it would also ensure that any especially valuable genes would be retained, and similarly, that the survivors would not be saddled, just by chance, with a disproportionate number of defectives. But of course, there are far fewer orangutans. Imagine that one local population consists of merely a few hundred, broken up into smaller groups of a dozen or so. Any one of these groups may contain a high frequency of some disadvantageous genes or a low frequency of especially well-suited ones, simply by the luck of the draw.

A variant of genetic drift is the so-called founder effect or its close ally, the "bottleneck principle," whereby a population is established based on a small number of colonists ("founders"), or after high mortality eliminates all but a few survivors (which passed through a kind of bottleneck). In either case, the result is a very small and thus unrepresentative sample of what existed previously, and whereas the large population likely reflects the action of natural selection, such a small sample may not.

In *The Song of the Dodo*, which explores the phenomenon of extinction in island populations, nature writer David Quammen gives a nice analogy. Imagine that you have a drawer full of socks, most of them black, brown, or blue, but one pair is flamingo pink. Packing your clothes for a trip, in a hurry and in the dark, you grab a handful of socks. The chances are that the socks in your luggage will be reasonably appropriate for your purposes; it is always possible, however, that you will end up with one or both of those unusable pink atrocities. On the other hand, it is also possible that if your trip takes you to Las Vegas, those flamingo beauties will be just the thing—but are unlikely to have made it through your sock-grabbing bottleneck.

The Pingelapese people of one of the Caroline Islands in the South Pacific evidently passed through a kind of bottleneck when a severe typhoon in 1780 killed all but nine men, from whom the current population is largely descended. Just by chance, these "founders" evidently happened to have a high frequency of a gene leading to what is now known as Pingelapese blindness, a very rare genetically based illness leading to color blindness, severe myopia, and cataracts. There is no reason to think that genes for Pingelapese blindness are adaptive or that they conferred any advantage when the typhoon struck. It appears that the gene was given an evolutionary head start simply because the population was small and among the survivors were a disproportionate number of gene carriers. Currently, between 4 and 10 percent of the Pingelapese population suffer from this condition, which inspired the book *Island of the Blind*, by neurologist Oliver Sachs.

There are other factors contributing to the less-than-perfect outcome of evolution; call them situational as opposed to genetic. We've already considered some consequences of the fact that evolution must deal with historical contingencies (hence it produces gill slits in mammal embryos, an overly narrow human birth canal, etc.). The past can also restrict the future, in seemingly nonadaptive ways. For example, it would probably be adaptive for elephants to fly, or at least to hover 30 feet or so above the ground, happily grazing on leaves that would otherwise be out of reach. But there are no real-life Dumbos, simply because all living things labor under the constraints of their evolutionary past. Birds have been selected for extremely light weight as part of their adaptations for flight, including bones that are virtually hollow. Elephant bodies have undergone selection for something very different, namely, the ability to grow very large, with bones that are both dense and immense. Once elephants began traveling down their particular evolutionary road (on foot, rather than by wing) they could no longer turn back and become birds, any more than they could climb trees or play the piano, even if it might be adaptive, or optimal, for them to do so.

Human beings and mammals in general might similarly be evolutionarily blessed if they had a third eye (perhaps in the back of their heads, to warn about ambushes from behind), or gills (think of the drownings that would be prevented). There is nothing intrinsic to being a mammal that precludes gills, a third eye, or a third arm for that matter. It's just that such major structural innovations are not available to animals that carry with them the baggage of hundreds of millions of years of evolution based on lungs and a bilaterally symmetric body plan. We, like all other living things in their own way, are prisoners of our past, which means that we are often stuck in suboptimality.

Another obstacle to optimality is imbedded in the fact that much of life is a zero-sum game, in which winners imply losers (thus, the sum of winning and losing is zero). As a result, the success of an evolutionary competitor often occurs at the expense of another's failure. The one that fails is, in a sense, suboptimal, but such an outcome is also unavoidable, if only as the flip side of the winner's success. For example, despite all the marvelous antipredator adaptations of the

Thomson's gazelles of Tanzania's Serengeti Plain, these animals are often killed by leopards. When a leopard wins, a gazelle loses, and vice versa. Most of the time, in fact, gazelles win and leopards lose, simply because the cost of losing is greater for a gazelle (it loses its life) than for a leopard, which only loses a meal. For gazelles, it would be evolutionarily optimal if they could run so fast that no leopard would ever catch them. But gazelles are unlikely to evolve into such safety, if only because natural selection carries within it a built-in counterbalance: faster, more alert gazelles select for faster, more alert leopards, because in a world of quick-moving gazelles, only quicker-moving leopards get to survive and reproduce. Similarly, highly tuned leopards select for hard-to-kill gazelles, because in a world of efficient leopards, only more efficient gazelles are likely to succeed. At the same time, whenever a leopard takes off after a gazelle, there will be a winner and a loser.

The final obstacle is increasingly important and widespread: evolution cannot anticipate the future. Indeed, it cannot even plan for the present. It can only respond to the past, equipping animals and plants with adaptations that served their ancestors well, based on those traits that worked once upon a time. The key, then, is the extent to which the present and the future resemble the past. Insofar as environments change slowly, adaptations are likely to be reasonably good, or at least adequate. But the more rapidly things change— the more the present and future depart from those past environments to which evolution is attuned—the more suboptimal, or even maladapted, living things will be.

Here's an example. Musk oxen live on the northern tundra, looking like great shaggy buffalo. For eons, they have defended themselves from their most lethal predators, wolves, by a simple but highly effective strategy. They form a dense circular array, with the adults facing outward like spokes of a wheel and the young calves protectively huddled inside. As a result, an attacking wolf, no matter the direction of his or her assault, must face a musk ox head-on, contending with the formidable combination of sharp horns and bony forehead plate, backed up by a thousand pounds of angry arctic pot roast. But in recent decades, things have changed. Instead of facing wolves, musk oxen are challenged by a new danger: human hunters, armed with high-powered rifles and riding snowmobiles. The animals respond as they always have, forming their age-old defensive circle, and are easily shot. It would be far better for them to spread out and run away, but they stick to the only response with which evolution has equipped them. Maybe in a few thousand years, if the species survives and if human hunting continues unchanged, musk oxen will have evolved a new and more effective response. But for now their defensive behavior is clearly outmoded and maladaptive, given the extent to which human beings have changed their world.

The predicament of the musk ox is dramatic but not unique. Current environments are filled with manmade innovations, including climate change, pollution, habitat rearrangement, and the rapid introduction of species into areas they had not previously occupied. In short, things are changing very rapidly. The future, it is sometimes said, is not what it used to be. This is increasingly true, and not simply because it is different; after all, the future has always been different. Our point is that because of human intervention, the future is *more different*, by which we mean that there is a greater disconnect between past and future environments than perhaps at any time in the history of life, except for immediately after worldwide calamities such as a major meteor impact. For many living things, the "time is out of joint," and their adaptations are more likely than ever to resemble those of the musk ox, which is currently threatened with extinction.

Evolutionary Ethics?

When it comes to oxymorons, "evolutionary ethics" ranks up there with "creation science." And yet, people have repeatedly tried to derive moral principles from evolution. Not surprisingly, they haven't won many converts, but still the enterprise persists. Like some nightmare of the undead, evolutionary ethics lurches along, foul-smelling and covered with flies but refusing to lie down or go away.

It has been said that for every complex problem there is a solution that is elegantly simple—but wrong. Evolution by natural selection is an elegantly simple solution to the question of why life is as it is, with the added virtue of being right. But please don't look to it for ethical guidance. Evolution cannot tell us how things ought to be or how we ought to live. Not that people haven't tried. "Social Darwinism" flourished in the late nineteenth and early twentieth centuries, espoused especially by the wealthy and the powerful, since it seemed to offer a "natural" justification for the oppression of the weak by the strong. After all, they claimed, the poor and the downtrodden were merely getting what they deserved, and the excesses of laissez-faire capitalism, colonialism, racism, aggressive war, and other horrors were simply human beings acting out "the survival of the fittest."

This is a grave blunder, in several ways. First, it confuses the actual mechanism of evolution—differential reproduction—with physical, technological, and economic superiority. Evolution rewards success in projecting genes into the future, not success in beating up one's rivals.* More important, evolutionary ethics confuses a scientifically valid account of a natural process with an ethically dubious proposition: that what *is* is therefore right. Evolution explains life, just as gravity explains weight, but at the same time, evolution is no more a touchstone for how we ought to behave than is gravity, or oxidation, or the Big Bang.

Many of us are biased—understandably so, in most cases—toward things that are "natural," including foods, lifestyles, environments, and so forth. But the fact that something is natural—as in natural selection—does not make it laudable. Nothing is more natural than a virus, but even the most ardent social Darwinists would probably not consider smallpox a paragon of ethical propriety. Right or wrong is a human judgment, and whereas evolutionary principles can doubtless help people understand the origin of their own human nature—maybe even including their inclinations to judge things ethical or unethical—no amount of evolutionary insight can actually tell us what is wrong and what is right.

We would go further and suggest that for all its importance and the fascinating insight it offers into how the living world is put together, evolution by natural selection is, if anything, downright evil, at least by the moral standards we find appealing. Admirers of Mussolini during the 1930s used to comment that he made the trains run on time. For all its inadequacies, its failure to achieve perfect optimality, evolution by natural selection is remarkably good at making the biological trains run on time. But the technical efficiency of fascism and Nazism could not wipe away their ethical depravity. So, too, with evolution.

Evolution is not malevolent. Indeed, it is downright benevolent toward whatever works, ruthless toward what falls short. It is simply *a*moral, a mindless, sometimes blundering trial-and-error process

---

* *To be sure, sometimes evolutionary success is achieved via aggressiveness, but there are also cases in which excessive animal aggressiveness results in reduced fitness, because, for example, violent individuals are more likely to be killed or injured, or to neglect their offspring while pursuing their rivals.*

that often rewards the selfish, the violent, and those who plan only for today (even if that means extinction tomorrow). Natural selection has no compunction about producing parasites that consume their living prey from the inside, or rewarding individuals who brutalize their fellows if that is what it takes to yield reproductive advantage. It is less lurid, in many cases, than Tennyson's "nature red in tooth and claw" but is far too bloodthirsty and ethically vacant to serve as a decent model for anyone's morality.

"Earth can not count the sons she bore:" wrote Edna St. Vincent Millay in "The Return." "The wounded lynx, the wounded man/ Come trailing blood unto her door;" We are not so sure about the poem's next line: "She shelters both as best she can." It is not the earth that protects the wounded lynx or man, but the efforts of lynx and man to shelter themselves or, more rarely, each other. Natural selection—Millay's Earth—is wildly profligate, spewing forth countless numbers of sons and daughters, not unlike Napoleon, who once bragged, "You can never defeat me, I spend fifty thousand men a month."

We can never defeat natural selection. Even in a human-dominated world, it continues to sift and winnow among the genetic possibilities offered by human, halibut, or hickory tree. It produces results, but so did Dachau, the Manhattan Project, and Genghis Khan. Evolution is beautiful, but it is a cold, unfeeling, utterly amoral beauty. It has no room for mercy, compassion, right or wrong, or indeed anything except insofar as it gives a boost, however slight, to genes jousting to get ahead. It is overwhelmingly important, but as something to understand, not to emulate.

As Ezra Pound wrote in his poem "The Unmoving Cloud":

"The clouds have gathered

And the rain falls and falls,

The eight ply of the heavens are all folded into
    one darkness,

And the wide, flat road stretches out.

SOCIOBIOLOGY:

Gene Machines in Action

"Descended from monkeys?" the good wife of the bishop of Worcester ejaculated (to use a perfectly good nineteenth-century word). "My dear," she went on, "let us hope that it isn't true! But if it is true, let us hope that it doesn't become widely known!" Well, it *is* true, and moreover, it has become widely known.*

Although controversial in its time, and even disputed today—at least among the flat-earth, "creation science" crowd—the reality of human descent from "monkeys" is no longer in serious dispute. Even the Catholic church has made its peace with evolution, including human evolution. Oddly enough, however, the ghost of the good bishop's wife is still with us; or perhaps it is the ghost of Tycho Brahe.

Brahean Thinking

Tycho Brahe was a renowned sixteenth-century astronomer, whose intellectual contortions provide a metaphor for a pervasive human penchant. Brahe worked during a troubled but exciting time for astronomy, when the widely

---

Actually, the truth is a bit more shaded. We are closely related to monkeys, descended—along with them—from early primates.

held, Earth-centered model of the solar system was becoming untenable. Unmistakable evidence showed that the known planets—Mercury, Venus, Mars, Jupiter, and Saturn—circled the sun. We can just imagine a sixteenth-century antecedent of the bishop's wife decrying this unseemly development, piously hoping that, if nothing else, it could be hushed up. We can also imagine how disconcerting these facts of planetary movement were to Brahe, who was both a devout Christian and a hardheaded empirical scientist.

The creative Brahe was up to the challenge, however. He proposed an inventive solution: a new astronomical model in which the five planets indeed went around the sun—as observation demanded—but with the whole business (the sun plus its five satellites) obediently circling the Earth!

We call this "Brahean thinking," and it deserves to be, if not honored, then at least recognized, for it rears its head in many situations. Basically, Brahean thinking induces people to accept, grudgingly if need be, awkward and discomfiting facts (what the facts demand), while at the same time retaining the core of their beliefs (what their bias demands). In Tycho Brahe's case, the core belief was that the solar system if not the whole universe was geocentric, not heliocentric. In the case of many people considering evolution, the core belief is that at some crucial, fundamental level, human beings are not really animals after all. Thus, today's educated people are likely to accept the reality that evolution is a historical fact, underlying all life. But in their heart of hearts, they are also inclined to relegate evolution to dusty museums, to tuck it away with dinosaurs, Charles Darwin, perhaps the gill slits of human embryos and the bones of "cave men." It does not apply to their own lives. They accept the peripheral facts of evolution, while retaining their core belief: that human beings are so important, so special, so *different*, that at a fundamental level, the deepest evolutionary reality—continuity—doesn't really apply to *Homo sapiens*. In short, Brahean thinking applied to evolution leads to the conviction that although evolution is real, it doesn't really matter!

It is hard to overstate the fervor with which many people demand that despite all evidence—from the DNA within our cells, to the structure and function of organs and organ systems, to our phylogenetic history and place in the ecological scheme of things—we somehow exist outside the natural world. The premise of this book, that human beings are perfectly good mammals, to be understood by looking in that mirror called science, is widely anathema. Nowhere, it seems, is this resistance more intense than when someone has the effrontery to point out that our mammalness includes not only biochemistry, physiology, and anatomy, but behavior, too.

In his poem "The White-Tailed Hornet," Robert Frost caught some of the outrage that results from identifying human beings with animals:

> As long on earth
>
> As our comparisons were stoutly upward
>
> With gods and angels, we were men at least,
>
> But little lower than the gods and angels.
>
> But once comparisons were yielded downward,
>
> Once we began to see our images
>
> Reflected in the mud and even dust,
>
> 'Twas disillusion upon disillusion.
>
> We were lost piecemeal to the animals,
>
> Like people thrown out to delay the wolves.

For others, including Charles Darwin, the connection is not only true but exalting. Here is the final paragraph of *The Origin of Species*:

> It is interesting to contemplate a tangled bank, clothed with many plants of many kinds, with birds singing on the bushes, with various insects flitting about, and with worms crawling through the damp earth, and to reflect that these elaborately constructed forms, so different from each other, and dependent upon each other in so complex a manner, have all been produced by laws acting around us. . . . Thus, from the war of nature, from famine and death, the most exalted object which we are capable of conceiving, namely, the production of the higher animals, directly follows. There is grandeur in this view of life, with its several powers, having been originally breathed by the Creator into a few forms or into one; and that, whilst this planet has gone cycling on according to the fixed law of gravity, from so simple a beginning endless forms most beautiful and most wonderful have been, and are being evolved.

The most conspicuous descendants of Darwin today are the sociobiologists (sometimes known as evolutionary psychologists, when they turn their attention specifically to human behavioral tendencies).

Their self-appointed task is to follow the light of evolutionary biology right into that stronghold of human conceit: our minds and our behavior. The word "sociobiology" had floated about obscurely in the scientific literature before being revived by Harvard University's Edward O. Wilson, who used it to title a large and impressive book, a masterful synthesis of evolutionary biology, ecology, and animal behavior that was published in 1975.

Sociobiology draws from a wide range of sources, seeking to point out unifying principles that make sense of social behavior in species ranging from honeybees to human beings. When it comes to living things "below" the rank of *Homo sapiens*, the battle has long been decided: sociobiology won, hands down. The ultimate victory in science goes to any approach that works, that suggests interesting explanations, furthers inquiry, and—almost miraculously—attracts adherents simply because it does a good job revealing that elusive but corny prize known as Truth.

But when it comes to the citadel—human beings themselves—the outcome is less one-sided than when applying sociobiologic thinking to other animals. From here on, when we refer to sociobiology, we mean human sociobiology or evolutionary psychology, since in its application to nonhuman animals, sociobiology is no longer in debate. The adverse reaction to evolutionary thinking applied to human beings has been predictably fierce, and not only out of fear that we are being "lost piecemeal to the animals."

Howls from the Critics

Hell hath no fury like someone whose treasured worldview is being overthrown. And this applies to scientists no less than anyone else. Social scientists (anthropologists, psychologists, sociologists) have been particularly irate, largely because their worldview emphasizes the overwhelming importance of learning, culture, and social traditions. Human specialness, they argue, derives from what we teach one another and ourselves, especially our complex symbolic systems, including language. In short, there is a powerful bias in favor of human specialness, according to which sociobiologists do their species a disservice by emphasizing those human traits that unite *Homo sapiens* to the rest of the natural world. Far better, more important, more interesting, and more gratifying, argue the critics, to focus on those traits that uniquely distinguish our species. Even renowned evolutionary biologist Julian Huxley warned against "nothing butism," the erroneous notion that just because human beings are animals, they are nothing but animals.

Huxley's point is well-taken. Whatever else we are, human beings are indeed the product of learning, culture, social norms, an extraordinary penchant for symbols, complex language, and elaborate thought processes. But the sociobiologic perspective also turns Huxley's argument around, pointing out that even though human beings are all of the above, it is equally erroneous to presume that they are *nothing but* the product of their learning, their culture, their social norms, and so forth. People, argue the sociobiologists, are also the product of evolution by natural selection, perfectly good mammals whose biological nature is revealed in their behavior no less than in their fossils, physiology, or physique.

For sociobiologists, in short, there is such a thing as human nature, just as there is harvest-mouse nature or hyacinth nature. Carl Sandburg said it beautifully in his poem "Wilderness," although perhaps overplaying the aggressive side of things. For Sandburg, there is a "menagerie" inside every person, but at the same time there is "something else."

> There is a wolf in me . . . fangs pointed for tearing gashes . . . a red tongue for raw meat . . . and the hot lapping of blood—I keep this wolf because the wilderness gave it to me and the wilderness will not let it go.

> There is a hog in me . . . a snout and a belly . . . a machinery for eating and grunting . . . a machinery for sleeping satisfied in the sun—I got this too from the wilderness and the wilderness will not let it go.

> There is a baboon in me . . . clambering-clawed . . . dog-faced . . . yawping a galoot's hunger . . . hairy under the armpits . . . ready to snarl and kill . . . ready to sing and give milk . . . waiting—I keep the baboon because the wilderness says so.

> Oh, I got a zoo, I got a menagerie, inside my ribs, under my bony head, under my red-valve heart—and I got something else: it is a man-child heart, a woman-child heart: it is a father and mother and lover: it came from God-Knows-Where: it is going God-Knows-Where—

> For I am the keeper of the zoo: I say yes and no: I sing and kill and work: I am a pal of the world; I came from the wilderness.

Let's consider some specific objections to sociobiology.

"Sociobiology is a doctrine of genetic determinism that deprives people of their free will."

First, no science can deprive people of their free will, or indeed of anything else, although applications (rather, misapplications) of science can sometimes do so. If people possess free will, sociobiology cannot take it away; if they lack it, sociobiology cannot provide it.

Second, genetic *determinism* is not at the heart of sociobiology. Genetic *influence* is. The two are quite different: whereas blood type, for example, is genetically determined, tendencies to behave aggressively, parentally, romantically, and so forth appear to be genetically influenced, which simply means that certain genetic combinations predispose people to behave in one direction or another. And a predisposition, inclination, or tendency is a far cry from a guaranteed certainty from which there is no escape.

Finally, it may be even more erosive of freedom and dignity if human behavior were entirely a function of learning, cultural traditions, and social norms, as social-science traditionalists often claim, because in that case, people are simply a pale reflection of what they have experienced, instead of bringing something (their "nature," derived from evolution) to the encounter. Moreover, it would be a totalitarian's dream if human beings were simply molded by their experiences, for then any tyranny—if it possessed the appropriate behaviorist technology—could sculpt obliging subjects. But if people really exist in their own right, as creatures with inherent inclinations, then they are likely to resist such manipulation. And human rights, inherent in being of the species *Homo sapiens*, might not be far behind.

"Sociobiology is a political and social doctrine that is conservative, if not downright reactionary, since it supports the status quo as the way things naturally are and thus how they ought to be."

This confuses *explaining* something with *excusing* it. To be sure, there is something appealing about the "natural" and the "organic," but some of the most natural things are also among the most unpleasant: gangrene, for example, or syphilis. When virologists study HIV, they are not justifying it. And when sociobiologists seek to unravel the underlying basis of male violence, polygyny, female-biased parental care, and so forth, they are seeking insight, not moral authority. Ironically, Karl Marx was so impressed with *The Origin of Species* that he proposed dedicating his first edition of *Das Kapital* to Darwin! After all, evolution is a doctrine of change, even of revolution, demonstrating that the living world is in flux due to competition, as opposed to the Special Creation of an unvarying natural or social order. When *The Origin of Species* first appeared, it certainly wasn't seen as conservative or supporting the status quo; far from it. The royal houses of Europe, recently shaken by the revolutions of 1848, were in fact terribly frightened that evolution, by demonstrating the naturalness of change, might provide ammunition for social revolution. Either way, however, there is no necessary connection between how things are (a statement of physical and biological fact) and how they ought to be (an ethical judgment).

"Sociobiology is racist."

Given the terrible history of misused genetics—from the hijacking of racial theory by the Nazis to the intellectual thuggery of *The Bell Curve*—this worry is, sadly, understandable. But it seems misplaced. Thus, sociobiology says virtually nothing about the differences among human groups but a lot about their underlying similarities. A sociobiologic view of human behavior focuses on what some anthropologists call "cross-cultural universals," those behavioral traits that people share, regardless of such superficial differences as cultural tradition, language, or skin color. If anything, sociobiology offers a potential antidote to racism.

"Sociobiology is sexist."

If sexism means the identification of male-female differences, then sociobiology is indeed sexist—but so are anatomy, physiology, embryology, biochemical genetics, and so on. Some of sociobiology's most cogent insights explain why men and women, boys and girls are different. This is a far cry, however, from valuing the two sexes differently or suggesting social policy; rather, it aims to illuminate the biology that underlies human reality. Interestingly, the "sexism" of sociobiology is compatible with an increasingly influential approach, known as "difference feminism," which glorifies certain male-female differences, such as the purported female penchant for social relationships over hierarchy and for nurturance over violence. Sociobiology promises to help us understand some of these differences. What, if anything, people choose to do about them is their business, a matter of ethics and social policy, not science.

"Sociobiology cannot be tested."

Actually, it can and is being tested constantly. In a sense, the immense diversity of human cultures and social systems is itself a huge, ongoing scientific experiment. One thing is held constant: *Homo sapiens*, the same from Greenland to New Guinea, from Manhattan to Tierra del Fuego, while other "things" are varied: namely, social and cultural rules, languages, and so forth. If, in such cases, universal patterns exist—and they do—it is only reasonable to attribute these commonalities to that one thing that the great human experiment on planet Earth also holds constant: namely, the biological nature of human beings. The existence of consistent, shared behavior patterns among all people is itself, therefore, prima facie evidence for a consistent, shared biology. Such evidence is even more suggestive when the shared behavior patterns are in directions predicted by evolutionary biology.

"Sociobiology relies on genetic assumptions that have no foundation."

There is some truth here, in that sociobiology is based on the principle that behavior is influenced by genes, most of which have not yet been clearly identified. The behaviors in question can be diverse and complex, including altruism, sexual inclinations, parenting tendencies, and so forth. In recent years, research has in fact uncovered genetic inclinations for a remarkable and increasing range of human behavior, from thrill-seeking to homosexuality to disease susceptibility.

But even though the evidence supporting genetic influence on a remarkably precise array of things is overwhelming, it still isn't necessary to postulate the existence of genes "for" everything under the sun. All that is needed is for individuals carrying certain genes to behave differently, on balance, from individuals carrying other genes—which is not the same as presuming that the behavior in question is rigidly "encoded" by DNA.

Take thrill-seeking. It is now evident that people carrying a particular genetic variant—which influences metabolism of the neurotransmitter dopamine—are statistically more likely to engage in high-risk behavior. Sociobiologists can therefore speak about genes "for" bungee jumping without literally implying that there are genes for bungee jumping! (Rather, the ranks of bungee jumpers are statistically more likely to include such gene carriers than are couch potatoes; the same would apply to high-angle ice climbers compared with people who would rather play shuffleboard.) Similarly, sociobiologists have been able to achieve stunning insights by thinking about genes "for" altruism versus selfishness, without necessarily specifying the exact pattern of DNA or its precise biochemical effects.

"Sociobiology promises more than it delivers. By focusing on people's genetic and evolutionary components, it leaves a lot out."

Sociobiology is more like a telescope than a microscope, since its power derives from seeing larger patterns rather than revealing the details of fine structure. To be sure, it is possible to learn about stars without telescopes; people did it for hundreds, presumably thousands of years before Galileo. And Galileo, interestingly, was unable to persuade certain clerics even to look through his telescope: if God had meant people to view the stars with powerful lenses, they argued, we would have been born with them. Others were disappointed when they couldn't see God's footprint on Jupiter, or the North Star, or use their new devices to unravel all the secrets of the visible universe. Telescopes haven't answered all astronomical questions; on balance, they may even have raised more questions than they have satisfied. All to the good. There is no doubt that telescopes have added to the sum total of human knowledge, although they aren't the be-all and end-all of such knowledge, or even of astronomy. The same can be said of sociobiology. It doesn't explain everything about the behavior of *Homo sapiens*, but it provides at least a revealing glimpse of one of this planet's most interesting mammals.

Much has already been heard from sociobiologists and their kindred spirits, evolutionary (or Darwinian) psychologists. There are, in addition, sociobiologically inclined anthropologists, sociologists, economists, political scientists, and legal scholars, but—as might be expected in disciplines that are notoriously nonbiological and often downright antibiological—they are in the minority. The likelihood, nonetheless, is that sociobiology will continue to advance its science in the coming years. Moreover, because of its novel insights into *Homo sapiens*, it will probably continue to catch the public eye. This makes it especially important to avoid likely misunderstandings, two in particular.

One of these involves the kind of questions asked. The physical sciences are typically unconcerned with "why" questions, devoting themselves exclusively to "how." Physicists don't ask *why* there are negative and positive charges; they ask *how* they function. They usually don't ask *why* light travels at 186,000 miles per second; they ask *how* this impacts the material world. Biologists, too, are largely concerned with answering "how" questions: previous chapters in this book have dealt with how DNA works, how viruses cause disease, how cells operate, how neurons create thought, how energy is processed, and so forth.

On the other hand, sociobiology usually focuses on *why*. Why are males typically more aggressive than females? Why does altruism occur? Why is step-parenting so difficult? To be sure, these questions can be answered in "how" terms, with reference to hormones, physiological or anatomical mechanisms, prior experience, and so forth. But this ignores an underlying question: Why is the body sensitive to, say, testosterone, producing a particular effect . . . rather than some other outcome? For example, it is not sufficient to say that males are more aggressive than females "because of" testosterone, when in many species, testosterone increases female aggression as well. The question then becomes: Why do males produce more testosterone? Which is really: Why are males, on balance, more aggressive than females?

Sociobiology's drive to address this sort of question introduces an evolutionary bias: namely, what is the adaptive significance of the phenomenon at issue? Thus, when sociobiologists pose "why" questions about such things as male versus female aggressiveness, they are less concerned with the immediate mechanisms that generate an effect than with the underlying evolutionary reason for its existence. When confronted with a sociobiologic explanation for something, it is quite reasonable to ask, "But *how* is this achieved?" At the same time, sociobiologists' concern with *why* over *how* is not a weakness but simply a difference.

The second common source of misunderstandings is over the kind of answers expected. Many people think that a sociobiologic interpretation of human behavior will be wholly "genetic," denying the importance of culture, learning, and other environmental influences. Given the emphasis sociobiologists place on genes and evolution, this is understandable but an error nonetheless. In fact, one of the most important insights gained in recent decades by students of behavior, genetics, and development is that no characteristic of any living thing—whether structure, physiological mechanism, or behavior—is produced by either genes or environment acting alone. For example, if a person is 5 feet 6 inches tall, it is absurd to assert that 3 feet of her height is due to genes and 2 feet 6 inches to her nutrition. Every inch of her stature is attributable to her genes and her experiences, acting together. So it is with behavior. (Despite this assertion, the discerning reader will note that the various sociobiologic concepts discussed in the rest of this chapter do in fact lean almost exclusively on genetic presumptions! This is not inappropriate, since the overwhelming bulk of social-science research and speculation weighs in on the environmental side, making the sociobiologic contribution a welcome counterbalance.)

As described in Chapter 8, biologists generally agree that evolution works most powerfully at the lowest levels of selection, especially that of genes. Much of the unique power of sociobiology comes from this focus. As British zoologist Richard Dawkins emphasizes, genes are deeply "selfish," which is to say they are concerned with promoting their own success . . . actually, not *their own* success so much as that of identical copies of themselves, contained in other bodies. From the sociobiologic perspective, that is *all* they are concerned with. That is, genes have evolved to maximize their "fitness," which is their success in projecting copies of themselves into the future. And bodies are the means whereby they achieve this end. The various ideas of sociobiology ultimately come down to this: insights into the "why" of behavior, derived from the proposition that, whether they know it or not, living things in general—and genes in particular—go about their lives seeking to maximize their fitness.*

Now we briefly review some of the central findings of human sociobiology, as we look at our own species, not so much in the mirror as through the telescope of evolution.

Altruism: Paradox and Paradigm

For biologists, altruism has long been a paradox. Evolution, after all, is fundamentally selfish, since, as we have seen, natural selection is competitive: genes as well as individuals joust with one another, and success crowns the victors. The problem is deceptively simple. How can natural selection produce and maintain behavior that is, by definition, self-defeating? It is crucial to realize that an altruist isn't just nice to someone else but is self-denying at the same time. There is no great evolutionary mystery, for example, about "nice" behavior by which everyone gains: if both recipient and donor are better off as a result, then natural selection could readily favor such niceness, from the donor's perspective no less than the beneficiary's. Such cooperation used to be known as symbiosis and is now called mutualism.

But altruists ends up *worse* off as a result of their act. While ethicists are delighted by altruistic deeds, evolutionists are intrigued, even disbelieving, because any genetic tendency to behave altruistically should incur the wrath of natural selection. Genes that predispose their carriers to do something detrimental to themselves should eventually disappear, replaced by alternative genes that whisper, "Look out for number one."

---

* *This is just a verbal shorthand. Genes needn't consciously try to do anything; rather, those that succeeded were more fit, so the ones that predominate are those that act in a particular way, regardless of motivation.*

When evolution was thought to operate "for the good of the species," biologists paid little attention to altruism, which was taken for granted: of course individuals are inclined to help others, even at substantial cost to themselves, because the species as a whole is better off as a result! But with the focus shifted to individual and gene benefit, altruism seemed puzzling indeed.

The most exciting and persuasive answer was provided by British geneticist William D. Hamilton, who published his findings in a now-classic paper in 1964. The gist of his work is deceptively simple, one of those rare and remarkable insights that, like Newton's discovery of gravity, are obvious in retrospect . . . that is, after someone else makes them! Hamilton's insight was this: genes promote the success of other genes—identical copies of themselves—by contributing to the success of bodies that house them. These bodies are often one's children, in which case the means of promotion and success is called reproduction. But reproduction is only a special case of the more general phenomenon by which genes are projected into the future via the success of *any* bodies that carry them. Accordingly, genes and individuals are really concerned about something more inclusive than baby-making. In the long run, natural selection favors genes that induce their bodies to do whatever is most likely to assist identical copies of those genes. For their part, genes don't run around naked; rather, they exist inside bodies, which, in fact, they created for their own benefit. Under the right conditions, genes can promote themselves by getting those bodies to value their kin more than their skin.

As Robert Browning might have put it, "How do I love thee? Let me count thy genes!"

Incidentally, the fact that people don't spend much time counting one another's genes doesn't mean that they don't behave with regard to the outcome. To take an extreme case, a rock doesn't know very much about gravity but is nonetheless inclined to behave as gravity dictates. Birds, similarly, need not consciously know the benefits of migration in order to migrate, any more than plants require knowledge of the seasons in order to flower in spring rather than winter. With bird migration, plant flowering, or human altruism, it is enough that individuals who behave adaptively—that is, whose genes gain an advantage as a result—will be giving a head start to those genes, so that over time, populations will tend to be comprised of individuals tending to behave in this manner. All this regardless of whether they can count, calculate, think, or in any way acknowledge the reality of natural selection.

In English, we have a word for individuals with a definable probability of carrying genes already present in a target individual: *relatives*.

The "closer" the relative, the higher that probability. Recall Hamilton's insight: natural selection rewards not baby-making as such but rather the propagation of genes. This can be done by reproducing and/or by helping others, with those "others" valued in proportion as they are more closely related. One offspring is thus "worth" one half of one's self, or equivalent to one full sibling. A grandchild and a niece or nephew are equivalent, genetically, and two grandchildren (or two nieces or nephews) are "worth" one child, and so forth.*

Several decades before William Hamilton shook up the sociobiological world—one might even say initiated it—the renowned biologist J. B. S. Haldane was asked if he would give his life for a brother. "No," he reputedly answered, after scribbling on the back of a napkin, "but for three brothers, or nine cousins."†

How can this explain the paradox of altruism? By showing that it isn't a paradox at all! Genes for altruism can prosper, even though they are *less* successful (are "selected against") within the body of an altruist, so long as they are *more* successful overall. This can happen if they increase the chances that other genes, identical to themselves, are selected for among genetic relatives that benefit from the altruistic behavior. Another way of looking at it: what appears to be altruistic, at the level of individuals, can actually be selfish at the level of genes. The paradox melts away.

Here is an animal example. Prairie dogs give alarm calls when they spot a coyote. By drawing attention to itself, the alarmist becomes vulnerable while benefiting others who hear his call. So, alarm-calling genes (and thus, alarm-calling itself) should disappear. But if alarm callers alert other prairie dogs, and if these others are likely to be relatives, then alarm-calling genes can be ahead in the long run. They are, in a sense, selfish, since what appears to be altruistic is actually a means of aiding themselves.

Examples of this sort are now commonplace, so that when biologists are confronted with a case of seeming altruism, their first thought is to look for benefits derived by the altruist's relatives. Food sharing, for example, or social tolerance generally, or cooperative defense

---

*A grandchild's genes have a one-fourth probability of being present in oneself; similarly for a niece or nephew. Two grandchildren (2 × ¼) are genetically equivalent to one child, each of whose genes has a one-half probability of existing in the parent.

† A brother, like a child, is genetically equivalent to one half of oneself. So three brothers "equal" 3 × ½, which exceeds one's own genetic worth. Similarly for nine cousins, since one cousin is genetically equivalent to one eighth of one's self, and 9 × ⅛ is greater than one.

against predators: these behaviors often reveal a predictable kin bias. Hence the widely used expression "kin selection."

For another example reflecting some of life's complexity, consider mating behavior among lions. Typically, there is more than one male in a pride, although mating is not democratic. The most dominant lion gets to do most of the breeding, which would not appear to profit the fitness of subordinate males—especially when there are four or more males, and those at the low end of the pecking order are unlikely to breed at all. What's in it for them? Why do they contribute to the pride in other ways, if they are not going to breed? The answer seems to be that subordinate individuals tend to be related to the dominant, reproductively successful males, often as brothers or half-brothers. As a result, even though the subordinates may never get to be fathers, they may succeed, via kin selection, in being uncles, with a surprisingly high "inclusive fitness" of their own.

Now make sense of these facts. Male lions are willing to join a small pride even if unrelated to the other males, whereas they will only join a large pride if it consists of their genetic relatives. Here is a likely reason: in a small pride, low-ranking individuals have a greater chance of mating; they don't need a kin-selected payoff. In a large pride, on the other hand, the subordinates are pretty much guaranteed to be reproductively shut out, and in such cases, not surprisingly, they cooperate only with other males who are kin. From the perspective of a subordinate male lion, it probably isn't evolutionarily worthwhile to join a large pride, in which you won't get to breed, unless there is the payoff that comes from being genetically related to the successful breeders.

You might say that in such cases subordinate male lions are helping dominant males to reproduce. Dominants profit by increasing their direct fitness; subordinates, via kin selection. Similar patterns have been reported for many other animals, especially birds, giving rise to numerous examples of "helpers at the nest." Ornithologists had long been puzzled by healthy young adult birds often forgoing their own reproduction and instead assisting other—typically older—adults to raise a family. This "helping" is now revealed to be genetic selfishness in disguise, since most helpers are actually assisting their parents to rear siblings for themselves, thus receiving a kin-selected benefit.

Incidentally, helping of this or any other sort does not preclude a bit of old-fashioned direct benefit as well. For helpers at the nest, the prospect exists that by hanging around the parental homestead, they increase the chance of later inheriting some of the family real estate, on which they will reproduce—directly—perhaps with assistance from a new generation of helpers, which they, in turn, may have helped to raise. Similarly, subordinate lions may be in line to replace the dominants, or they may just be making the best of their situation.

Cases of reproductive helping provide the most dramatic examples of kin selection in action and some of the most convincing evidence for Hamilton's concept of inclusive fitness, which "includes" both direct (parental) and indirect (kin-selected) components. The social insects, notably bees, wasps, and ants, occupy center stage in this drama. Many of these animals consist of large numbers of female workers, so designated because they are helpers at the hive, where they show an extreme form of nonreproductive altruism. Unlike subordinate lions or young adult helping birds, both of which are at least capable of reproducing directly if the opportunity arises, workers among the eusocial—"truly social"—insects are sterile. They have essentially given up the option of having their own offspring, laboring instead for the reproductive success of their mother, the queen. Why do workers accept such a high degree of reproductive restraint?

Hamilton, an entomologist as well as a geneticist, points out that these insects share an unusual biological trait: their genetic system is "haplo-diploid." This means that males are haploid, produced from unfertilized eggs, whereas females (including the queen as well as the workers) are diploid, having developed "normally" from fertilized eggs. As a result, each worker female enjoys a 75 percent probability that a gene present in herself is also present in her sisters, whom she will help to rear, as compared with a 50 percent probability that her genes would be passed on to her own offspring if she were to reproduce normally. This is because her haploid father produces sperm without meiosis. He therefore confers identical sperm to every worker, making them all identical with respect to the paternal half of their genes. To this is added the usual contribution of 50 percent identity through their mother, who provides the other half of the diploid genotype. (Complete identity at half the genotype plus 50 percent identity at the other half—50 percent of one half equals 25 percent—yields a 75 percent genetic similarity between sisters.)

Some of these sisters will go on to become successful reproducing queens, thus rewarding the workers' reproductive restraint. The bottom line is that because of their unusual haplo-diploid genetic system, females of the eusocial insects do more to promote their genes by staying home and helping to rear sisters, to whom they are related by 75 percent, than if they were to attempt to have offspring of their own, to whom they would be related by 50 percent. Once again, an extreme form of altruism at the level of bodies and behavior is revealed to be selfishness at the level of genes.

What about human beings? Not surprisingly, kin selection applied to *Homo sapiens* promises to open a vast new realm of research and potential insight as well as controversy. Among humanity's strongly suggestive cross-cultural universals, one of the most prominent is favoritism toward relatives, commonly known as nepotism (from the Latin *nepotem*, referring to "nephew"). To be sure, sometimes people dislike their relatives, if only because they know their kin better than anyone else and familiarity often breeds contempt. But inclusive-fitness theory and the biology of altruism allow us to understand *why* it is that people around the world are so closely involved with their relatives and why it is that despite the inevitable squabbles and disagreements, people are invariably biased in favor of their kin.

Armies of anthropologists have struggled to unravel the complex terminology whereby people identify their relatives and the rules by which they favor them. Although the fine print differs from one group to another, three basic patterns are clear. First, human beings are involved, almost obsessed, with situating themselves within a network of kin. There isn't a society on earth that does not in some way identify mothers, fathers, in-laws, grandparents, even complex and more distant relatives such as cousins, nieces and nephews, aunts and uncles, and so forth. Second, although these socially identified patterns don't always correspond exactly to the underlying genetic realities, they are remarkably close, with any deviations occurring at the periphery, among "distant relatives" (as seen in the Western concept of "distant cousins"). Third, this worldwide penchant for locating oneself within a kin-biased network is not an idle pastime. It has consequences that are as consistent as the phenomenon itself: closer relatives receive preferential treatment.

Not surprisingly, there have been objections to this "biologizing" of human behavior. (No one seems to object, by contrast, to the biologizing of lion or honeybee behavior.) In the forefront is the argument that complex human interactions are "socially constructed," which is undeniably true. The crucial question is whether they are *wholly* constructed by arbitrary societal whim, or if biology also whispers behind the scenes. Those arguing for the former point to exceptions that seem to work against kin selection. For a notable example, let's consider so-called mother's brother societies, which comprise nearly one third of human arrangements. In these cases, fathers do not provide the bulk of paternal care; rather, it is typically an uncle—

the mother's brother—who acts like the father. This clearly violates inclusive fitness theory. Or does it?

Sociobiologists have taken up the challenge, with results that are sobering for those who claim that human beings, alone among mammals, are somehow exempt from biology. A key point is that "father" is a biological as well as a social construct, and the two need not coincide. In short, a mother's husband may or may not be the father of her child. It is well within the realm of biological possibility that her egg(s) were fertilized by someone else. In such a case, the mother's husband may have no kin-selected interest in that child, whereas by contrast, the mother is guaranteed a 50 percent genetic investment, and her brother is therefore assured to be at least a half-uncle.*

Add to this the fact that avunculate societies, in which the mother's brother is the primary "father," exist disproportionately among those in which extramarital affairs are relatively frequent. It is precisely in such circumstances that husbands should be less paternal than uncles. And they are. Although not quite a case of the exception proving the rule, this can only enhance confidence that human beings are influenced by gene-oriented considerations of altruism and selfishness, whether they know it or not.

Reciprocity

Supposedly, there are many ways to skin a cat. There are also many ways to achieve altruism. Shared genes is one. Another is known as reciprocity, or "you scratch my back, I'll scratch yours."

Just as human beings had resonated to the tune of kin selection long before William D. Hamilton explained it scientifically, reciprocity has an ancient and honorable history, predating the sociobiologic revolution. Gifts, Christmas cards, and dinner invitations are typically reciprocated. Otherwise, a kind of unspoken social contract is violated, and a clear message conveyed: such a one-sided relationship is not likely to continue. As with kin selection, now widely seen to underlie nepotism, the expectations of reciprocity had long gone unexamined by biologists. (Indeed, a useful generalization emerges: whatever seems so obvious as to be taken for granted is likely to have an underlying biological basis.)

---

* Half because just as the child may not be the offspring of the purported father, brother and sister also might not share the same father, although they can be guaranteed the same mother.

What Hamilton did for nepotism, Robert L. Trivers (then a graduate student at Harvard, now a professor at Rutgers) did for reciprocity. He showed not only that it can be fully biological but that it can be selected for, even among individuals who do *not* share genes. To appreciate the issue, it is necessary, as with altruism generally, to point out why reciprocity is a dilemma and how this dilemma can be resolved.

If reciprocity were a simple matter of simultaneous transactions for mutual benefit—with no one losing and everyone winning, as in the case of mutualism—then no one would be surprised; the same holds for altruism based on kin selection. But what if the altruist and beneficiary do not share genes?

The idea of reciprocity is deceptively simple. Evolution can promote altruism from A toward B even in the absence of shared genes, so long as there is sufficient chance that at some time in the future, B will reciprocate and help A. Genes would still be profiting, but these reciprocity genes needn't be identical or derived from a common ancestor. Trivers showed, however, that this is more easily said than done, because of B's temptation to cheat: to take aid from A and then renege when it comes to paying back. Admittedly, both A and B would be ahead in the long run if they cooperated, but B might be even more ahead by gathering the benefit of A's altruism and not giving anything in return.

Trivers pointed out that the situation is similar to one that has long fascinated students of mathematical game theory: the prisoner's dilemma. In this idealized situation, two individuals have the choice of cooperating, of cheating while the other cooperates, or of both cheating at the same time. Depending on how the various payoffs are arranged, a genuine dilemma can be produced. Each individual, not knowing what the other will do but fearful that the partner/opponent may cheat and leave him a sucker with the lowest payoff of all, will be tempted to cheat. That way, if the partner cooperates, the cheater gets the highest payoff, and if the partner cheats, at least a fellow cheater avoids the lowest, sucker's payoff. The dilemma is that since both individuals come to the same conclusion, both wind up with a punishing outcome: cheating and being cheated. Admittedly, it's not as bad a payoff as being a sucker (cooperating when the partner cheats), but it's a lesser return than both would receive if they could somehow figure out how to cooperate; in our case, mutual cooperation means reciprocity.

Robert Axelrod, a mathematically oriented political scientist, has found that there is at least one way out of the prisoner's dilemma, notably a strategy of "tit-for-tat," by which a player attempts to cooperate but pays back noncooperation in kind. Nonetheless, reciprocity, for all its tempting logic, remains strangely rare in the natural world.

Probably the clearest example involves food sharing among—of all things—vampire bats. These otherwise unsavory creatures live on the edge of potential starvation. Because of their rapid metabolism, they cannot go more than seventy-two hours without a blood meal (typically obtained from horses or cows, not human victims), yet success varies unpredictably from one night to the next. In addition, vampire bats spend their days roosting in small social groups within which everyone knows everyone else, but which, interestingly, are not composed of close relatives. A distended belly makes it obvious who fed well the previous night. Hungry bats sidle up to well-nourished ones and, by a variety of begging movements, make their need clear. A well-fed altruist will typically respond by regurgitating some precious blood, which is eagerly consumed. A key observation is this: the beneficiary, when striking it rich sometime in the future, is likely to reciprocate, regurgitating blood for the benefit of the initial altruist.

Vampire bats seem to possess most of the prerequisites for squelching the temptation to cheat and achieving reciprocity. First, since they live in small groups and are capable of identifying one another, recipients can be expected to do their part when the tables are turned. And second, opportunities for payback are relatively frequent, while altruistic regurgitation contributes greatly to the success of the recipient without costing the donor too dearly (when foraging vampires are successful, they typically obtain more than they need, so sharing isn't very costly).

Many other cases of animal reciprocity have been proposed, notably among porpoises, baboons, chimpanzees, and some birds.

We've already seen that selfish benefit may often combine with a kin-selected payoff to generate multiple pressures for altruism among genetic relatives. Similarly, there is every reason for reciprocity and kin selection to co-occur, with kin cooperating and returning favors. It also seems likely that among all living things, the champion reciprocator is *Homo sapiens*. As a species, human beings meet many of the vampire prerequisites. Among other things, we have very large brains, especially important, perhaps, when it comes to establishing stable systems of reciprocity because of the subtle temptation to cheat. These temptations, in turn, may well generate powerful pressures to distinguish likely cheaters from noncheaters, to assess motivation, and to recall past obligations, as well as past performance. It has even been proposed that the benefits of successful reciprocating as well as of cheating may have exerted some of the selective pressures originally responsible for the rapid evolution of the large human brain in the first place. (Significantly, perhaps, vampires have the biggest-known bat brains.)

Reciprocity is fascinatingly complex. For example, it seems likely that natural selection has favored a tendency to deceive as well as to detect deception. And since the most effective liars are those who believe their own misrepresentations, it is possible that human beings have also been endowed with an adaptive tendency for *self*-deception, of which, by definition, they are not only unaware but especially inclined to disavow! It is an important sociobiological principle that the brain has not evolved to provide an objectively accurate portrayal of the world, either "out there" or "in here," but rather to maximize the long-term reproductive success of the body that carries it and the genes that help construct it. That's what it's *for*.

There seems little doubt that reciprocity exerts a powerful role in human affairs. There is also little doubt, however, that for all the cogency of kin selection and/or reciprocity, some examples of human altruism appear to defy either sociobiologic principle. Take, for example, disinterested heroism on the battlefield or even the more mundane generosity of giving money to one's National Public Radio station, or donating to a blood bank. In these cases it is possible as well as tempting to "cheat," taking advantage of other people's contributions to public radio, or getting blood if needed, without having given any. Many people succumb to such temptations, and perhaps that is answer enough. But sociobiology is not without an alternative explanation for those who donate without expectation of either reciprocity or kin-selected payoff: people are exceedingly sensitive to "third-party" or "reputation" effects. It is sometimes said that character is what you do when no one is watching; the sad truth is that it makes a big difference to most people whether someone is watching. For many individuals to "do the right thing" in the absence of kin benefits or potential reciprocity, it helps greatly if there is a prospect of receiving some sort of social recognition, which can eventually translate into greater fitness.

And so, we have medals and public acclaim for heroes and philanthropists. Genuine, disinterested altruism may in fact be as rare in *Homo sapiens* as in other species. One study, for example, showed that people are significantly more likely to donate blood if they receive a small insignia pin in return. Although it is unlikely that human beings possess a gene for blood donation, it is entirely plausible that we have been outfitted with adaptive psychological mechanisms that incline us to respond positively to situations that result in being identified as generous, as someone to trust and rely on. (Imagine a vampire bat proudly displaying a lapel pin that says "I gave at my roost.")

Gametes, Gonads, Genes, and Gender: Why Men and Women Are Different

A small number of interesting exceptions aside, most of us—and most other living things, too—can readily be identified as either male or female. But surprisingly few people understand the basis for these two fundamental categories. In Chapter 5 we briefly reviewed the defining qualities of sperm and eggs. Here we look into some of the implications of that basic difference when it comes to the behavior of the two sexes. A distinction is sometimes made between "sex" and "gender," with the former limited strictly to biology and the latter, to social roles attributed to one sex or the other, such as aggressiveness, nurturance, sexual coyness, and so forth. The strong implication is often added that such gender roles are socially constructed as opposed to biologically imposed, and that accordingly they are arbitrary and readily neutralized or even reversed.

Any fair-minded look at history as well as current events must conclude that women have gotten, and by and large still get, a rotten deal: typically overworked, underpaid, overlooked, undereducated, overrepresented among the poor, underrepresented in the higher ranks of government, commerce, academia, and the professions. In large measure, such gross unfairness has indeed been socially constructed, and most sociobiologists have no desire to further such outrages. Biology has often been used as a justification for society's malfeasance. But at the same time, a fair-minded look at the human species would suggest that although gender differences are often constructed by society, they are frequently built upon a foundation of genuine biological distinctions between male and female.

In the next few pages we review some of these distinctions as emphasized by sociobiologists, focusing on just four: sexual inclinations, violence, parenting, and some aspects of the human body. Don't be distracted if you can think of exceptions to the generalizations to come, for generalizations are just that: statements that are *generally* true, and which are not disproved by the occasional exception. Men, for instance, are taller than women. This is a valid, biologically based generalization, despite the fact that some women are very tall and some men very short. Women also have generally higher-pitched voices than men, although altos sing lower than countertenors. And whereas some women have sexual appetites that are more often associated with men, as we will see, there is a predictable and generally consistent sexual pattern that distinguishes men from women.

To understand male-female differences, evolutionary psychologists and sociobiologists turn especially to the crucial distinction between eggs and sperm. Each egg is a literal mother lode of metabolic investment, immeasurably more valuable than a single sperm. Among birds, for example, it is not uncommon for a female's eggs to comprise 15 to 25 percent of her body weight; no comparable investment is made by males in their sperm. Even among mammals, the egg-sperm disparity in metabolic investment is huge, despite the fact that mammalian eggs are tiny compared with those of birds, reptiles, and even most fish. Thus, a mammal's egg is only the beginning of a lengthy, expensive parental commitment by any female whose gametes might become fertilized. The young will be nourished within the mother's body, and then, following birth, an even greater investment of maternal resources is initiated: lactation, the distinguishing characteristic of mammals, involves several times more expenditure (in calories alone) than gestation. During all this time, biology makes no comparable demand on the male.

Not only are sperm cheap and easily replaced, but since they are produced in the hundreds of millions at a time, there is a reasonable chance that a relatively small number of matings will succeed in fertilizing the comparatively few eggs produced by any one female. Consider a cow elk or elephant—or human being for that matter. In most cases, she will produce one fertilizable egg at a time. Once inseminated, her attention and effort (physiological as well as behavioral) is likely to be directed toward insuring that her zygote, then embryo, then fetus, then offspring is as successful as possible. Males, too, are concerned with success, but theirs is more likely achieved by obtaining access to additional females—and, not coincidentally, their eggs and all the parental investment and potential evolutionary success they represent. So while cow elk and elephants occupy themselves with quality, bull elk and elephants are largely concerned with quantity: notably, sexual opportunities, pure and simple.

Sexual Inclinations

Several aspects of male-female differences in sexual inclinations should now be discernible. For one, males generally are sexually pushy and aggressive. After all, they invest little in any one mating, and accordingly, they have little to lose if they engage in an amorous encounter that turns out to be a "mistake." By contrast, females, with a comparatively large investment in every reproductive event, tend to be shy, sexually reticent comparison shoppers. This applies to human beings no less than to most other mammals.

It seems no coincidence, for example, that throughout the world, men will pay for sexual encounters with women, and not vice versa. In a very real sense, women already bring a valuable resource to any sexual encounter: their precious promise of parental investment. Based on their biology alone, women have something that men want, and typically, men are willing to pay for, whether the payment be in flowers, goats, sacks of rice, or charge cards. Similarly, the huge worldwide pornography industry exploits the relatively low male threshold for sexual stimulation as well as variety.

Researchers consistently find that men are focused on sex, women on romance. When, in an experiment, male and female college students were accosted by attractive members of the opposite sex who asked, "Would you go to bed with me tonight?" more than 75 percent of the men said yes, but not one woman did. When couples were asked, separately, to imagine what kind of "two-timing" behavior by their partner they would find most upsetting, significantly more men indicated that they would be most troubled if their female partner had a sexual relationship with another man, whereas women were most upset by signs that their male partner was romantically involved with someone else. For these women, it was not the prospect of their partner's sex act with someone else that was so threatening, but indications of waning affection (and by implication, lessened probability of helping provide future resources and assistance). Studies of singles ads find, similarly, that men consistently shop for indications of physical attractiveness and sexual availability, whereas women look for maturity, sincerity, indications of wealth and reliability.

Males also have a lower threshold for sexual excitation. Walk through a marsh in the American South during the spring, and you are likely to find a bullfrog clinging to the toe of your boot, sometimes one on each! At mating time, such males will attempt to mate with almost anything and have even evolved a special call that says,

"Let go, I'm a male, you dummy!" Not surprisingly, female bullfrogs are considerably more choosy, bestowing their eggs only on a male who meets rather strict criteria. Meaning no disrespect to either sex, the sociobiologic point is that men are recognizably like male bullfrogs, and women like female ones.

Perhaps the most impressive evidence for the connection between male-female differences in parental investment and corresponding differences in sexual behavior comes from study of those (very few) animal examples in which the usual pattern is reversed and males actually invest more than females. Among certain South American wading birds known as jacanas, males construct the nests, incubate the eggs, and care for the young. Sure enough, female jacanas are malelike. They are bigger than the males, more brightly colored, and more pushy. Large, aggressive female jacanas vigorously defend territories against other, comparable females, and within these territories they will mate with several small, sexually coy, and retiring males, who are "Mr. Mom" to the max.

Another dramatic example of sex-role reversal comes from the so-called Mormon cricket (actually a katydid and of no particular religious persuasion). Each male produces a large glob of proteinaceous glop known as a spermatophore, which is transferred to the female as part of courtship, along with his sperm. The male's investment in his spermatophore is considerably larger than the female's in her eggs. Once again, as predicted by parental-investment theory, females of this species solicit males for sex, while males are reticent and choosy, reluctant to expend their valuable spermatophore on just any courting female. "A guy can't be too careful," one can hear the male Mormon cricket repeat to himself. "Those females just want one thing."

These are all exceptions, and whereas they don't prove any rule, they consistently support the sociobiological insight that individuals of the sex that invests less—usually male—tend to be sexually pushy, seeking access to individuals of the sex investing more—usually female. There are no comparable exceptions among mammals, all of whom show high levels of female parental investment and notable male sexual pushiness.

Continuing the pattern, males of most species are generally stimulated by the prospect of sex with a new partner; females, much less so. Although women may also appreciate sexual novelty, the evidence is clear that by and large, the sexual dynamic is quite different for males and females, at least as measured by sexual responsiveness as such. Thus, women are much more likely to experience orgasm during sex with a loved and trusted partner; men, by contrast, are likely to be aroused by the mere prospect of sexual novelty for its own sake, a phenomenon dubbed the "Coolidge effect."

The story goes that President Calvin Coolidge and his wife were separately touring a model farm. Mrs. Coolidge noticed an impressive rooster surrounded by many hens and, in response to her question, was assured that this one animal was able to service all the hens. "Mrs. Coolidge wanted you to be told that this rooster mates many times each day," the guide later mentioned to the president.

"Always with the same hen?" asked Silent Cal.

"No, sir!" replied the guide.

"Please tell *that* to Mrs. Coolidge," said the president.

The Coolidge effect operates as follows: Place a male and a sexually receptive female together, and after a while, sexual activity declines. Then introduce a new female. The male shows renewed sexual interest, which, again, declines over time. For most males, familiarity may not breed contempt, but rather, diminished excitement. Even the most jaded male sexual appetite tends to perk up, however, at least temporarily, at the prospect of a sexual encounter with a new partner. This has been confirmed for nearly all mammals, including cattle, horses, pigs, sheep, goats, and human beings. Nothing comparable applies to female mammals, including women. Given the fundamental reproductive biology of maleness compared to femaleness, such a difference makes perfect sense.

Violence

For every Bonnie, there are about a hundred Clydes. For every Lorena Bobbit (famous for cutting off her husband's penis), there are about a hundred O. J. Simpsons (who, if not a murderer, is an admitted wife beater). When it comes to violence, the difference between males and females seems enormous. Like the proverbial intelligent fish who, if asked about its environment, probably would not describe it as "wet," people take certain things for granted, like the fact that men are overwhelmingly more violent than women. It is part of the ocean in which we swim. The exact numbers differ from country to country, but within any society, men are about ten times more likely to commit violent crimes. Given the immense personal and social cost of violence, this is something that cries out to be understood. Given what we now know about the biology of males and females, it is also something that cries out for an evolutionary interpretation.

Once again, the train of evidence harkens back to eggs and sperm, and the resulting fact that one male can fertilize many females, and that in such cases, males are under strong evolutionary pressures to try. First, it is worth noting that the great majority of mammals are in fact polygynous, with males seeking and obtaining multiple mates. Monogamy among mammals is exceedingly rare, in all likelihood because of the disproportionate payoff between males and females in having multiple partners. Thus, if a male mammal changes from monogamist to bigamist, he is likely to double his fitness; a trigamist, in turn, may enhance his fitness threefold, and so forth. By contrast, females gain comparatively little by increasing their number of mates, since female reproductive success is limited by the success of their offspring, whereas male reproductive success is limited by their success in gaining access to females. And so, harem formation is the sought-after state of affairs—from the male perspective. But since there are roughly equal numbers of males and females, the success of any one male comes at a proportionate cost to the others.

Consider any polygynous species, such as elk or human beings. If one male is able to keep, say, ten females, then his success comes at the expense of nine other males, who are reproductively excluded. These disappointed peripheral males typically fight with one another and periodically seek to overthrow the harem master and take his place. The result is that in polygynous species males are selected to be larger, often armed with lethal weaponry such as tusks, antlers, or fangs, as well as a violent temperament. Males, in short, are under evolutionary pressure to be physically and to do behaviorally whatever it takes to outcompete their fellows. This is the sociobiologic explanation for why males are more violent. As Gilbert and Sullivan's Princess Ida put it, "Darwinian man, though well-behaved, at best is only a monkey shaved."

Many people would like to think that human beings are well-behaved (although often they aren't) and also naturally monogamous, which almost certainly they aren't. Statistically, human monogamy is unlikely; out of nearly four thousand species of mammals, only a handful are monogamous. A Martian biologist visiting Earth would have little doubt that *Homo sapiens* belongs to the mammalian majority, being mildly polygynous. Our Martian would find a species with the following characteristics. Males, on average, are larger and heavier than females. Males are consistently more violence-prone. Females become sexually mature before males: this is the standard pattern among polygynous species, in which individuals of the harem-keeping sex delay entering the dangerous rough-and-tumble arena of reproductive competition until they are large enough, strong enough, tough enough, and smart enough to have some chance of success, or at least a reasonable prospect of not being killed or seriously injured. Since in most cases female-female competition is comparatively subtle, less head-to-head and more concerned with maneuvering for the success of their offspring, there is no comparable reason for individuals that are "kept" to delay breeding. Females who are ready to breed can usually do so.

Prior to the cultural and marital homogenization that came with Western colonialism and missionary zeal, polygyny was the prevalent system among *Homo sapiens*. About 85 percent of human societies were preferentially polygynous, compared with fewer than 15 percent monogamous, and less than one tenth of 1 percent polyandrous (in which a woman has multiple husbands).* Examined objectively, the polygyny of *Homo sapiens* must be considered proved. And examined sociobiologically, the implications are diverse, among them male-female differences in sexual proclivities, in physical traits, and in behavioral tendencies, including but not limited to the dangerous male penchant for violence.

Despite the attention given sexual harassment and wife battering, most male violence is directed at other men. Once again, this is the typical mammalian pattern, fully predicted by polygyny, which sets the stage for male-male competition in a variety of forms. Disputes and disagreements among men—especially young, hormonally driven, and socially insecure men—have a disconcerting tendency to escalate to violence. Arguments and misunderstandings over, say, an offhand comment or an ill-considered action may appear trivial and thus inappropriate reasons for violence. But in the previous eras that established human behavioral inclinations, such encounters may well have had consequences for social status and, therefore, evolutionary success.

It is also worth noting that whereas male-male violence is likely to have covert sexual motivation (that is, over social status), male violence toward women frequently carries more direct sexual overtones. Thus, cases of male-female assault or homicide are disproportionately associated with accusations of infidelity. And the most

---

* *It is probably noteworthy that in the very few cases of human polyandry, a woman's husbands are often brothers, a situation that is remarkably similar to that of lions or other cases of multiple male breeding systems, and that conforms closely to the expectations of kin selection.*

sexual of violent male-female crimes—rape—is widespread among animals, where it is usually a behavior of socially and sexually marginalized individuals (those excluded, resentful bachelors left in the wake of polygyny). The sociobiological assessment of human rape is still in its infancy; at present, it seems likely that rape is a violent, misguided behavior, overwhelmingly male, whose roots can be at least partly illuminated by considering its likely evolutionary basis.

Parenting

"The heart of woman's oppression," wrote radical feminist Shulamith Firestone, "is her childbearing and childrearing roles." In a sense, she is absolutely correct. Where Ms. Firestone got it wrong, as sociobiologists see it, is in her assumption, widely shared, that these roles are determined by social forces alone and that women are necessarily strong-armed into childbearing and childrearing by churlish men.

Sociobiologists and evolutionary psychologists emphasize that when it comes to parenting, there are biological factors at work, notably confidence of genetic relatedness. This is not necessarily conscious, self-aware confidence (although in some cases that, too, is at issue), but rather a deep undercurrent of biologically generated inclination similar to the widespread tendency to favor one's genetic relatives (kin selection and nepotism) or to do unto others as they have done unto you (reciprocity).

For many living things, and all mammals, reproduction requires more than getting one's gametes fertilized. It means behaving parentally, which we may describe subjectively as "loving" but objectively as providing defense, advice, assistance, and material resources. There are costs to such provisioning, but living things willingly pay them, because of "love," which for biologists is a route to achieving fitness. Just as natural selection would operate against true altruism, it would also operate against parental behavior directed toward offspring who are not one's own, that is, who do not represent one's own genetic heritage.

Here we come to another of those fundamental bedrock dichotomies of the natural world, comparable in its male-female significance to the great divide of egg versus sperm: internal versus external fertilization, more widely known as "Mommy's babies, Daddy's maybes." The simple, stunning reality is that whenever fertilization takes place inside an animal's body, a guaranteed asymmetry is established, in that individuals of one sex can be absolutely certain that they are genetically related to the offspring that eventually emerge from their body, whereas the impregnating others can only take their mates' word for it.

There is not a single species of mammal in which males do as much parenting as females. And of course, there is not a single species of mammal in which males have anything like the females' confidence of being related to those offspring. Interestingly, paternal care is not unusual among animals, notably certain fishes and amphibians, in which fertilization is *external*; in such cases, males typically defend a territory, squirting out their sperm immediately after females have extruded their eggs. Among externally fertilizing species, males and females can be about equally confident that the offspring are in fact theirs, and in fact, males and females are about equally likely to assume the parenting role when fertilization does not convey an asymmetry in genetic confidence.

But all mammals partake of internal fertilization, and significantly, among all mammals, it is the females that lactate. Once again, as with obvious and thus unexamined phenomena such as nepotism or male violence, the fact that male mammals don't nurse their young is usually taken for granted. But why don't they? Evolution has contrived far more elaborate and seemingly impossible feats than males making milk. Moreover, after the mother has undergone pregnancy plus the stress of giving birth, it would appear only fair for the father to step in at this point and share the physiological burden. But this ignores the fact that female mammals "know" they are related to their newborn, whereas male mammals do not. (And also that evolution is concerned not with what is fair but with what maximizes fitness.)

Among animals generally, a sure way to reduce the likelihood that males will participate in child care is to reduce the likelihood that they have fathered the offspring in question. And vice versa. Females that engage in "extra-pair copulations" typically lose some degree of parental assistance from their mate but often gain assistance, or at least tolerance, from the possible father. Similarly, when monogamous birds—which normally practice biparental care—are experimentally induced to establish "stepfamilies" in which the male has not fathered the offspring in question, such males consistently refrain from provisioning or defending their stepchildren. And among mammals, paternal behavior, rare as it is, is essentially limited to those very few species that are also monogamous: in such cases, the males have a high confidence of being the fathers.

Men can be very good fathers, and indeed, *Homo sapiens* is a species in which male parental investment is often substantial. Compared with most other mammal species, human fathers are unusual in how much time they spend with their children and how much, on average, they contribute. Nonetheless, there is no society on earth in which fathers are as paternal as mothers are maternal. If this enormous gap were solely due to arbitrary cultural conventions, then there should be as many cases of primarily father-oriented human societies as mother-oriented ones. Or at least some.

As with so many other aspects of male-female differences, the gender gap when it comes to parenting is fraught with various interpretations. Thus, it is interesting that in *The Second Sex*, even so liberated a woman as Simone de Beauvoir did pay homage (perhaps excessively) to the specialness of pregnant women: "Against the light of the mind they oppose the fecund darkness of life; against the clarity of consciousness, the mysteries of inwardness; against productive liberty, the weight of this belly growing there enormously without human will." It is possible (and maybe particularly tempting for a childless woman like de Beauvoir) to idealize the biology of femaleness, especially of reproduction. "The mother," she wrote, "is the root which, sunk in the depths of the cosmos, can draw up its juices; she is the fountain whence springs forth the living water, water that is also a nourishing milk."

Alternatively, when the uniquely female biology of reproduction has not been celebrated and even adored, it has often been devalued even as it has been romanticized. Witness the infamous Nazi ideal of women's roles: *Kinder, Kirchen, und Kuchen* ("children, church, and kitchen"). From one extreme to the other, either women have been supposed to possess a mystical "knowing," unfathomable in its depth, or they have been widely despised as brutishly organic compared to men, who alone were thought to possess the cerebral qualities that distinguish animal from truly human.

Ironically, in emphasizing the role of women as childbearers and nurturers, some radical feminists have converged with strange allies: right-wing ideologues. In the United States today the radical right uses and abuses biology to argue against government-sponsored child care and even affirmative action, by claiming that such programs undermine "family values," according to which women are breeders and little else. Up to a point, they are correct. Women *are* biologically designed to be childbearers and nurturers, but this is only a part of what they are (just as being pushy competitors is only part of what men are). Such a single-minded view of gender roles is sure to undermine anyone's advancement, whether woman or man.

Our point is that there appears to be a tendency from either end of the political spectrum to go overboard, so emphasizing the biological uniqueness of women that they are diminished by being relegated to a debased, "lower" realm of earthiness and reproduction. There is a downside to giddily adoring women as uniquely connected through their biology to the "fecund darkness of life" and thus to deep spiritual verities inaccessible to men. Such exaggeration, no less than the more obvious put-down of women as mere breeding stock, can readily degenerate into sexist oppression and stereotyping.

There is no question that men can be excellent fathers, just as women can be poor mothers, or excellent CEOs. And if, as biology seems to incline them, there are differences between men and women in those ways, this does not mean that such differences cannot be overcome, or alternatively, exaggerated. These would be very human decisions, entirely appropriate to complement, or confront, our very human biology.

Finally, we need to address part of the sociobiology of parenting that appears to be strongly influenced by confidence of genetic relatedness and *not* fundamentally driven by male-female difference: stepfamilies. Stepfamilies are increasingly common, and by and large, they function well. But it is a dramatic and unavoidable fact that for a child, living with a nonbiological parent is the highest of all identified risk factors for child abuse or neglect. In committing themselves to going against evolutionary inclinations to care for one's own offspring and not someone else's, stepparents enter a challenging realm. In some cases they are unable to rise above the siren song of their genes, whispering within them like a small evolutionary devil perched on their shoulder.*

Stepparenting is a troubled phenomenon in the animal world, too. The most disturbing cases involve outright infanticide, commonly observed when a newcomer male replaces an existing harem master. The newly ascendant harem keeper is, in a sense, a stepparent, and with distressing perseverance, he often kills those infants sired by his predecessor. Since they are not his, in the cold calculus of natural selection, their murder is no loss to his fitness. Under these conditions, in fact, infanticide actually promotes his genes, since the victims' mothers then stop lactating, begin ovulating, and are available to mate with him and produce his progeny.

---

* Stepparenting should be distinguished from adoption, which also involves care of unrelated children, but which does not carry an increased risk of abuse or neglect. Like stepparenting, adoption also involves a conscious decision to care for someone else's child as though it is one's own, but (1) adoption typically occurs when individuals are unable to reproduce biologically, which, if feasible, is nearly always the preferred choice, and (2) a stepparent almost always becomes affiliated with one or more stepchildren because of a chosen connection with the child's parent, not with the youngster(s) per se.

Don't misunderstand: most human stepparents do a good job. Certainly, they are neither abusive nor infanticidal, although they are often stressed in ways that go beyond the experience of biological parents. Our point is that when it comes to shedding light on a pressing human problem, a little sociobiology seems to go a long way.

Bodies

For sociobiologists and evolutionary psychologists concerned with unraveling biology's contribution to the various gender gaps, gametes (and, to a lesser extent, confidence of genetic relatedness) are where it's at. Beards and breasts, penises and vaginas are secondary at best. Nonetheless, there are evolutionary insights to be gained when it comes to examining the human body.

For one, we have already mentioned the universal reality that male bodies are bigger and heavier than female bodies; this is consistent with *Homo sapiens'* primitive penchant for polygyny. For another, there is the equally universal observation that although men are stronger than women when it comes to gross muscular capacity, women exceed men in a more significant measure of biological "strength": namely, life span. In certain cases, this difference is in fact socially constructed, such as susceptibility to lung cancer, in which women's rates began approaching those of men once it became socially acceptable for women to smoke. But just because male-female differences in death and disability are often attributable to male-female differences in behavior does not mean that in many cases these differences do not also have underlying evolutionary causes. Thus, greater male susceptibility to hypertension, atherosclerosis, and coronary artery disease may well be associated with job-related and other competitive stresses. At the same time, this is consistent with biologically influenced differences, insofar as men are more likely to engage in stressful, competitive activities *because* they are men. For example, accident rates are consistently higher for men than for women, and violent death is primarily a risk factor for men. Both reflect male-male competition. If human beings were biologically polyandrous, with competitive females inclined to maintain husband harems, it is likely that men would generally outlive women.

Some interesting speculation and research have centered around the shape of male and female bodies. Earlier we discussed the mystery of women, unique among mammals, sporting prominent nonlactating breasts. One possibility is that fatty mammary tissue serves as a means whereby women deceive men as to their milk-producing capacity. Continuing this line of thought, researchers have found evidence of men counterattacking, in the widespread male preference for the classic hourglass female figure: by preferring women who have narrow waists relative to their hips, men may unconsciously increase the probability that their chosen mates are not already pregnant, have not recently been pregnant, and are not simply padding their bodies with adipose tissue. To this, women may have counter-counterattacked by employing adipose tissue selectively, distributing it specifically on the breasts and thighs.

By contrast, sociobiologists have expended relatively little effort unraveling women's role in evolutionarily sculpting the male body. It is clear, however, that taller men are universally preferred over shorter ones: another predictable consequence of polygyny. As expected in a biologically harem-forming species, male "heightism" is not paralleled by comparable discrimination against petite women. Beards and baldness, however, remain mysterious.

As to genitalia, there is, sadly, relatively little to say about the human vagina, or that of mammals generally. It is a muscular tube connecting the uterus with the outside, suitable for transporting babies and menstrual blood one way, penises and sperm the other. Whether, as Freud suggested—and we deny—women's psychology is a particularly "dark continent" compared with that of men, it is certainly true that women's genitalia have received less attention (or at least, less interpretive, analytic, and scientific scrutiny) than the male equivalent.

"Some are fancy on the inside," Mr. Rogers, of children's television fame, used to sing. And "some are fancy on the outside." But in fact, a woman's genital anatomy is fancy on the outside, too! And every bit as worthy of praise and pride as a man's. The clitoris, for example, may be the Rodney Dangerfield of anatomic structures, usually getting very little respect. Yet it may be unique as the only body organ designed for just one reason: sexual gratification.

French psychoanalyst Luce Irigaray claims that women are uniquely fancy, with a distinct psychology because of the physical structure of their bodies, especially their genitals: "A woman 'touches herself' constantly without anyone being able to forbid her to do so, for her sex is composed of two lips which embrace continually." Of course, women can no more feel this "continual embrace" than they (or anyone else, for that matter) can sense the erotic undulations of a periodically emptying gallbladder, but it makes for an intriguing image, even if, like penis envy, it is utter nonsense.

We now offer a brief (dare we say short?) treatise on the penis. For nonbiologists, this overrated appendage is the crucial difference between men and women. Men have it; women don't. Even self-proclaimed feminists wind up paying special heed to the penis, albeit to revile rather than admire. "There is a far more effective weapon than brute strength for accomplishing male domination over females," writes one scholar. "This weapon is . . . the penis. . . . As feminists and others have been arguing for some time now, phallo-centrism and patriarchal power are inseparable."

The truth is quite otherwise. Even the most ferociously tumes-cent penis is trivial compared to just the forearm of a petite woman. It may be a blow to the male ego, but human penile anatomy just isn't very impressive. It is ineffective as a digging implement, a spear, or a club: nobody, to our knowledge, has ever been beaten over the head with one. Sure, penises can do damage, but only because their posses-sors are generally larger, stronger, faster, and more violent than those equipped with a vagina.

Although the really significant male-female differences are not found in their genitalia, genitals keep popping up, at least in the imagination of theorists. More recently, sociobiologists, too, have begun to take notice. In certain insects as well as snails and other invertebrates, vaginas are long and complex, often with elaborate labyrinthine twists and turns guarded at their openings by an array of flanges, flaps, sphincters, siphons, and so forth. Nonetheless, we believe that it is no slight to females to point out that in general, evo-lution has been even more inventive in creating penises than vaginas, probably because of the powerful biological pressures operating on males to insure that their sperm get priority treatment. It is only through a quirk of embryology that males urinate through the penis. The real work of this organ is depositing sperm where they are most likely to win the evolutionary prize.

Seen in the context of other species, the human penis isn't much to crow about. In the black-winged damselfly, a common streamside insect of the eastern United States, females mate with more than one male. Each male black-winged damselfly sports a specialized penis outfitted with lateral horns and spines, not unlike a scrub brush. Copulating male damselflies use their penis to clean out from 90 to 100 percent of their predecessor's sperm before depositing their own. One specialist in insect reproduction commented that the vari-ous valves, levers, awls, corkscrews, scoops, spines, and other peculiar-ities of these penises comprise a veritable Swiss Army knife of mechanical structures designed to introduce sperm while removing or superseding that of competitors.

Then there are the many species of mammals—including dogs, rodents, seals, and walruses—that possess a bacculum, or penis bone. Presumably, this confers a mating advantage of some sort; no one seems to know what.

The human penis, by contrast, is relatively unadorned. Compared with a man's body size, however, it is huge. Sociobiologists do not agree as to why. Some think the comparatively long penis of *Homo sapiens* might help introduce sperm deep into the vaginal tract, an adaptation to human bipedalism and upright posture that has caused the female's pelvis to be somewhat tilted from the angle found in most other primates.

Testicle size, however, follows a relatively straightforward pattern. Taken as a percentage of total body weight, human testicles are about twice as large as orangutans' and fully five times larger than gorillas'. But in this dubious realm of anatomic competition, the overwhelming champion is the chimpanzee, whose testes are more than three times larger than men's and fifteen times those of gorillas. Why would evolution favor smaller testes in some primates, larger ones in others? The reason seems clear: gorillas—and, to a lesser extent, orangutans—rely on body size and strength to win access to females. A dominant silverback male gorilla who drives away potential competitors by virtue of his physical strength is pretty much guaranteed sole access to the females in his troop. Therefore, his testicles need only be large enough to make sufficient sperm to fertilize the females he has accumulated.

Chimpanzees, on the other hand, live in multimale groups, in which many males often copulate with a single estrous female. Compared with gorillas, male chimpanzees compete less with their bodies and more with their balls. To be sure, male chimps physically compete with one another as well; thus there is considerable sexual dimorphism in body size. But they leave much of the competition to their sperm: the moist confines of the chimpanzee uterus and Fallopian tubes have become a miniature Roman Coliseum, within which tiny spermatic gladiators vie to outfox, outrun, or simply overwhelm the opponents' sperm. Under these conditions, he who produces the most sperm is likely to win.

Whales, incidentally, show a similar pattern. The world's largest animal, the blue whale, has testicles that weigh "only" about 200 pounds, to go with a body of 150 tons or more. The world's largest testicles are found in the right whale, whose body is only about half the size of the blue whale but whose testes can exceed 9 feet in length, weighing in at 1,000 pounds. The reason? Right whales mate promiscuously, not unlike chimpanzees, so that many males copulate with the same female. They are also the only whale species in which unambiguous instances of rape have been reported.

It should be clear at this point that genitalia (of either sex) warrant no mystical significance, and moreover, that the much-ballyhooed male physical "superiority" is limited to such grosser qualities as the ability to run faster, lift heavier weights, and so forth. As we have already described, this constellation of male traits generally goes a long way in conferring social status, along with the prospect of direct physical and social domination. Indeed, this may be why such traits as size and strength are generally associated with glory and renown—at least, in the minds of men.

Parent-Offspring Conflict

Some of the most interesting ideas are counterintuitive. The sun, we all *know* intuitively, moves around the earth. The earth itself, of course, is flat. And as for parents and offspring, it seems as intuitively obvious as the progress of the sun through the sky that what's good for a parent is good for the child. After all, parent and child have the same evolutionary interest, since children are the primary vehicles for achieving the parent's fitness. But here again, intuition fails.

The primary insight in this case comes from the work of Robert Trivers, of reciprocity renown (he also formulated the most coherent treatment of parental-investment theory). As with other sociobiologic concepts, notably altruism and kin selection, the key is to focus on genes rather than individuals. Parents are indeed keenly interested in their children, just as children are interested in themselves (and sometimes in their parents, too). But the concerns of parents and offspring are not identical, because parents and offspring are not genetically identical. Any gene present in a child has a one-half probability of being present in its parent. To put it another way: a child is twice as interested in itself as in its parent. (Ditto for each parent, at least if he or she is capable of reproducing again.)

The result is parent-offspring conflict. Take weaning, for example. A nursing infant is likely to want parental investment from the mother in the form of milk. And initially at least, the mother is equally inclined to provide it. The fitness of both parties is served by the same behavior, so there is no conflict . . . yet. But as time passes, the mother, like most mammal mothers, becomes increasingly interested in having another child, and this in turn is facilitated by weaning the infant. After all, mothers generally experience "lactational amenorrhea" as their reproductive system shuts down in the course of nursing so as to avoid diluting their precious parental investment.

The babe in arms, however, is likely to see things differently. Since it is entirely related to itself but only one-half related to its mother, it devalues her needs, inclinations, and overall fitness relative to its own. And so there is a predictable period of disagreement, during which the mother's fitness is maximized by discontinuing nursing and seeking to invest in the next child, while the current offspring wants to keep things as they are. The weaning conflict that results has been abundantly described for many mammals, although—prior to sociobiology—not interpreted this way.

Things are eventually resolved. After all, think of the adults you know: every one, we suspect, has been weaned. But this doesn't necessarily mean that the mother has won; rather, as time goes on, the interests of parent and child once more coincide. As the child grows, its need for milk decreases, while the mother's payoff for weaning is likely to increase. At a certain point, therefore—and in spite of the twofold difference in fitness assessment between parent and child—both parties agree to discontinue nursing.

In addition to conflict over the time course of investment, a similar analysis suggests parent-offspring conflict over the amount of investment—food, time, money, et cetera—conveyed from parent to offspring at any given instance. It also leads to other predictions. For example, parents and offspring are expected to disagree about the behavior of each offspring toward its brothers and sisters. This is because parents are equally related to each of their offspring, while by contrast, any given child is more closely related to itself than to any of its siblings, just as it is more closely related to itself than to its mother. Its "coefficient of genetic relationship" is 1 with itself and only .5 with a sibling; similarly, it is .5 with any of its own eventual offspring, compared with .25 with any of its siblings' offspring.

The sociobiologic prediction, therefore, is that children should be less inclined to share, to play nicely with, or to be otherwise altruistic and solicitous toward their brothers and sisters than their parents would like—more precisely, one half as much.

In the course of resolving these and other evolution-inspired conflicts, parents and offspring can be expected to employ different tactics. For example, parents can take advantage of their greater size, experience, control of resources, and physical power. But children are not altogether helpless. Although, as Trivers has pointed out, an infant cannot fling its mother to the ground and nurse at will, it can employ psychological tactics that emphasize—and likely exaggerate—its neediness. Since parents ultimately derive an evolutionary benefit from the success of their children, and moreover, since their own information as to their offspring's neediness is bound to be less precise than that of the offspring themselves, the stage is set for offspring to manipulate their credulous parents, even as parents might seek to force their interests on seemingly helpless children.

All this may seem a uniquely cheerless view of parenting, and doubtless it is not the entire story. But neither is the standard view of sociologists and developmental psychologists, namely, that any rough spots in the parent-offspring connection result from the inevitable difficulties encountered in the process of socializing primitive and untutored juveniles, who simply have not yet learned where their best interests actually reside. An interesting derivative from parent-offspring conflict theory is that the traditional view from the social sciences, in which parents are thought simply to have the child's best interests in mind, is precisely the kind of poppycock that adults would be expected to promote! By contrast, the sociobiologic perspective suggests that parent and child—more precisely, genes within parent and child—all have their own agendas, and that parent-offspring behavior, just like behavior generally, takes place within a complex, shifting arena of individuals and genes struggling to get ahead, and sometimes cooperating, sometimes competing toward that end.

The goal of sociobiology is to help make sense of behavior, employing an evolutionary perspective. It does not try to explain everything. It is worthwhile insofar as it is useful, providing stimulating ideas and suggesting new ways of looking at ourselves. If sociobiology prospers—and we suspect that it will— it will likely reorient part of the human self-image, connecting *Homo sapiens* yet more to the rest of the living world. We think this will be a gain rather than a loss. In any event, it should be possible for even nonsociobiologists to understand what is going on and to judge the outcome for themselves.

CONCLUSION:
Digging for Water

"Just as people find water wherever they dig, they everywhere find the incomprehensible, sooner or later." So wrote Georg Christoph Lichtenberg, eighteenth-century physicist and satirist. Two centuries later it isn't at all clear that people will find water *wherever* they dig (especially with the worldwide shrinkage of underground aquifers). Nor is it certain that wherever we dig in the realm of life, we shall find the incomprehensible. The point of this book, in fact, has been just the opposite: that the living world *is* comprehensible, just as our own part in that world is undeniable.

Post-*Sojourner*, there is much about Mars that remains unknown, but being uncomprehended does not make it incomprehensible. It isn't terribly long ago that reputable astronomers entertained the notion that there were canals on the red planet, for example. From our current vantage, this is clearly fantasy. But Mars has not become any less intriguing for becoming known, just as the moon didn't lose any appreciable panache once it became undeniably made of rock rather than green cheese.

We hope that the same thing applies to life back here on Earth. The alternative to mystery isn't triviality or mundaneness. Rainbows are no less beautiful when we understand that they are produced by white light separated into its spectral components, just as life is not rendered tedious when seen as structured by nucleic acids rather than "protoplasm." Or when evolution is seen as the ultimate force behind that structuring. Or when human beings are seen as inextricably connected to the whole business.

There will be more *Sojourners*, reporting more and newer information to planet Earth and its inhabitants. At the same time, there will be lots more sojourning and pathfinding right here on Earth, more careful gazing at that strange, yet familiar mammal in the mirror, and maybe there will be future editions of this book, reporting more and newer information about the life around us and in us. Perhaps there will be a chapter on how AIDS, or cancer, or heart disease was finally overcome. Perhaps there will be a report on how neuroscience and consciousness fit together like a hand in a glove; if so, the synthesis would be about as momentous as physicists' long-sought unification of relativity and quantum theory. Perhaps our understanding of evolution and the fine tuning of ecological systems will have advanced to the point at which predictions will be totally reliable and information will be available upon which ecosystem-preserving decisions can be based.

Perhaps this book itself will become obsolete, not just too simple but even out-and-out wrong, like theories about canals on Mars or lunar gorgonzola. Maybe then, perfectly good mirror-gazing mammals such as yourselves will look back in surprise that you ever believed such elementary foolishness in the first place. We certainly hope so.

RECOMMENDED READINGS

1. Key to Life: The ABCs of DNA

Aldridge, Susan. *The Thread of Life: The Story of Genes and Genetic Engineering.* Cambridge, England: Cambridge University Press, 1996. A very effective book about many new advances in genetic engineering and biotechnology, although written pre-Dolly.

Frank-Kamenetskii, Maxim D. *Unraveling DNA: The Most Important Molecule of Life.* Reading, Mass.: Addison-Wesley, 1997. A well-written and definitive guide to DNA in all its guises, in an easy-to-understand and clear format.

Gonick, Larry, and Mark Wheelis. *The Cartoon Guide to Genetics.* New York: HarperCollins, 1991. A very easy-to-read, amusing, and scientifically accurate genetics text written in the form of a comic book; does an excellent job of clarifying confusing concepts, while expanding greatly on the genetics coverage in this chapter and offering entertaining anecdotes about early scientific theories of heredity.

Lyon, Jeff, and Peter Gorner. *Altered Fates: Gene Therapy and the Retooling of Human Life.* New York: W. W. Norton, 1996. A thoughtful look at both the promises and possible harmful effects of gene therapy and its predecessor, the Human Genome Project; brings together the stories of many of the scientists who have been influential in the field and raises questions about where future research may be heading.

Watson, James D. *The Double Helix: A Personal Account of the Discovery of the Structure of DNA.* New York: New American Library, 1991. A now-classic tale written by one of the founders of molecular biology describing the race to determine the structure of DNA; captures the drama and fun of science.

2. Tiny and Troublesome: Virus Pirates and Other Nasty Little Guys

Garrett, Laurie. *The Coming Plague.* New York: Penguin, 1994. A dramatic account of many of the new and dangerous diseases that threaten us. A call to arms in the style of Rachel Carson's ecological masterpiece, *Silent Spring*, warning about future threats of emergent diseases.

Klitzman, Robert. *Trembling Mountain: A Personal Account of Kuru, Cannibals, and Mad Cow Disease*. New York: Plenum, 1998. A personal account of the discovery and exploration of *kuru* in the Fore tribe of Papua New Guinea. An autobiography and coming-of-age story, combined with the history of this perplexing disease.

Oldstone, Michael B. A. *Viruses, Plagues, and History*. Oxford, England: Oxford University Press, 1998. A description of the new diseases that threaten humanity and the history of their discovery; emphasizes the triumphs of science and scientists rather than the dangers associated with the diseases; in the spirit of, but more modern than, Paul De Kruif's classic *Microbe Hunters*.

Olshaker, Mark, and C. J. Peters. *Virus Hunter: Thirty Years of Battling Hot Viruses around the World*. New York: Anchor, 1998. A personal description of the challenges and adventures of being a virus hunter; the author describes in clear detail the diseases he has "chased" and tries to quell much of the worry over them by describing how contained they actually are—for the most part.

Preston, Richard. *The Hot Zone*. New York: Doubleday, 1994. A true story about Ebola virus and a near-miss outbreak in Reston, Virginia, that reads like a thriller.

3. Cell-Sized: Building Blocks of the Body

Burnet, Frank MacFarlane. *Integrity of the Body: A Discussion of Modern Immunological Ideas*. Cambridge, Mass.: Harvard University Press, 1962. Although the subtitle is no longer accurate, this is a first-rate account of the excitement of early immunology, by one of the pioneers in the field.

Davies, Kevin, and Michael White. *Breakthrough: The Race to Find the Breast Cancer Gene*. New York: John Wiley & Sons, 1996. A clear account of the search to find the "breast cancer gene," including insights into the competition between pharmaceutical companies and academic researchers.

Hall, Stephen S. *A Commotion in the Blood: Life, Death and the Immune System*. New York: Henry Holt, 1997. An up-to-date description of some of the prospects, promises, and heartbreaks of cancer immunotherapy.

Rensberger, Boyce. *Life Itself: Exploring the Realm of the Living Cell*. Oxford, England: Oxford University Press, 1997. A report from the front lines of cell biology, including topics such as cell regeneration, cell movement, and even the definition of life.

Waldholz, Michael. *Curing Cancer: Solving One of the Greatest Medical Mysteries of Our Time*. New York: Simon & Schuster, 1997. A people-centered account, focusing on "cancer genes" and their detection.

4. The Brain and Behavior: On the Matter of Mind

Calvin, William. *How Brains Think: Evolving Intelligence Then and Now*. New York: Basic, 1997. An eminent neurophysiologist and writer explains the evolution of intelligence, from the inside out.

Crick, Francis. *The Astounding Hypothesis: The Scientific Search for the Soul*. New York: Touchstone, 1995. The famed DNA researcher describes the evidence for a materialistic interpretation of human experience, with special emphasis on information processing in the visual system.

Damasio, Antonio. *Descartes' Error: Emotion, Reason and the Human Brain*. New York: Avon, 1995. A neurosurgeon meditates on the connectedness of mind and body and of reason and emotion.

Ledoux, Joseph. *The Emotional Brain: The Mysterious Underpinnings of Emotional Life*. New York: Touchstone, 1998. A comprehensive description of the complex systems underlying emotion in the human brain.

Restak, Richard. *Brainscapes: An Introduction to What Neuroscience Has Learned about the Structure, Function and Abilities of the Brain*. New York: Hyperion, 1996. A neurologist discusses basic brain biology, including a guided tour from the micro to the macro; also includes a strong description of various imaging techniques.

5. Relevant Reproduction: Sex and the Human Animal

Kolata, Gina Bari. *Clone: The Road to Dolly and the Path Ahead*. New York: William Morrow, 1998. A science journalist surveys the fundamental research as well as the personalities leading to the famous sheep, and offers solid speculation about what the future may hold.

Money, John. *The Adam Principle: Genes, Genitals, Hormones and Gender*. Prometheus, 1993. Informative essays that are especially strong on the question of sex hormones, genes, and sexual identity.

Planned Parenthood Federation of America. *All about Birth Control: A Personal Guide*. Three Rivers Press, 1998. The title says it well: solid, personal advice, including the pros and cons of the major methods.

Ridley, Matt. *The Red Queen: Sex and the Evolution of Human Nature*. New York: Macmillan, 1993. A fine account of some of the evolutionary riddles posed by sex, emphasizing that sexual reproduction generates a never-satisfied striving for genetic diversity.

Vaughan, Christopher. *How Life Begins: The Science of Life in the Womb.* New York: Dell, 1997. Lavishly illustrated depiction of human embryology, backed by scientifically valid information.

6. The Energetic Life: Food, Fuel, and Fat

Fraser, Laura. *Losing It: America's Obsession with Weight and the Industry That Feeds on It.* New York: E. P. Dutton, 1997. A personalized expose of the weight-loss industry and its misrepresentations, this book may be rather liberating for people hooked on and disappointed by the latest diet fad.

Kirschmann, Gayla, and John D. Kirschmann. *Nutrition Almanac.* New York: McGraw-Hill, 1996. A reference work for the general reader that covers advances in nutritional research, including good material on herbal medicine.

Klein, Richard. *Eat Fat.* New York: Pantheon, 1996. A work of literature, history, and sociology that espouses a contrarian view (which we do not share) regarding the "benefits" of fat; worthwhile if only because it is so provocative.

Schmidt-Nielsen, Knut. *Animal Physiology: Adaptation and Environment.* New York: Cambridge University Press, 1990. Although a textbook, this is quite readable, offering a very strong comparative approach to basic processes including nutrition and digestion as well as circulation, respiration, salt and temperature balance, and more.

Thomas, Paul, et al. *Weighing the Options.* National Academy Press, 1995. A solid description of various weight-loss programs, plus practical and scientifically valid tips on how to evaluate them.

7. Ecology: The Cloud in the Paper

Cohen, Joel. *How Many People Can the Earth Support?* New York: W. W. Norton, 1995. A remarkably reasoned yet readable depiction of the practical issues faced by demographers, ecologists, agriculturalists, and everybody else as we enter the twenty-first century with an ever-growing human population.

Gradwohl, J., and R. Greenberg. *Saving the Tropical Forests.* Washington, D.C.: Island Press, 1988. Describes the problems facing tropical forests while extolling some local successes in averting their destruction.

Leopold, Aldo. *A Sand County Almanac*. New York: Oxford University Press, 1987. Among the great classics of ecological writing, as poetic as it is scientifically valid, by the founder of the field of wildlife management.

Quammen, David. *The Song of the Dodo*. New York: Scribner, 1996. One of our best nature writers describes his personal odyssey to experience and understand the role of islands and of diminishing populations in the lives and future of endangered species.

Wilson, Edward O. *The Diversity of Life*. Cambridge, Mass.: Harvard University Press, 1992. An informed clarion call for appreciating and protecting the world's biodiversity, by a modern master of ecology and evolutionary biology.

8. Evolution: The Road Stretches Out

Dennett, Daniel C. *Darwin's Dangerous Idea*. New York: Simon & Schuster, 1995. An effective rebuttal to many misconceptions about modern Darwinism. Written by a noted and occasionally controversial philosopher, it also provides some thought-provoking suggestions about evolutionary issues.

Lewin, Roger. *Bones of Contention: Controversies in the Search for Human Origins*. Chicago: University of Chicago Press, 1997. An engrossing account of modern paleoanthropology and the people behind it, including current ideas, debates, and personal conflicts.

Mayr, Ernst. *The Growth of Biological Thought*. Cambridge, Mass.: Harvard University Press, 1982. An admirable review of evolutionary thinking, especially about the role of natural selection in producing distinct species, by one of the founders of the "modern synthesis" of genetics and natural selection, who is also an accomplished historian.

Weiner, John. *The Beak of the Finch*. New York: Knopf, 1994. A very readable description of evolution during "real time," focusing on the research by Peter and Rosemary Grant.

Williams, George C. *The Pony Fish's Glow*. New York: Basic Books, 1997. A delightfully accessible, cogently argued gem of a book, touching on basic themes of adaptation, by one of the towering figures in modern evolutionary biology.

9. Sociobiology: Gene Machines in Action

Barash, David P., and Judith Eve Lipton. *Making Sense of Sex*. Washington, D.C.: Island Press, 1997. Uses many examples from animal as well as human behavior to point out the implications of sociobiology for understanding male-female differences, especially in sexual behavior, violence, parenting, childhood, brains, and bodies.

Dawkins, Richard. *The Selfish Gene*. Oxford University Press, 1989. A modern classic; although not especially concerned with applying sociobiology to *Homo sapiens*, it is an eloquent argument in favor of the "gene-centered" thinking that has given sociobiology much of its intellectual momentum.

Pinker, Steven. *How the Mind Works*. New York: W. W. Norton, 1997. Develops and explains many of the important concepts in evolutionary psychology and is particularly strong in showing the fruitfulness of an evolutionary approach in illuminating processes of human thought.

Ridley, Matt. *The Origins of Virtue*. New York: Viking, 1996. Does an excellent job of recounting evolutionary aspects of altruism and ethics, including some delightful treatments of semimathematical paradoxes (although we disagree with the author's effort to draw politically conservative lessons from his material).

Wilson, Edward O. *On Human Nature*. Cambridge, Mass.: Harvard University Press, 1978. A wide-ranging, thoughtful, and controversial classic of human sociobiology, written by the renowned biologist who first brought the discipline into the popular and scientific limelight.

INDEX